Multiscale Modeling of Cancer

An Integrated Experimental and Mathematical Modeling Approach

Mathematical modeling, analysis, and simulation are set to play crucial roles in explaining tumor behavior and the uncontrolled growth of cancer cells over multiple time and spatial scales. This book, the first to integrate state-of-the-art numerical techniques with experimental data, provides an in-depth assessment of tumor cell modeling at multiple scales. The first part of the text presents a detailed biological background with an examination of single-phase and multi-phase continuum tumor modeling, discrete cell modeling, and hybrid continuum-discrete modeling. In the final two chapters, the authors guide the reader through problem-based illustrations and case studies of brain and breast cancer, to demonstrate the future potential of modeling in cancer research. This book has wide interdisciplinary appeal and is a valuable resource for mathematical biologists, biomedical engineers, and clinical cancer research communities wishing to understand this emerging field.

Vittorio Cristini is Professor of Health Information Sciences and Biomedical Engineering at the University of Texas, and of Systems Biology at the MD Anderson Cancer Center, Houston. He is also Honorary Professor of Mathematics at the University of Dundee, Scotland. Professor Cristini is a leading researcher in the fields of mathematical and computational biology, complex fluids, materials science, and mathematical oncology. He has published several chapters in books and over 60 journal articles. His research has been supported by various institutions, including the US National Science Foundation, the National Institutes of Health, the Cullen Trust for Health Care, and the US Department of Defense.

John Lowengrub is Chancellor's Professor of Mathematics, Materials Science, and Biomedical and Chemical Engineering at the University of California, Irvine. He has published over 80 journal articles along with several book chapters. Professor Lowengrub is a leading researcher in the fields of mathematical and computational biology, mathematical oncology, complex fluids, and materials science. His research has been supported by various institutions including the National Science Foundation, the National Institutes of Health, and the State of California.

This past decade in the field of cancer research can be described as a decade of data generation. The next step in the war on cancer is figuring out what to do with the data. Cristini and Lowengrub have put together a remarkable treatise on ways to interpret these data and start modeling cancer behavior. This book presents a multi-scale approach toward understanding the behavior of cancer – such an important beginning to a critical field.
Professor David B. Agus, M.D., *University of Southern California Keck School of Medicine*

Cristini and Lowengrub have leaped over an obstacle previously assumed to be insurmountable: how to model complex biological systems without precise measurements of every event. Biologists are struggling to produce such measurements, which could take centuries to accomplish. In an unprecedented tour-de-force through a number of peer-reviewed publications synthesized in this book, Cristini and Lowengrub found a way to reliably predict biological behavior, thereby circumventing the need for precise measurements to generate predictive modeling. Their method assumes that the physical laws of nature are obeyed by biology – what could be more simple? By making this assumption, they have generated useful and predictive models of cancer progression that apply to pathological diagnosis. Such predictive modeling will be a powerful tool for diagnosis, prognosis, and treatment in the next decade.
Professor Elaine L. Bearer, MD., *University of New Mexico Health Sciences Center.*

This is a comprehensive and authoritative account . . . by leading experts in the field. The work brings powerful new ideas and tools of mathematical and computational modeling to the field of cancer research.
Professor J. Tinsley Oden, *The University of Texas at Austin*

This is a wonderful book covering most of the literature that has appeared in the last ten years on cancer modeling. It covers both theoretical and experimental aspects, drawing a strong link between them, and describes all phases of tumor growth, from the avascular to the vascular phase through the angiogenic process. It presents both discrete and continuous models, with the aim of linking them in line with the current thought that important insights into the complexity of tumour growth can only be reached by closely relating the phenomena occurring at the sub-cellular, cellular and tissue scale. Though at present multiscale models can be considered still in their infancy, this book gives a lot of ideas on how such models could be developed. For this reason, the book is of great value for young researchers who want to devote their attention to this crucial aspect of mathematical modelling in medicine in general.
Professor Luigi Preziosi, *Politecnico di Torino, Italy*

If you've ever wondered what mathematical and computational modeling can do for the field of cancer research, this book is a key to finding compelling reasons to integrate theory and experiment to unlock the mysteries of tumor growth and invasion. *Multiscale Modeling of Cancer: An Integrated Experimental and Mathematical Modeling Approach* tells the complex and dynamic story of tumorigenesis, vascular tumor growth, and invasion from the latest mathematical perspective, in less than 300 pages. I'm recommending this book as a must-read for all of my graduate students and postdocs!
Professor Trachette Jackson, *University of Michigan*

Multiscale Modeling of Cancer

An Integrated Experimental and Mathematical Modeling Approach

VITTORIO CRISTINI

University of Texas

JOHN LOWENGRUB

University of California, Irvine

CAMBRIDGE
UNIVERSITY PRESS

CAMBRIDGE UNIVERSITY PRESS
Cambridge, New York, Melbourne, Madrid, Cape Town, Singapore,
São Paulo, Delhi, Dubai, Tokyo, Mexico City

Cambridge University Press
The Edinburgh Building, Cambridge CB2 8RU, UK

Published in the United States of America by Cambridge University Press, New York

www.cambridge.org
Information on this title: www.cambridge.org/9780521884426

First published 2010

Printed in the United Kingdom at the University Press, Cambridge

A catalog record for this publication is available from the British Library

Library of Congress Cataloging in Publication data
Cristini, Vittorio, 1970–
Multiscale modeling of cancer : an integrated experimental and mathematical modeling
approach / Vittorio Cristini, John Lowengrub.
    p.   ;   cm.
Includes bibliographical references and index.
ISBN 978-0-521-88442-6 (hardback)
1. Cancer – Research – Mathematical models.   I. Lowengrub, John.
II. Title.
[DNLM: 1. Neoplastic Processes.   2. Models, Biological. QZ 202 C933m 2010]
RC267.C75   2010
362.196′994 – dc22   2010017932

ISBN 978-0-521-88442-6 hardback

VC: This work is dedicated to my wife, Jennifer, and to my children Giovanni, Gabriella and Tizita, to my mother and father, and to the excellent scientists I have been fortunate to train, without whom this work would not have been possible.

JL: This work is dedicated to my wife, Elizabeth, my children Catherine, David, Collette, Hillary and Mark, my parents Mort and Carol, to whom I am much indebted, and to my students and post-doctoral fellows, from whom I have learned more than I have taught.

Finally, we dedicate this work to the health care professionals and patients who bravely do battle every day on the front lines of the war against cancer.

Contents

List of contributors

Yao-Li Chuang, Ph.D.

Yao-Li Chuang, who holds a Ph.D. in physics from Duke University, is a post-doctoral researcher at the University of Texas Health Science Center in Houston. He specializes in nonlinear dynamical systems, with an emphasis on the statistical and numerical analysis of discrete many-body problems and their continuum limit. His areas of interest include the formation and stability of patterns arising in ordinary and partial differential equation systems, with a recent focus on the high-performance scientific computing of Cahn–Hilliard equations applied to tumor growth modeling.

Mary E. Edgerton, M.D., Ph.D.

Mary Edgerton, who holds an M.D. from the Medical College of Pennsylvania and a Ph.D. in biophysics from the University of East Anglia (United Kingdom), is an Associate Professor at the M. D. Anderson Cancer Center. Dr Edgerton's research focuses on the discovery of mechanisms in cancer genesis and progression. She has worked on the development of mathematical models for computer simulations of the spread of ductal carcinoma in situ and on the role of cell motility in extensive disease. She researches methods for the analysis of gene-expression array data for pathway discovery. She has published articles on the application of these methods to lung cancer and breast cancer profile data. Dr Edgerton has also worked on the development of integrated information platforms for tissue acquisition, clinical annotation, and molecular profiling.

Hermann B. Frieboes, Ph.D.

Hermann Frieboes, who completed a Ph.D. in biomedical engineering at the University of California, Irvine, is a postdoctoral researcher at the University of Texas Health Science Center in Houston and works in collaboration with the Department of Mathematics at the University of California, Irvine. He specializes in applying computational and experimental cancer modeling techniques to predictive oncology. His interests include the development of quantitative approaches to the study of tumor growth and treatment, furthering the interaction of basic research with clinical and translational programs and exploring the use of nanotechnology in disease diagnosis and treatment.

Fang Jin, Ph.D.

Fang Jin, who holds a Ph.D. in chemical engineering from Johns Hopkins University, is a postdoctoral researcher at the University of Texas Health Science Center in Houston and works in collaboration with the Department of Mathematics at the University of California, Irvine. He specializes in scientific modeling and toolset design in the simulation and visualization of multiscale problems. His areas of interest include surface tension induced morphology changes in the micro-size environment, angiogenesis network topology modeling in the human body, and the modeling of tumor growth with discrete–continuum multiscale interchange.

Xiangrong Li, Ph.D.

Xiangrong Li completed his Ph.D. in mathematics in 2007 from the University of California, Irvine. He is now a postdoctoral researcher at the Center for Mathematical and Computational Biology at the University, Irvine. His research focuses on tumor modeling and moving-interface and free-boundary problems.

Paul Macklin, Ph.D.

Paul Macklin, a lecturer at the University of Dundee, was formerly an assistant professor at the University of Texas Health Science Center in Houston. He completed his Ph.D. in applied and computational mathematics at the University of California, Irvine, and specializes in discrete and continuum biomathematics and computation with particular emphasis on cancer. He has recently focused on improving the state-of-the-art in patient-specific multiscale cancer modeling and enjoys the fusion of mathematics, computer science, biology, the physical sciences, and medicine that is necessary to describe and calibrate biophysical systems.

Steven M. Wise, Ph.D.

Steven Wise completed his Ph.D. in engineering physics at the University of Virginia and is now an assistant professor of mathematics at the University of Tennessee. He specializes in mathematical biology, computational and applied mathematics, and computational materials science and has helped to push forward the state-of-the-art in nonlinear multigrid solution techniques using adaptive meshes.

Preface

In the past several decades there has been significant progress in understanding and identifying the causes of cancer and in developing effective treatment strategies. Nevertheless, a cure remains frustratingly elusive. At its most essential level, cancer involves the abnormal growth and spread of tissues within a body. Yet each cancer is unique, based on the tissue in the body where it originates and the particular person who has it. While molecular mechanisms and cell-scale dynamics governing tumor cell migration and proliferation are well studied from a biological perspective, cancer progression actually involves events that occur at multiple time and spatial scales. What occurs at the nano-scale of molecules and micro-scale of cells affects the behavior of tissue at the centimeter-scale – and vice versa. In order to better understand these multiscale linkages, mathematical modeling, analysis, and simulation have been employed to study tumor behavior. The complex shapes and invasive behavior of tumors requires a nonlinear approach, meaning that effects at various physical scales within the tissue do not necessarily influence each other additively. Hence, the combination of events may yield a response greater or less than of each component, depending whether there is synchrony. The application of such computational models in the clinical setting, however, is still in its infancy.

In this book we outline recent advances in the field of mathematical modeling and the simulation of cancer, particularly with respect to multiscale, nonlinear, computational models that integrate theory and experiment. We present state-of-the-art numerical methods as tools for analyzing the nonlinear behavior predicted by the models. The book focuses on the challenging problem of developing models that connect intratumor molecular and cellular properties with critical tumor behaviors such as invasiveness and clinically observable properties such as morphology. In this context we discuss the incorporation of experimental and clinical data into predictive mathematical and computational models. The interactions between cellular proliferation and adhesion and other phenotypic properties are reflected in both the surface characteristics of the tumor–host interface and the invasive characteristics of the tumors. These cellular and molecular properties are influenced by the cellular genetics and by microenvironmental conditions such as oxygen deprivation (hypoxia). This close connection between tumor morphology and the underlying cellular and molecular dynamics is of fundamental importance, in that the cellular dynamics that give rise to various tumor morphologies also control its ability to invade. This allows the observable properties of a tumor, such as its morphology, to be used both to understand the underlying cellular physiology and to predict subsequent invasive behavior.

There are significant challenges in the multiscale and multidisciplinary modeling of cancer. These include the development of realistic models, the achievement of numerical solutions, and the incorporation of experimental and clinical data. This book offers three novel features to address these needs. First, we present and critically evaluate state-of-the-art mathematical models calibrated with experimental results. We demonstrate how experiments are used to determine functional relationships between the phenotypic variables and parameters of the models and the microenvironmental and molecular agents that affect tumor progression and invasion. Second, we evaluate patient-specific calibration protocols in a multiscale modeling framework spanning the cell-scale to the tissue-scale. Third, we present the state-of-the-art numerical algorithms that are indispensable tools for studying nonlinear predictions of the mathematical models. It is our sincere hope that the presentation of the material in this publication will further the goal of the eventual clinical application of multiscale modeling to cancer patients.

About the cover

Cover illustration: the computer simulation of a growing tumor and the corresponding host vascular response are shown in images representing predicted morphologies at 20, 40, 80, and 110 days since inception. The inner necrotic region is shown in dark red, surrounded by a layer of viable proliferating cells in blue. The blood-conducting vessels are indicated as thicker red lines, while sprouting (non-conducting) vessels are shown as thinner orange lines.

Acknowledgements

We acknowledge support from the US National Science Foundation through the Division of Mathematical Sciences. We also acknowledge partial support from the National Institutes of Health, the Cullen Trust for Health Care, and the US Department of Defense. We are greatly indebted to David Agus, Elaine L. Bearer, Alex Broom, Kim-Anh Do, Robert Gatenby, Jahun Kim, Mauro Ferrari, Sandeep Sanga, Giovanna Tomaiuolo, Wei Yang, and Xiaoming Zheng for their contributions to the research presented in this book.

Notation

Sets and set operations

$\emptyset$	Empty set
$x \in A$	x is contained in the set A
$x \notin A$	x is not contained in the set A
$A \cap B$	The intersection of sets A and B
$A \cup B$	The union of sets A and B
$A \subset B$	Set A is a subset of set B
$A \subseteq B$	Set A is a subset of set B or identical to B
$A \setminus B$	Set A after subtraction of subset B, i.e., $x \in A$ such that $x \notin B$
$A \times B$	Cartesian product of sets A and B: $\{(a, b)$ such that $a \in A$ and $b \in B\}$
$\bigcup_{i=1}^{N} A_i$	Union of sets A_i for $i = 0$ to N

Number systems and operations

$\mathbb{N}$	Natural numbers $(0, 1, 2, \ldots)$
$\mathbb{Z}$	Integers $(\ldots -2, -1, 0, 1, 2, \ldots)$
$\mathbb{Q}$	Rational numbers
$\mathbb{R}$	Real numbers
$\sum_{i=0}^{N} a_i$	Summation of a_i for $i = 0$ to N
$\forall$	"for all"

Vectors, matrices, and tensors

$\mathbf{v}$	Vector $\mathbf{v}$
$\mathbf{v} \parallel \mathbf{w}$	$\mathbf{v}$ is parallel to $\mathbf{w}$
$\mathbf{v} \perp \mathbf{w}$	$\mathbf{v}$ is perpendicular to $\mathbf{w}$
$|\mathbf{v}|$	Length of $\mathbf{v}$
$\mathbf{I}$	Identity matrix
$\mathbf{A}^T$	Transpose of a matrix (tensor) $\mathbf{A}$
$\mathbf{A}^{-1}$	Inverse of a matrix (tensor) $\mathbf{A}$

Topology

$[a, b]$	Closed interval of real numbers x satisfying $a \leq x \leq b$
(a, b)	Open interval of real numbers x satisfying $a < x < b$
$B_r(\mathbf{x})$	Open ball of radius r centered at $\mathbf{x}$, i.e., $\{\mathbf{v}$ such that $|\mathbf{v} - \mathbf{x}| < r\}$
$\partial\Omega$	Boundary of a domain Ω

Differentiation and integration

∇f	Gradient of f
$\nabla \cdot \mathbf{v}$	Divergence of $\mathbf{v}$
$\nabla^2 f$	Laplacian ($\nabla \cdot \nabla$) of f
$\partial f / \partial x_i$	Partial derivative of $f(x_1, x_2, \cdots)$ with respect to (w.r.t.) x_i
Df/Dt $= \partial f/\partial t + \mathbf{u} \cdot \nabla f$	Advective derivative of $f(\mathbf{x}, t)$ in a velocity stream $\mathbf{u}$
$\dfrac{\delta f}{\delta \varphi}$	Variational derivative of a functional f w.r.t. a variable φ
$\int_a^b f(x)dx$	Integral of f on $a \leq x \leq b$
$\oint_\gamma f(x)dx$	Line integral of f over the curve γ

Special functions

$\mathbf{1}_A(\mathbf{x})$	Characteristic function satisfying $\mathbf{1}_A(\mathbf{x}) = \begin{cases} 1 & \text{if } \mathbf{x} \in A, \\ 0 & \text{if } \mathbf{x} \notin A \end{cases}$
$(x)_+$	Positive part of x satisfying $(x)_+ = \max(x, 0)$
$\mathcal{H}(x)$	Heaviside step function satisfying $\mathcal{H}(x) = \begin{cases} 0 & \text{if } x < 0, \\ 1 & \text{if } x \geq 0 \end{cases}$
$\delta(x)$	Dirac delta function
δ_{ij}	Kronecker delta function satisfying $\delta_{ij} = \begin{cases} 1 & \text{if } i = j, \\ 0 & \text{if } i \neq j \end{cases}$
$[x]$	Nearest integer to a real number x

Probability and statistics

$\Pr(X)$	Probability of the event X
$\Pr(X|Y)$	Conditional probability of the event X given the event Y
$\langle x \rangle$	Mean of the measurable quantity x
$\text{Ex}[X]$	Expected value of the random variable X
$\text{Var}[X]$	Variance (standard deviation squared) of the random variable X

Part I

Theory

1 Introduction

In this book we describe recent efforts to model tumor growth and invasion using an interdisciplinary approach that integrates mathematical and computational models of cancer with laboratory experiments and clinical data. The aim of these efforts has been to provide insight into the root causes of solid tumor invasion and metastasis, to aid in the understanding of experimental and clinical observations, and to help design new, targeted, experiments and treatment strategies. The ultimate goal is for modeling and simulation to aid in the development of individualized therapy protocols that minimize patient suffering while maximizing treatment effectiveness. In order to achieve this objective, mathematical and computational models are needed that quantify the links of three-dimensional tumor-tissue architecture with the growth, invasion, and underlying micro-scale cellular and environmental characteristics. This approach requires a multiscale modeling framework that is capable of linking the molecular and cell scales directly to the patient data.

There are many ways to evaluate the progression of these efforts. Here we follow the major stages that progressively incorporate the complexity of the tumor environment: (i) modeling of avascular tumors *in vitro* and *in silico* to assess stages of tumor growth; (ii) interactions between a tumor and its *in vivo* microenvironment; (iii) modeling of vascularized tumors *in silico* to assess angiogenesis and vascular growth; (iv) modeling of vascularized tumors *in vivo* and *in silico* to assess tumor progression in the body. We describe in detail biologically founded approaches that employ multiscale mathematical models of tumor progression in two and three dimensions. This enables the study of how molecular-scale phenomena regulating cell-proliferation, migration, and adhesion forces, including those associated with the genetic evolution from lower- to higher-grade tumors, may generate, in a predictable and quantifiable way, heterogeneous cell activity and oxygen and nutrient demand across the tumor mass, thus determining its morphology and degree of invasiveness.

Through the years numerous mathematical models have been developed to study the progression of cancer (e.g., see the reviews by Adam [5], Chaplain [120], Bellomo and Preziosi [61], Moreira and Deutsch [475], Bellomo *et al.* [59], Swanson *et al.* [649], Araujo and McElwain [34], Mantzaris *et al.* [452], Friedman [236], Ribba *et al.* [566], Quaranta *et al.* [553], Hatzikirou *et al.* [307], Nagy [490], Byrne *et al.* [96], Fasano *et al.* [209], van Leeuwen *et al.* [668], Roose *et al.* [573], Graziano and Preziosi [291], Friedman *et al.* [237], Sanga *et al.* [584], Deisboeck *et al.* [169], Anderson and Quaranta [29], Bellomo *et al.* [60], Cristini *et al.* [146], and Lowengrub *et al.*

[425]). Most models fall into two broad categories based on how the tumor tissue is represented: discrete cell-based models and continuum models. Although the continuum and discrete approaches have each provided important insight into cancer-related processes occurring at particular length and time scales, the complexity of cancer and the interactions between the cell- and tissue-level scales may be elucidated further by means of a multiscale (hybrid) approach that uses both continuum and discrete representations of tumor cells and components of the tumor microenvironment. Thereby biological phenomena from the molecular and cellular scales are coupled to the tumor scale (e.g., the recent work by Kim *et al.* [382], Bearer *et al.* [57] and Lowengrub *et al.* [425]). Such an approach can, for example, capture transitions from collective to individual behavior and combine the best features of continuum and discrete models.

Continuum tumor models are based on reaction–diffusion equations describing the tumor cell density (e.g., [22, 532, 680]), the extracellular matrix (ECM), matrix degrading enzymes (MDEs) (e.g., [82, 83, 123, 316]), and concentrations of cell substrates such as glucose, oxygen, and growth factors and inhibitors (e.g., [73, 118, 121, 303, 315, 442, 495]). Classical work [292, 293] used ordinary differential equations to model tumors as a homogeneous population, as well as partial differential equation models confined to a spherical geometry. In the case of avascular tumors, growth has been modeled as a function of cell substrate concentration, usually oxygen. More recent work has incorporated cell movement through diffusion (e.g., [121, 607, 608]), convection (e.g., [101, 157, 532, 679, 682]), and chemotaxis or haptotaxis (e.g., [453, 532, 607]). Cell proliferation, death, and pressure have also been considered (e.g., [21, 40, 59, 78–80, 92–95, 100, 101, 103, 104, 149, 164, 220, 222, 258, 262, 318, 335–338, 352, 399, 413, 429, 532, 536, 574, 589, 632, 636, 657, 673, 681]). Linear and weakly nonlinear analyses have been performed to assess the stability of spherical tumors to asymmetric perturbations (e.g., [34, 95–97, 99, 100, 104, 121, 149, 273, 416]) in order to characterize the degree of aggression. Various interactions with the microenvironment, such as nutrient- or stress-induced limitations to growth, have also been studied (e.g., [18, 20, 22, 35, 37, 38, 352, 574]). The models may account for observations of the stronger cell–cell interactions (cell–cell adhesion and communications), the high polarity, and the strong pulling forces exchanged by cells and the ECM [127–129, 235]. Extracellular matrix reorganization by tumor cells [235] has been incorporated, and various degrees of dependence of the cells on signals from the matrix have been modeled. The models are typically single-species (e.g., single-phase tumors), treating the tumor or, more generally, biological tissues as fluid (e.g., [80, 81, 97, 99, 100, 103, 119, 241, 293]), elastic or hyperelastic (e.g., [17, 20, 35, 248, 249, 257, 328, 352, 462, 604, 673]), poroelastic (e.g., [574]), viscoelastic (e.g., [46, 382]), or elasto-viscoplastic (e.g., [23]). More recently, multiphase models (e.g. using mixture theory) have been developed to simulate multiple solid-cell species and extra- or intracellular liquids (see below). Theoretical nonlinear analyses of the various single-phase tumor models mentioned above have also been performed (e.g., [55, 86, 150–160, 176, 209, 236, 238–240, 242, 243, 490, 520, 608, 652, 653, 675, 701, 706, 707, 723, 724]).

Building upon the formulation of classical models [100, 101, 293, 460], a break-through nonlinear simulation of a continuum tumor model was provided by Cristini *et al.* in 2003 [149], who studied solid tumor growth in the nonlinear regime using boundary-integral simulations in two dimensions to explore complex morphologies. This work demonstrated that non-necrotic tumor evolution could be described by a reduced set of two parameters that characterize families of solutions. One parameter describes the relative rate of mitosis to the relaxation mechanisms (cell mobility and cell–cell adhesion). The other describes the balance between apoptosis (programmed cell death) and mitosis. Both parameters also include the effect of vascularization. The results revealed that tumor growth can be divided into three regimes, associated with increasing degrees of vascularization: low (diffusion-dominated), moderate, and high. Critical conditions exist for which the tumor evolves to nontrivial dormant states or grows self-similarly (i.e., in a shape invariant way). Away from these critical conditions evolution may be unstable, leading to invasive fingering into the external tissues and to topological transitions such as tumor breakup and reconnection. This work identified for the first time the concept of tumor "diffusional instability" in the low vascularization regime as a mechanism for invasion. While previous work [97, 99, 100] had demonstrated that steady-state avascular symmetric tumors could be unstable, the results of Cristini *et al.* [149] showed that instability during growth can allow tumors to grow indefinitely, bypassing the symmetric steady state. This idea has since been studied in a variety of different configurations using a number of different models. Interestingly, the shape of highly vascularized tumors was predicted to remain compact and without invasive fingering, even while growing unboundedly. This suggests that the invasive growth of highly vascularized tumors is associated with vascular anisotropy and other inhomogeneties in the microenvironment (e.g. cell-substrate inhomogeneities, elastic anisotropy). The self-similar behavior described above leads to the possibility of controlling tumor shape control and of the release of tumor angiogenic factors by restricting the tumor volume-to-surface-area ratio.

Expanding on the idea of "diffusional instability," subsequent work [116, 145, 230, 436, 437, 620, 722] has developed the hypothesis that, through heterogeneous cell proliferation and migration, microenvironmental substrate gradients, e.g., of cell nutrients or the extracellular matrix, may drive tumor invasion through morphological instability with separation of cell clusters from the tumor edge and infiltration into surrounding normal tissue. Tumor morphology would be determined by the competition between heterogeneous cell proliferation caused by spatial diffusion gradients, driving shape instability, and invasive tumor morphologies, and stabilizing mechanical forces, e.g., cell–cell and cell–matrix adhesion. Following these ideas, the stability of avascular tumor growth has also been investigated in discrete models [25, 273, 538].

To investigate further the stability of avascular tumors, parameter-based statistics providing input to the mathematical model were obtained from *in vitro* glioblastoma tumors [230]. Employing a linear stability analysis of the model from Cristini *et al.* [149], these results predicted that tumor spheroid morphology would be marginally stable. In agreement with this prediction, unbounded growth of the tumor mass and invasion of the environment were observed *in vitro*. The mechanism of tumor invasion was characterized as recursive sub-spheroid component development (i.e., the formation of "buds") at the tumor viable rim and separation from the parent spheroid. Computer simulations of the mathematical model closely resembled the morphologies and spatial arrangement of tumor cells from the *in vitro* model. Simulations and *in vitro* experiments provided further evidence [145] that morphological instability could be suppressed *in vivo* by spatially homogeneous oxygen and nutrient supply because normoxic conditions act both by decreasing gradients and increasing cell adhesion and, therefore, the mechanical forces that maintain a well-defined tumor boundary. Taking into account the effect of the microenvironment, it was also found that tumor morphological stability could be enhanced by improving the nutrient supply [437].

Recently, multiphase (mixture) models have been developed that are capable of describing detailed interactions between multiple solid cell species and extra- or intra-cellular liquids (see Chapter 5 for references). Vascular tumor growth has been studied in three dimensions [57, 229], and the simulation results compare well with clinical tumor data. In particular, in [57] a biologically founded multiphase model was applied to identify and quantify tumor biologic and molecular properties relating to clinical and morphological phenotype, and to demonstrate that tumor growth and invasion are predictable processes governed by biophysical laws and regulated by heterogeneity in phenotypic, genotypic, and microenvironmental parameters. This heterogeneity drives the migration and proliferation of more aggressive clones up the cell substrate gradients within and beyond the central tumor mass, while often also inducing loss of cell adhesion. The models predict that this process triggers a gross morphologic instability that leads to tumor invasion via individual cells, cell chains, strands, or detached clusters infiltrating into adjacent tissue and producing the typical morphologic patterns seen, e.g., in the histopathology of brain cancers such as glioblastoma multiforme. The model further predicts that the different morphologies of infiltration correspond to different stages of tumor progression regulated by heterogeneity.

This mathematical and computer modeling provides evidence that tumor morphogenesis *in vivo* may be a function of marginally stable environmental conditions caused by spatial variations in cell nutrients, oxygen, and growth factors. A properly working tumor microvasculature could help maintain compact non-infiltrating tumor morphologies by means of minimizing the oxygen and nutrient gradients. Controlling the environmental conditions by decreasing spatial gradients may benefit treatment outcomes whereas current treatments, and especially anti-angiogenic therapy, may trigger microenvironmental heterogeneity (e.g., local hypoxia), thus causing invasive instability. Indeed, the mathematical models show that the parameters that control the tumor mass shape also control its ability to invade [149, 230]. Thus, tumor morphology may serve as a predictor of invasiveness and treatment prognosis.

The theoretical models also provide an explanation for the highly variable outcome of anti-angiogenic therapy in multiple clinical trials [145]. Anti-angiogenic therapy may promote morphological instability, leading to invasive patterns even under conditions in which the overall tumor mass shrinks. Thus, therapeutic strategies focused solely on the reduction of vascular density may paradoxically increase invasive behavior. Anti-angiogenic strategies may be more consistently successful when aimed at "normalizing" the vasculature [342] and when combined with therapies that increase cell adhesion [528], so that morphological instability is suppressed and compact non-invasive tumor morphologies are enforced. This could be done, for example, through anti-invasive therapy [468, 476]. Further, by quantifying the link between the tumor boundary morphology and the invasive phenotype, multiscale mathematical modeling provides a quantitative tool for the study of tumor progression and diagnostic and prognostic applications. This establishes a framework for monitoring system perturbation towards the development of therapeutic strategies and obtaining correlations to clinical outcome for prognosis.

The outline of this book is as follows. In Part I, we focus on the theory and the numerics. In Chapter 2, we review basic cancer biology as a background to the modeling. In Chapter 3, we present continuum modeling and the incorporation of biologically relevant parameter values into multiscale models of tumor growth and invasion. We describe a basic model founded on classical work and then expand it to include vascularization. In Chapter 4, we evaluate the theory of stability, including different regimes of growth and linear analyses. We then consider in Chapter 5 multiphase modeling to simulate multiple cell species and include the effects from tumor cell chemotaxis as well as tumor-induced vascularization in three dimensions. Chapter 6 presents the modeling of discrete cells by evaluating an agent-based cell model with applications to cancer. Chapter 7 explores the modeling of tumor invasion by using a hybrid continuum–discrete multiscale framework. Chapter 8 introduces the numerical schemes used to solve the model equations.

In Part II, we look at specific case studies. Chapter 9 describes the multidisciplinary modeling of tumor growth and invasion through the incorporation of experimental and clinical observations into the parameters of the tumor model. Chapter 10 considers the application of agent-based modeling to breast cancer. The Conclusion following Chapter 10 introduces patient-specific modeling with a prototype of the multiscale modeling framework by calibrating the molecular or cell scale directly to patient data and upscaling to calibrate the continuum scale.

2 Biological background[1]

With P. Macklin

In this chapter, we present some of the key biological concepts necessary to motivate, develop, and understand the tumor models introduced in this book. We introduce the molecular and cellular biology of noncancerous tissue (Section 2.1) and then discuss how this biology is altered during cancer progression (Section 2.2). The discussion may be more detailed in some areas than is necessary for the models that we present; the intention is to offer a sample of the rich world of molecular and cellular biology, helping the reader to consider how these and other details may need to be incorporated in the work of cancer modeling. For greater depth on any of the topics, refer to such excellent texts as [12] for molecular and cellular biology and [384] for cancer cell biology.

2.1 Key molecular and cellular biology

We focus upon the molecular and cellular biology of epithelial cells, the stroma, and the mesenchymal cells that create and maintain the stroma (Section 2.1.1). Specific and often anisotropic adhesive forces help to maintain tissue architecture (Section 2.1.2). Epithelial and stromal cells have the same basic subcellular structure (Section 2.1.3) and share much in common. They progress through a cell cycle when preparing to divide, can control their entry into and exit from the cycle, and can self-terminate (apoptose) when they detect irreparable DNA errors or other damage (Section 2.1.4). Their behavior is governed by a signaling network that integrates genetic and proteomic information with extracellular signals received through membrane-bound receptors (Section 2.1.5). Sometimes cells respond to signaling events by moving within the stroma or along the basement membrane (Section 2.1.6). In pathologic conditions leading to hypoxia, cells can respond through a variety of mechanisms or can succumb to necrosis; in some cases, the necrotic cellular debris becomes calcified (Section 2.1.7).

2.1.1 Tissue microarchitecture and maintenance

The *epithelium* is composed of sheets of tightly adhered epithelial cells that cover organ surfaces and often perform specialized functions. The epithelium is supported by the *stroma*, a loose connective tissue. The main component of the stroma is the *extracellular*

---

[1] This introduction to cancer biology updates and expands the original exposition in [431].

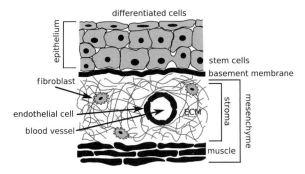

Figure 2.1 Typical tissue structure showing epithelium separated from the stroma by a basement membrane.

matrix (ECM), a scaffolding of fibers (collagen, elastin, fibronectin, etc.) embedded in a mixture of water and glycoproteins. The ECM is secreted and maintained by *stromal cells*, specialized mesenchymal cells that can freely move within the stroma as they maintain the tissue; fibroblasts are the primary stromal cells in loose connective tissue (epithelial stroma). The stroma is interlaced by blood vessels, nerves, and lymphatic vessels, and it may rest on an additional layer of muscle or bone, depending upon the location. A thin, semi-permeable *basement membrane* (BM, a specialized type of ECM) separates the epithelium from the stroma. See Figure 2.1.

This complex tissue structure is maintained by careful regulation of the cell population and a specific balance of adhesive forces. These processes are often tied together through cell signaling. For further information on tissue and organ structure, see [219] and [12] and the references therein.

Population dynamics

Each cell type population must be regulated by balancing proliferation and apoptosis. When a differentiated cell dies, a *somatic stem cell* may divide either symmetrically into two new stem cells or asymmetrically into a stem cell and a *progenitor cell*. The progenitor cell either further divides or terminally differentiates into a specific cell type and then migrates or is pushed to the correct position and assumes its function. This process is tightly regulated by intercellular communication via biochemical signals (growth factors) and mechanics; stromal cells help to maintain this signaling environment [423, 493, 725]. Each cell's response to the microenvironment is governed by surface receptors that interact with an internal signaling network. We note that stem-cell dynamics are not fully understood; see the excellent overviews in [71, 725].

Epithelial cell polarity and adhesion

Epithelium can be broadly classified as *simple* or *stratified* on the basis of its cell arrangement. In simple epithelium, cells are arranged in a single layer along the basement membrane. The cells are *polarized*, with a well-defined base adhering to the BM and an apex exposed to the *lumen* (e.g., a cavity in an organ); the apical side of the cell is often used to release secretory products. The epithelial cells adhere tightly to one another along their nonapical, nonbasal, sides. See Figure 2.2, left. In stratified epithelium, a

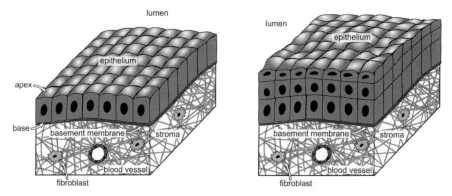

Figure 2.2 Simple (left) and stratified (right) cuboidal epithelium.

single cell layer adheres to the BM (as in simple epithelium), with additional layers above. The cells in the upper layers adhere to the layers above and below them and tend to be flattened. See Figure 2.2, right. Overall, the careful orchestration of cell–BM and cell–cell adhesion helps determine the tissue geometry [301, 350, 685]. In fact, heterogeneities in the balance of cell–cell and cell–BM adhesion can lead to epithelium invagination [400], folding [672], and other nontrivial geometries [633]. The molecular mechanisms of adhesion are further explored in Section 2.1.2. More information on epithelial cell polarization can be found in standard biology texts, such as [12].

Interaction between cell adhesion and population dynamics

Cell adhesion and population dynamics are, in fact, linked to one another. Epithelial cell cycle progression and proliferation are controlled in part by cell–cell adhesion: when an epithelial cell is in (adhesive) contact with many neighbors, its cell cycle and proliferation are suppressed. This helps to maintain the epithelial cell population by reducing proliferation when the epithelium is fully populated, and by increasing proliferation near gaps in the epithelium (e.g., due to apoptosis) [142, 301, 685]. Hence, cell–cell contact-dependent proliferation helps to prevent overproliferation. This theme is discussed further in Section 2.1.5.

Cell populations are also controlled by contact with the extracellular matrix and basement membrane. Polarized epithelial cells often become apoptotic after losing adhesive contact with the BM [245, 276, 330, 640, 677]; this specialized type of apoptosis, termed *anoikis*, helps prevent overproliferation of unattached cells into the lumen [162]. The ECM also plays a major role in regulating stromal cells [276]. For example, ECM-bound proteoglycans control the proliferation, differentiation, and apoptosis of bone marrow stromal cells [68] and integrin ligands in the ECM regulate endometrial stromal cells [598].

2.1.2   Cellular adhesion and cell sorting

Adhesion is essential to multicellular arrangement and motility: cell–cell, cell–ECM, and cell–BM adhesion are responsible for maintaining the tissue arrangement, while cell–BM and cell–ECM adhesion are essential for traction during motility.

Adhesion

Cells can exhibit both homophilic and heterophilic adhesion. In *homophilic* adhesion, adhesion receptor molecules on the cell surface bond to identical ligands (a receptor's "target" molecules) either on neighboring cells (in cell–cell adhesion) or in the microenvironment (in cell–ECM or cell–BM adhesion). This is the mode of E-cadherin-mediated cell–cell adhesion in epithelial cells, including carcinoma [521]. In *heterophilic* adhesion, surface adhesion molecules of one type bond to unlike ligand molecules in the extracellular matrix, on the basement membrane, or on neighboring cells. Cell–ECM and cell–BM adhesion are heterophilic between integrin molecules on the cell surface and ligands such as laminin and fibronectin in the microenvironment [90]. Heterophilic cell–cell adhesion is also observed, for example in T-cell lymphocytes via immunoglobulin–integrin bonds [427, 630, 654].

Cell adhesion and cell sorting

While epithelial cell–cell adhesion is generally homophilic and mediated by E-cadherin, other cadherins complicate the picture. For example, E-cadherin binds with the greatest strength and specificity to E-cadherin, but it can also bind to N-cadherin [521] and certain integrins [361]. Hence the mixture of adhesion molecules on the two cells' surfaces (and the specificity and kinetics of the bonds between the molecules) will determine the strength of their adhesion. Adhesive differences between cell types can lead to self-sorting behavior based upon adhesion gradients, which contributes to epithelial cell organization in tissues [551]. Such cell sorting has been observed experimentally [41].

2.1.3   Subcellular structure

A cell is composed of a well-defined nucleus containing the cell's DNA, surrounded by cytosol (the liquid in the cell) and enveloped in a bilipid cell membrane. The cytoplasm contains organelles that carry out the cell's functions, such as the mitochondria (which synthesize adenosine triphosphate (ATP) from glucose and oxygen to provide energy to the cell) and endoplasmic reticulum (which provides ideal conditions for protein synthesis, folding, and transport), all supported by a *cytoskeleton* of microtubules and actin polymer fibers. See Figure 2.3. The bilipid membrane separates the cell from the microenvironment. It is permeable to the passive diffusion of small molecular species such as oxygen and glucose, actively pumps other molecular species (e.g., potassium and sodium) to maintain the cell's internal pH and chemical composition, and is impermeable to other, larger, molecules such as growth factors. Embedded in the membrane are a variety of macromolecules that pump smaller molecules (e.g., potassium) against gradients, exchange mechanical forces with the extracellular matrix, basement membrane, and other cells, and transmit microenvironmental information to the cell interior.

2.1.4   Cell cycle, proliferation, and apoptosis

Cell division is regulated by a highly regimented series of stages known as the *cell cycle*. In the first stage in the cell cycle, G1 (gap 1), the cell physically grows, proteins are

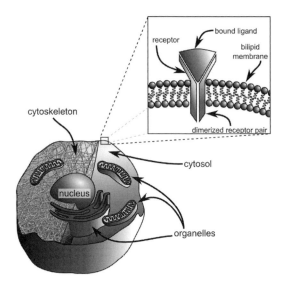

Figure 2.3 A eukaryotic cell: a bilipid cell membrane contains the cytosol, nucleus, and organelles, all supported by a cytoskeleton. Inset: Membrane-embedded receptors transmit microenvironmental information to the cell interior.

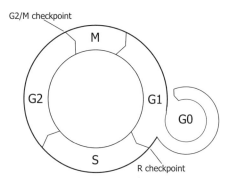

Figure 2.4 The cell cycle.

synthesized, new organelles are constructed, and the cell prepares for DNA replication. In the following S (synthesis) phase, the DNA is copied and, in the G2 (gap 2) phase, final preparations are made within the cell nucleus for the division of the cell. In the final M (mitosis) phase, the two copies of the DNA are separated and incorporated into two nuclei (*mitosis*), and the cytoplasm and the organelles are divided into two daughter cells (*cytokinesis*). See Figure 2.4.

The cell cycle contains numerous checkpoints that allow the cell to check for and repair DNA damage, as well as to control or halt cycle progression. At the R (restriction) checkpoint, late in the G1 phase, the cell either commits to division (and progresses to the S phase) or exits the cell cycle (and enters the G0 *quiescent state*) [70, 717]. Most noncancerous somatic cells stay in this "resting" state because microenvironmental signals maintaining homeostasis have been received prior to the R checkpoint; after the

R checkpoint, cells are committed to division and are less responsive to environmental signals to halt the cycle [606].

There are numerous checkpoints in the S and G2 phases to detect and repair DNA damage (e.g., between G2 and M). See Figure 2.4. Cells that fail to repair DNA damage at such checkpoints induce apoptosis [138]. In the process, "executioner" proteins (Caspases) in the cytoplasm break down the organelles, degrade the cytoskeleton, and fragment the DNA. The cell shrinks, and the degraded cell contents are released as harmless (i.e., chemically inert) vesicles known as *apoptotic bodies*, which are ingested (phagocytosed) by specialized immune cells as well as neighboring epithelial cells [369, 389].

A cell's speed of cell cycle progression is regulated by the production and balance of internal chemical signals, principally *cyclins* and *cyclin-dependent kinases* (CDKs). Surface receptors help control gene expression levels through complex signaling pathways. In turn the gene expression pattern determines the production and balance of proteins (including cyclins and CDKs). Hence, cell cycle progression is regulated by a complex interaction between the cell's internal biomachinery and its surrounding environment [138].

2.1.5   Genetics, gene expression, and cell signaling

Oncogenes and tumor suppressor genes

The correct interpretation of growth and inhibitory signals is key to maintaining healthy tissues. If the cell receives both growth-promoting and growth-inhibiting signals, its behavior is determined by the balance of the signals and the resulting gene expression pattern. Two types of genes are particularly relevant to regulating cell proliferation. *Oncogenes* respond to or create growth signals and promote cell cycle progression. *Tumor suppressor genes* (TSGs) respond to inhibitory signals, retard or halt the cell cycle, ensure proper DNA repair, and may trigger apoptosis under certain circumstances. Cancer initiation, or *carcinogenesis*, starts with the malfunction of one or more of these types of genes [300].

Genetic mutations can cause overactivity in oncogenes and impair the function of tumor suppressor genes. Sometimes, a single uncorrected point mutation is sufficient to affect the function of an oncogene [449] or functionally neutralize a tumor suppressor gene [323]. In other cases, cell division errors (e.g., during the M phase) can create a mutant fusion oncogene, where the protein coding portion of an oncogene is mistakenly fused with the triggering portion of another, frequently expressed, gene. As a result, signals are "misrouted" to the oncogene, thus boosting its activity. See [394], which describes the activation of the *MYC* oncogene by translocation with an immunoglobulin gene.

Other errors during cell division may cause a daughter cell mistakenly to receive extra copies of an oncogene (e.g., [109, 117, 216]) or too few copies of a TSG. Because normal cells possess two copies of each TSG, both copies must be damaged for a total loss of function of the gene. (See the Knudson two-hit model [386, 387], which led to the first discovered TSG [244].) While the probability of independent mutations in both copies of the TSG is ordinarily small, a *loss of heterozygosity* (two damaged copies of the TSG are passed to a daughter cell) can significantly accelerate the process [500]. Furthermore, the loss of just one TSG copy can significantly impair the TSG's activity and increase the probability of completion of a multistep carcinogenesis pathway [555].

Changes in gene expression

Gene expression is essential to maintaining proper cell function. Recent research has examined the over- and underexpression of genes, rather than outright genetic damage, as a potential contributor to unchecked cell proliferation. Viral infections (e.g., human papillomavirus, which can induce cervical cancer [692]) and microenvironmental signals (e.g., hypoxia; see the first example below and references therein) can also induce changes in gene expression. Because gene expression patterns can be heritable, such changes can potentially affect a cell's malignant transformation (e.g., by disabling a tumor suppressor gene) in the same way as a genetic mutation [354]. Lastly, we note that the biochemistry of gene expression is very complicated and is beyond the scope of this introduction; see [190, 354, 355, 423] for more on this topic.

Cell signaling networks

Gene expression is controlled by cell surface receptors after activation by various signaling factors. Internal chemical species (e.g., oxygen) can also affect gene expression. The cell integrates such information with its genetic and proteomic state using a complex signaling network to determine its phenotype. Aberrant cell signaling is often implicated in cancer, making it a key topic in molecular and cellular cancer biology. We illustrate with a few examples.

Example: HIF-1α signaling
A cell's response to hypoxia (low oxygen levels) is a key example of how internal protein levels can affect gene expression without the need for additional receptor signaling. All cells create HIF-1α (a hypoxia-inducible factor), which is ordinarily degraded in the presence of oxygen [85, 246, 287, 599]. When a cell experiences hypoxia, HIF-1α accumulates and activates downstream "target" genes. Among targets of importance to cancer biology, HIF-1α upregulates motility and the secretion of angiogenic-promoting factors and anaerobic glycolysis (an inefficient metabolism by which glucose reacts with glucose rather than oxygen) downregulates cell–cell and cell–ECM adhesion and reduces sensitivity to apoptotic signals [13, 303, 540, 713]. We discuss the significance of this signaling pathway in cancer biology in Sections 2.1.7 and 2.2.2.

Example: EGF signaling
Epidermal growth factor (EGF) can bind to and subsequently activate EGF receptors (EGFRs). When two activated EGFRs bind to one another (dimerize), they can transmit signals leading to increased HIF-1α secretion, increased cell proliferation, increased cell motility, or reduced sensitivity to apoptosis. See Figure 2.5 and the excellent reviews in [132, 312, 502]. Malfunctions in this signaling process have been implicated in several cancers. For example, a mutant form of EGFR (HER2) commonly found in breast cancer

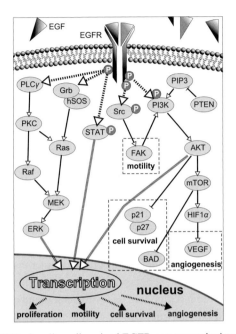

Figure 2.5 Simplified EGFR signaling: dimerized EGFR can transmit signals through a variety of molecular pathways that trigger proliferation, motility, or increased resistance to apoptosis.

is constitutively (i.e., permanently) active and does not require EGF binding for signaling activity; moreover, HER2 can bind to activated EGFR to provide a "shortcut" in the EGFR signaling cascade and thus increase EGFR signaling activity [132, 191]. In non-small-cell lung carcinoma (NSCLC), downstream targets of EGFR are often mutated, most notably a constitutively active form of K-ras that can function independently of upstream EGFR signals. Indeed, NSCLCs with K-ras mutations are generally resistant to therapies that target EGFR [192, 522]. Both these mutations effectively activate downstream targets of EGFR independently of receptor activity; i.e., the EGFR pathway switch is "stuck in the ON position," leading to excessive proliferation and other cancer-promoting activity.

Example: E-cadherin/β-catenin signaling

Some receptors have multiple simultaneous roles. E-cadherin mediates homophilic epithelial cell–cell adhesion (Section 2.1.2): the intracellular domain of E-cadherin binds to α-catenin (using β-catenin as an adapter protein) to couple mechanically an adhered cell to its actin cytoskeleton [188, 385]. Ligated E-cadherin also binds to β-catenin, which sequesters it at the cell membrane and prevents its downstream signaling. Unsequestered β-catenin would otherwise promote cell cycle progression by triggering the transcription of Cyclin D1, c-myc, and Axin2. Hence, E-cadherin not only plays a mechanical role in cell–cell interactions but also a signaling role, by inhibiting cell-cycle progression when it physically adheres to epithelial cells [69, 313, 428, 597]. This signaling pathway plays a key role in maintaining normal epithelial tissue microarchitecture

[142, 301, 685]; see Section 2.1.1. In many cancers (e.g., breast cancer [417]), the E-cadherin/β-catenin signaling pathway can be disrupted, leading to increased downstream oncogenic activity (e.g., increased cell cycle progression owing to Cyclin D1 overexpression [461]).

2.1.6   Cell motility

Motile cells demonstrate directed motion due to a complex interaction between cell signaling, their cytoskeleton, and adhesion with the ECM or BM. We describe here the key aspects of this process; more detail can be found in [143, 278, 403].

Gradients in microenvironmental signaling molecules (e.g., EGF) can be amplified by the multiple steps in signaling networks, leading to pronounced internal signaling gradients [377]. A key downstream effect of motility signaling is actin polymerization (the formation of linked chains of actin monomer that extend the actin cytoskeleton) and depolymerization (the spontaneous degradation of actin polymers). This process takes place within a thin region just below the cell membrane [357, 403]. Wherever polymerization exceeds depolymerization, there is net outward growth of the cell's cytoskeleton, which, in turn, deforms and extends the cell membrane. If net actin polymerization continues in a consistent direction, the cell forms a pseudopod (i.e., a "false foot") that extends from its leading edge into the microenvironment. Net actin depolymerization at the cell's trailing edge, along with internal microtubule activity, leads to cell contraction [143, 278, 403]. The signaling network creates and maintains this bias in actin polymerization. For example, dimerized EGFR can activate Src, which, in turn, can mediate the formation of Arp2/3-N-WASP complexes that nucleate actin polymerization; microenvironmental EGF gradients thus create internal polymerization gradients towards the cell's leading edge [143, 478, 593, 688, 689].

Cell motility requires a mechanical interaction between cell-membrane protrusions and the microenvironment. Individual cells may move through the stroma (in three dimensions) in an amoeboid motion by squeezing between ECM fibers (e.g., T-lymphocyte migration [700]) or by extending a slender, finger-like, pseudopod (a *filopodium*), which forms focal adhesions with the ECM to exert traction [403]. The latter, which occurs during cancer cell invasion of the stroma [234, 699], requires directed, coordinated, degradation of the ECM to create space for motion, and this is accomplished by the formation of tiny invadopodia on the filopodium surface that secrete proteases to degrade the ECM [173, 365, 684]. In other cases, cells may move along a surface by extending a sheet-like pseudopod (a *lamellipodium*) that focally adheres to the surface for traction [403]. This has been observed in Paget's disease of the breast (cancerous epithelial cells chemotax along the breast duct basement membrane towards the nipple [76]), in wound healing (keratinocytes crawl along the top of granular tissue [401]), and in fibrosarcoma metastasis (cancer cells crawl along lymph vessel walls [709]). Following membrane protrusion, nonamoeboid motility requires the release of integrin bonds along the cell's trailing edge and subsequent cell contraction, allowing net forward motion [403]. Directed cell motility also requires active intracellular transport of actin monomer [686], integrins [278], and other cytoskeletal components between the cell's trailing and leading edges [403].

2.1.7   Hypoxia, necrosis, and calcification

In Section 2.1.5, we discussed some of the cellular adaptations to hypoxia. Sustained hypoxia (as well as sustained hypoglycemia), such as that encountered in ischemic tissue [251, 383, 580] and in larger tumors [112, 213], can lead to ATP depletion and consequently cell death. This unplanned cell death is referred to as *necrosis*.

When a cell becomes necrotic, its surface ion pumps cease to function, resulting in osmosis of water into the cell, cell swelling, and subsequent bursting [49]. This differs from apoptosis, where the volume loss is orderly and the intracellular contents are contained in apoptotic bodies [49]. In necrotic cells, the remaining solid-cell fraction is generally not phagocytosed by the surrounding cells, as typically they themselves are also necrotic. In some cancers (e.g., breast cancer [638], liver cancer [308], ovarian cancer [628], and lymphoma [135, 334]) and other pathologic conditions (e.g., tuberculosis [52] and abscesses [396, 698]), necrotic tissue can undergo calcification: the solid cell components are replaced by calcium phosphate and/or calcium oxalate molecules that bond together to form calcite crystals that grow into hard *microcalcifications* [444].

2.2   The biology of cancer

Most simply stated, cancer occurs when defective genes cause cells to malfunction and interact with the body in an aberrant, hyperproliferative, manner (either by increased cell proliferation or reduced cell apoptosis). We now examine how the molecular and cellular biology previously introduced in Section 2.1 can break down, leading to cancer. Our discussion primarily focuses upon *carcinoma* (cancers arising from epithelial cells) rather than *sarcoma* (cancers arising from mesenchymal cells).

2.2.1   Carcinogenesis

Carcinogenesis is a multistage process, thought to begin with a genetic mutation or epigenetic alteration that overexpresses an oncogene or underexpresses a tumor supressor gene in one or a small number of cells. If the cell survives and the mutation escapes its DNA repair mechanisms, the cell (or its descendants) may, over time, acquire further mutations that enable it to ignore growth-inhibiting signals from its neighbors, bypass its internal controls and checkpoints, and form a colony of hyperproliferative aberrant cells. This accumulation of mutations may require years to progress but can be accelerated by exposure to carcinogens and other harsh, DNA-damaging, environmental effects.

Differentiated cells can only divide a limited number of times before reaching *senescence*: the point at which they permanently arrest in G0 or apoptose. Thus, differentiated cells alone cannot drive unlimited tumor growth without additional mutations to overcome senescence. Recent studies suggest that cancer may arise from mutated somatic

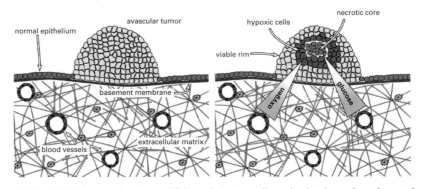

Figure 2.6 Left: initial avascular tumor. Right: substrate gradients lead to hypoxia and central necrosis.

stem cells rather than differentiated cells [56, 423, 605]. In this scenario the tumor is a mixed cell population, whose overgrowth is driven by a small subpopulation of cancer stem cells rather than by differentiated cells that have overcome senesence. With or without cancer stem cells, the result is the same at the multicell and tissue scales: a mass of cells that fail to respond to ordinary physiologic limits to their proliferation (Figure 2.6, left).

2.2.2   Avascular solid-tumor growth

Once a tumor has established a foothold in its host tissue, it begins an early period of growth as it becomes an in situ cancer. Epithelial cells are generally constrained by the basement membrane.

The limiting role of oxygen and nutrient diffusion, hypoxia, and necrosis

In this early stage of cancer, the tumor has no vascular system of its own, and so it must rely upon the host vasculature in the nearby stroma for crucial oxygen, nutrients, and growth factors; we refer to these generically as "substrates." Substrates diffuse from the surrounding vascularized tissue, enter the tumor, and are uptaken by proliferating tumor cells. This motion of substrates from external sources (the host vasculature) to internal sinks (the metabolically active tumor cells) causes substrate gradients to form within the tumor. Of particular importance is oxygen, which generally diffuses on the order of 100–200 μm into tissue before dropping to levels insufficient for cellular metabolism [112, 149, 213, 435, 437]. Interior tumor cells experience hypoxia and respond to their harsher microenvironment in a variety of ways (Section 2.1.5). Deeper within the tumor, oxygen and glucose levels drop to critcally low levels that cause the tumor cells to necrose. These dynamics are manifested as an outer tumor viable rim of proliferating cells, an interior band of hypoxic cells, and a central necrotic core. See Figure 2.6, right.

This affects the tumor mechanically. Prior to the formation of a necrotic core, proliferation throughout the tumor causes a net outward cell flux that expands the tumor

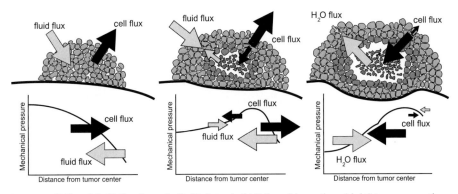

Figure 2.7 Cell and fluid flux in early (left), later (middle), and long-time (right) tumor growth.

(Figure 2.7, left). Simultaneously, the proliferating tumor cells absorb fluid from the interstitium to fuel their growth and eventual division, resulting in a net fluid flux into the tumor. Once a necrotic core has formed, cell lysis reduces the tumor cell volume and releases fluid that leaves the necrotic core and enters the proliferative rim interstitium. The subsequent reduction in mechanical pressure in the necrotic core redirects some of the viable rim cell flux towards the tumor interior (Figure 2.7, middle). As the tumor grows, the volume of its necrotic core increases, thus accentuating its cell volume sink effect. Once the tumor grows large enough, the cell flux resulting from proliferation balances with the fluid flux stemming from necrosis, leading to zero outward cell flux. This gives rise to a steady-state tumor spheroid (Figure 2.7, right).

2.2.3   Interaction with the microenvironment

As the nascent tumor grows in its host tissue, it interacts with the surrounding micro-environment in a variety of ways. It mechanically displaces and compresses the surrounding tissue, including the basement membrane (Figure 2.7, middle and right). The tumor degrades and remodels the extracellular matrix (ECM), both biomechanically and biochemically, by the secretion of enzymes such as matrix metalloproteinases (MMPs) that degrade the ECM. The degraded ECM, in turn, can release ECM-associated growth factors that fuel further tumor growth [644]. The degradation of the ECM by the MMPs increases the ability of the tumor to push into the surrounding tissue, both by reducing the mechanical rigidity of the surrounding tissue and by creating extra space for the growing tumor [325]. The combination of proliferation-induced pressure and proteolytic degradation of the surrounding tissue results in *tissue invasion*, the invasion of sheets or "fingers" of tumor cells into the surrounding tissue along the paths of least mechanical resistance. *Acidosis* (a decreased microenvironmental pH resulting from anaerobic glycolysis in hypoxic tumor cells) has also been hypothesized to play a role in tumor invasion, by inducing apoptosis in the surrounding normal epithelium, by giving invasive tumor cells a selective advantage over tumor cells that have not adapted to acidity, and by contributing to ECM degradation (due to proteases released by apoptotic cells) [208, 263–267, 279–282, 526, 624–627].

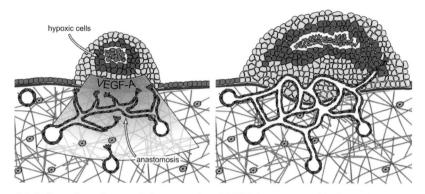

Figure 2.8 Left: angiogenic growth factors such as VEGF-A are secreted by hypoxic tumor cells, leading to angiogenesis. Right: the fresh nutrient supply allows for renewed tumor expansion.

There is recent evidence that tumors induce changes in gene expression in the nearby stroma, which helps to sustain tumor growth [326, 725]. For instance, carcinomas may release signaling molecules (e.g., IL-1β) that stimulate fibroblasts to secrete hepatocyte growth factor (HGF). The HGF, in turn, promotes tumor cell growth, decreases cell–cell adhesion, and increases MMP secretion [457]. Tumors may also alter gene expression in nearby noncancerous epithelial cells [333].

2.2.4   Vascular growth and metastasis

The next stage in cancer development can be viewed as a response to hypoxia. The ultimate result is *angiogenesis*, where the tumor induces endothelial cells to form a new vasculature that directly supplies the tumor with the nutrients enabling further expansion. Some of the same mechanisms as those responsible for angiogenesis play a role in *metastasis*, the spread of tumor cells to distant locations.

Angiogenesis

As discussed in Section 2.1.5, hypoxia inducible factors (e.g., HIF-1α) accumulate in hypoxic cells, which can trigger numerous downstream genetic targets. In particular, hypoxic cells secrete tumor angiogenic growth factors (TAFs) such as vascular endothelial growth factor (VEGF) [13, 362, 540, 713]. These TAFs diffuse outward from the hypoxic regions of the tumor and eventually reach nearby blood vessels. See Figure 2.8, left.

Blood vessels are composed of tightly connected squamous (flat and scalelike) endothelial cells that are surrounded by a basement membrane and other supporting cells, including smooth muscle cells and pericytes [424]. When the endothelial cells detect the TAF gradient emanating from the tumor, they secrete MMPs that degrade the basement membrane and extracellular matrix [46] (Figure 2.8, left). This allows the endothelial cells to migrate away from the blood vessel and toward the TAF source in the tumor. The leading endothelial cells are referred to as *sprout-tips*; immediately behind the sprout-tips, other endothelial cells divide, migrate, align, and form tubes of polarized

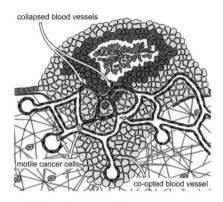

collapsed blood vessels

motile cancer cells

co-opted blood vessel

Figure 2.9 Invasive tumor growth into the stroma. The tumor grows to co-opt the neovasculature, leading to collapse of some vessels and renewed hypoxia.

endothelial cells surrounding a vascular lumen [491]. The vessels then link with one another to form a network of loops in a process called *anastomosis* (Figure 2.8, left). It can take on the order of 10 to 21 days for new vessels to form and connect to the parent vessels [46, 283, 486].

The end result is a *neovasculature* that provides the tumor with a direct supply of oxygen and nutrients. The configuration of the neovasculature is determined by the balance of pro- and anti-angiogenic growth factors, as well as by the mechanical pressures from the growing tumor and flow stresses within the nascent blood vessels [214, 285, 407, 554, 650]. The fresh nutrient supply allows a new stage of rapid tumor growth into the surrounding tissue (Figure 2.9).

Angiogenesis is not unique to tumor growth but is also a key part of wound healing, the menstrual cycle, and embryonic development [112, 213]. However, we note that tumor angiogenesis is pathological in nature, and the resulting vasculature is inefficient in a number of ways: the vessels are often "leaky" owing to the large gaps between endothelial cells; the newly formed vessels are not as stiff and rigid as mature vessels and may collapse when subjected to tissue stress (such as that created by rapidly growing tumors); the basement membrane around the new vessels may not be fully formed; some newly formed vessel walls may be composed of a mosaic of tumor and endothelial cells; and the tumor neovascular network tends to be much more tortuous than regular vascular networks [112, 217]. See Figure 2.9. This inefficiency may hinder drug delivery within tumors [340, 620] and also lead to the development of new hypoxic regions within the tumor and additional sessions of angiogenesis.

Tissue invasion and metastasis

A particularly damaging aspect of advanced cancer is *metastasis*, the spread of tumor cells to form secondary tumors in distant locations. Metastasis occurs most commonly in breast, prostate, and lung cancers [64], and it is estimated that over 90% of all deaths from solid tumors result from metastasis [298]. In spite of the great clinical importance of metastasis, it is poorly understood [360].

Metastasis is a complex phenomenon involving several mechanisms that are closely related to tissue invasion. Genetic instability, intrinsic limits (e.g., senescence), and extrinsic selective pressures (e.g., limited nutrients, immune system attacks) lead to competition within heterogeneous tumor cell populations and the eventual selection for pro-metastatic genes [298]. Hypoxia creates a strong selective pressure, leading to increasing internal HIF-1α levels in the tumor cells and the expression of genes responsible for increased motility, glycolysis, reduced response to apoptotic pathways, and increased production of MMPs [303]. These selective pressures also lead to increased expression of the genes responsible for locomotion [541]. As a result, tumor cells degrade the BM and ECM and invade the stroma, either individually, as small clumps of cells (emboli) or in cohort motion of sheets of cells linked by cell–cell adhesion [298, 487]. Eventually, invasive tumor cells can enter the vasculature or lymphatic system. See Figure 2.9.

For sarcomas (which already reside in the stroma), this is accomplished by proteolytic degradation of the ECM and BM surrounding the stromal vessels, followed by direct entry into the vessels. For carcinomas (which are separated from the stroma by the BM), entry into the vasculature could also be indirect via the lymphatic system [199]. The mesenchymally derived sarcoma cells move with built-in cellular machinery in a contractile manner: by first degrading the ECM on their leading edge, adhering to the ECM, and contracting, then by rebuilding the ECM on the trailing edge [669]; see Section 2.1.6. Epithelially derived carcinoma cells initially lack this locomotive ability, but mutations and altered gene expression can restore these locomotive mechanisms; the process is often referred to as the *epithelial–mesenchymal transition* (EMT) [541, 669].

Once the metastatic tumor cells have reached the vasculature, they circulate in the blood. Initially, survival of the circulating tumor cells is inhibited by the immune system, which kills most of the individual cells; emboli consisting of five to 10 cells are more likely to escape attack by the immune system [199]. Note that the complex role of the immune system is poorly understood; it may both promote and inhibit metastasis. Circulating tumor cells that do survive can eventually lodge in the capillary bed of distant organs; the most frequent destinations include the liver, lungs, and bones [64].

However, without further tumor–host interaction, the destination microenvironment will not support the newly arrived metastatic tumor cells. Different types of tumor cells tend to metastasize to specific tissues. This "seed and soil" idea, that only specific tissues are suitable to each tumor cell line, was first formulated by Stephen Paget in 1889 when studying breast cancer metastases [171, 484, 516]. The reasons for this are only now being elucidated, in an emerging area of cancer research. The theory is that tumors release cytokines, VEGF, and other chemical signals into the circulatory system that recruit progenitor and endothelial cells from the bone marrow and vasculature to assist in creating a *pre-metastatic niche*, a modified microenvironment in a distant host tissue that is suitable for tumor metastasis [298]. In the process, the chemical signals alter gene expression in the endothelial cells in capillary walls at the destination tissue; these cells then express additional adhesion molecules and secrete MMPs to degrade the basement membrane surrounding the capillaries [199, 314, 360]. The newly expressed adhesion molecules on the inner surface of the capillary bed improve the ability of the metastatic tumor cells to arrest at the destination, and the degraded BM assists in the extravasation of the tumor cells from capillaries into the destination tissue.

Once the metastatic tumor cells successfully invade the destination tissue, they secrete growth factors that induce additional changes in the new location. This growth is similar to the mechanisms of tissue invasion that were introduced earlier but with additional elements. Tumor-induced changes in the stromal cells cause them to degrade and remodel the matrix, even as the tumor cells are also secreting MMPs to degrade the matrix. Growth-promoting molecules that were previously sequestered in the ECM fuel further tumor growth [199]. With ample room to grow and a favorable microenvironment, these tumor cells can develop into secondary tumors. Because the metastatic tumor cells have been selected for their invasive phenotype, they are capable of expressing pro-angiogenic growth factors to initiate angiogenesis and enter vascularized growth. It is likely that the tissue specificity of this process is due to the combination and balance of cytokines and chemicals secreted by the tumors, which, in turn, depends upon the genetic makeup of the tumors [623]. It is thought that only a small fraction of the cells in the primary tumor have the ability to recruit the proper progenitor and endothelial cells to build a pre-metastatic niche [298].

The scientific understanding of metastasis is advancing rapidly, and the reader is encouraged to read the reviews [199, 298, 360, 518]. The reviews on bone metastases [64, 419] provide well-written concrete examples of the process, and they give an excellent overview of the latest developments in metastasis research.

2.3   Concluding remarks

In this chapter, we have presented a simplified overview of the major topics in biology that relate to cancer. Cancer modelers may wish to keep these topics in mind, as they study and extend the models presented in this book, and to explore the excellent references cited in this chapter and elsewhere to learn more about these biological themes in greater depth. In the following chapters of Part I we present state-of-the-art continuum, discrete, and hybrid models that incorporate a broad spectrum of the tumor progression and behavior presented in this chapter.

3 Continuum tumor modeling: single phase

With H. B. Frieboes

3.1 Introduction

We present a sharp-interface continuum model of tumor growth, in which the tumor is treated as a single-phase, or piecewise single-phase, biomaterial with sharp boundaries delineating the various tumor components (viable, quiescent, and necrotic cells). We present the model in order of increasing complexity, starting with avascular and vascular versions that incorporate varying degrees of microenvironmental heterogeneity and ending with a model that accounts for the nonlinear coupling between tumor growth and tumor-induced angiogenesis. In this chapter we focus on the presentation of the model. Analysis and numerical simulations are presented in Chapter 4.

3.1.1 Background

In larger-scale systems, continuum methods provide a good modeling strategy. The reader should refer to the Introduction (Chapter 1) for references and reviews on the continuum modeling of cancer. The approach draws upon principles from continuum mechanics to describe variables as continuous fields described by means of partial differential and integro-differential equations, and it allows for the development of fast numerical solvers. These models treat a tumor as a collection of tissue, for which densities or volume fractions of cells are described. Individual cells and other discrete elements are not tracked. The model parameters may describe volume fractions of various cell species as well as concentrations of cell substrates such as glucose, oxygen, and growth factors. On the one hand, the parameters at this macro-scale are somewhat easier to obtain, analyze, and control compared with those required for discrete models, which typically involve cellular and subcellular measurements (see Chapter 6). On the other hand, although continuum models are appropriate at the tissue scale, where gross tumor behavior can be quantified, they cannot represent individual cells and discrete events (e.g., epithelial-to-mesenchymal phenotypic transitions that lead to individual cell migration). This may be important when one is studying the effect of genetic, cellular, and microenvironment characteristics on overall tumor behavior.

A basic tumor growth model represents tumor cells proliferating in such a way that they form a small sphere-like structure without direct access to the vasculature. During this avascular growth, the tumor cells receive oxygen, nutrients, and growth factors via diffusion from the surrounding host tissue. This phase can be investigated by *in vitro* experiments, in which cancer cells are cultured in a three-dimensional geometry as tumor spheroids [378, 393, 479, 481, 647, 648, 676]. This can be done because certain cancer cell lines will self-organize by cell–cell adhesion into multicellular, roughly spheroidal, colonies. The outer cells tend to proliferate while the cells in the interior necrose (i.e., die) owing to insufficient oxygen and nutrients [648]. For example, the typical distance that an oxygen molecule diffuses before being uptaken by a cell lies in the range 100–200 μm. This limits the size to which a tumor spheroid can grow (to 1–2 mm in diameter) in experiments. A layer of quiescent (hypoxic) cells separates the necrotic core from the proliferating rim. Because of the three-dimensionality, the growth of multicellular spheroids is thought to be similar to that of *in vivo* avascular tumors. There is a significant amount of experimental data in the literature on the internal structure of multicellular spheroids and the spatiotemporal distribution of cell substrates (see the references above). Thus, the tumor spheroid is a good experimental system with which to test tumor predictions by mathematical models.

Greenspan [292, 293] developed one of the earliest continuum models, in which tumor growth is a function of the diffusion of cell substrates, as observed in previous studies (e.g., [88, 655]). McElwain and Morris [460] accounted for apoptosis and Adam [5] discussed the immune response. Byrne and Chaplain [97, 99, 101] studied the growth and stability of radially symmetric tumors without and with necrosis, as well as the effects of cell substrates and inhibitors. Chaplain [119] presented mathematical models of spherical tumor growth through the stages of avascular growth, angiogenesis, and vascularization, as well as pattern formation in cancer [120]. Friedman and Reitich [241, 242] and Cui and Friedman [156] studied non-necrotic vascularized radially symmetric and spatially patterned tumors modeled through a free-boundary problem, in which tumor growth was dependent on the level of diffusing cell substrates. See also [150, 152, 154, 238, 297, 652, 653, 701] for later extensions to a variety of different tumor models.

On the basis of these and other classical continuum tumor models (see Chapter 1), Cristini *et al.* [149] performed computer simulations of tumor growth beyond the limited capabilities of mathematical linear analyses and spherical geometries, thus enabling the nonlinear modeling of complex tumor morphologies. Using a new formulation of these classical models, they showed that tumor evolution could be described by a reduced set of two dimensionless parameters (related to mitosis rate, apoptosis rate, cell mobility, and cell adhesion), independently of the number of spatial dimensions. These parameters regulate the morphology and growth (invasiveness) of avascular and vascularized tumors. Critical conditions were predicted that separate compact, noninvasive, mass growth from unstable, fingering, infiltrative progression [149], thus suggesting that the mechanisms that control tumor morphology also control the tumor's ability to invade. This morphological instability provides a mechanism for invasion without angiogenesis and may allow the tumor to overcome diffusional limitations to growth by creating excess surface area that exposes more interior cells to oxygen and nutrient. Indeed, numerical simulations show that the tumor grows without bound by repeated sub-spheroid growth (budding), fingering and folding to create a complex shape. That is, morphological instability driven by microenvironmental gradients of nutrient and oxygen selects for locally higher cell proliferation. Tumors may thus escape diffusion-limited constraints without recourse to angiogenesis, as has been observed experimentally (e.g., [167, 230]). Recently, Li *et al.* [416] extended this work to arbitrary geometries in three dimensions.

3.2   Avascular tumor growth

3.2.1   Model without necrosis

We take Ω to be a computational tissue domain that contains two non-intersecting subdomains, the tumor $\Omega_T(t)$ and the external host tissue $\Omega_H(t)$, Σ to be the boundary between the tumor tissue and the host tissue, $\mathbf{n}$ to be the unit outward normal vector to Σ, and $\mathbf{x}$ to be the position in space. We let n denote the concentration of a vital cell substrate (e.g., oxygen or glucose). Since the diffusion rate of oxygen or glucose is much faster (on the order of seconds across the cell space) than the rate of cell proliferation (typically one cell per day), we may regard the cell substrates to be in a steady state for a given tumor morphology (e.g., [97, 149, 241, 293]). This can be represented as:

$$0 = D_T \nabla^2 n + \Gamma \quad \text{in} \quad \Omega_T, \tag{3.1}$$

$$0 = D_H \nabla^2 n \quad \text{in} \quad \Omega_H, \tag{3.2}$$

where D_T and D_H are the diffusion constants in the tumor tissue and the host tissue, and Γ is the rate at which cell substrates are added to Ω_T and is given by:

$$\Gamma = -\lambda_B(n - n_B) - \lambda n, \tag{3.3}$$

where the first term describes the source of substrates from the vasculature (λ_B is the blood–tissue transfer rate and n_B is the substrate concentration in the blood), and the second term describes the uptake of substrates by tumor cells.

Following [97, 149, 241, 292] and other works, we assume that the cell density is constant in the proliferating tumor domain. Therefore, mass changes correspond to volume changes, and mass and volume changes are equivalent. Note that apoptosis is modeled as a source of volume loss of the solid tumor component due to lysing; the lysing is assumed to occur instantaneously. Defining $\mathbf{u}$ to be the cell velocity, the local rate of volume change $\nabla \cdot \mathbf{u}$ is given by

$$\nabla \cdot \mathbf{u} = \lambda_p \quad \text{in} \quad \Omega_T, \tag{3.4}$$

where λ_p is the cell-proliferation rate and is given by

$$\lambda_p = bn - \lambda_A, \tag{3.5}$$

where λ_A is the rate of apoptosis and b is a measure of mitosis.

To determine the cell velocity, we use Darcy's law as the constitutive assumption (e.g., [97, 149, 241, 293]):

$$\mathbf{u} = -\mu \nabla P + \chi_n \nabla n \quad \text{in} \quad \Omega_T, \tag{3.6}$$

where P is the oncotic (solid) pressure, μ is a mobility that reflects the combined effects of cell–cell and cell–matrix adhesion, and χ_n is the coefficient of chemotaxis from regions of low to high cell substrates; chemotaxis is directed cell migration up gradients of soluble substances such as oxygen and nutrients. In contrast, haptotaxis describes directed cell migration along gradients of insoluble substances, e.g., up gradients of extracellular matrix molecules such as fibronectin.

On the tumor interface Σ, the boundary conditions are as follows:

$$[n] = 0, \quad [\mathbf{n} \cdot \bar{\mathcal{D}} \nabla n] = 0 \qquad \text{on} \quad \Sigma, \tag{3.7}$$

$$P = \gamma \kappa \qquad \text{on} \quad \Sigma; \tag{3.8}$$

here $[\cdot]$ denotes the jump in the value of a quantity from the inside to the outside of the interface, $\mathbf{n}$ is the vector normal to the interface, and $\bar{\mathcal{D}} = D_{\mathrm{T}} \mathbf{1}_{\Omega_{\mathrm{T}}} + D_{\mathrm{H}} \mathbf{1}_{\Omega_{\mathrm{H}}}$. Here $\mathbf{1}_S$ is the characteristic function on the set S, satisfying $\mathbf{1}_S = 1$ for $\mathbf{x} \in S$ and 0 otherwise. The pressure boundary condition (3.8) reflects the influence of cell–cell adhesion through the parameter γ, and κ is the local total curvature. Physically, Eq. (3.8) means that we approximate the surrounding tissue as infinitely biomechanically compliant, i.e. $P = 0$ outside the tumor. The more general case is discussed in Section 3.2.2 in the context of growth in inhomogeneous microenvironments.

We impose the following condition for cell substrates at the outer (far-field) boundary $\partial \Omega$:

$$(n)_{\partial \Omega} = n^{\infty}. \tag{3.9}$$

For simplicity, we assume that n^{∞} is constant.

The normal velocity $V = \mathbf{n} \cdot (\mathbf{u})_{\Sigma}$ of the tumor boundary is

$$V = -\mu \mathbf{n} \cdot (\nabla P)_{\Sigma} + \chi_n \mathbf{n} \cdot (\nabla n)_{\Sigma}. \tag{3.10}$$

Following [97, 149, 241, 292] and others, we assume that λ, λ_{B}, n_{B}, b are uniform. We denote $\lambda_{\mathrm{M}} = b n^{\infty}$ as the characteristic mitosis rate, $\lambda_{\mathrm{R}} = \mu \gamma L_{\mathrm{D}}^{-3}$ as the intrinsic relaxation time scale, and $B = (n_{\mathrm{B}} \lambda_{\mathrm{B}} / n^{\infty})(\lambda_{\mathrm{B}} + \lambda)$ as a measure of the extent of vascularization. Introducing the nondimensional length scale $L_{\mathrm{D}} = D_{\mathrm{T}}^{1/2} (\lambda_{\mathrm{B}} + \lambda)^{-1/2}$ and time scale $\lambda_{\mathrm{R}}^{-1}$, we define a modified concentration $\bar{\Gamma}$ and pressure $\bar{p}$ by

$$n = n^{\infty} \left[1 - (1 - B)(1 - \bar{\Gamma}) \right], \tag{3.11}$$

$$P = \frac{\gamma}{L_{\mathrm{D}}} \left(\bar{p} + G + (\chi - G) \bar{\Gamma} + A G \frac{\bar{\mathbf{x}} \cdot \bar{\mathbf{x}}}{2d} \right), \tag{3.12}$$

where

$$\chi = \frac{n^{\infty}(1 - B) L_{\mathrm{D}}}{\mu \gamma} \chi_n \tag{3.13}$$

is the nondimensionalized coefficient for chemotaxis due to cell substrates in a d-dimensional tumor ($d = 2, 3$), and $\bar{\mathbf{x}}$ is the nondimensional position in space. The parameters G and A measure the relative strengths of adhesion (cell–cell and cell–matrix) and apoptosis, respectively:

$$G = \frac{\lambda_M}{\lambda_R}(1 - B), \tag{3.14}$$

$$A = \frac{\lambda_A/\lambda_M - B}{1 - B}. \tag{3.15}$$

A large G value corresponds to low cell adhesion or aggressive proliferation. Dropping the bar notation, we obtain nondimensional equations for Γ and p:

$$0 = \nabla^2\Gamma - \Gamma \quad \text{in} \quad \Omega_T, \tag{3.16}$$

$$0 = D\nabla^2\Gamma \quad \text{in} \quad \Omega_H, \tag{3.17}$$

$$0 = \nabla^2 p \quad \text{in} \quad \Omega_T, \tag{3.18}$$

where $D = D_H/D_T$ (i.e., the diffusion constant outside the tumor is D times larger than that inside it).

The boundary conditions are given by

$$[\Gamma] = 0, \quad [\mathbf{n} \cdot \mathcal{D}\nabla\Gamma] = 0 \qquad \text{on} \quad \Sigma, \tag{3.19}$$

$$p = \kappa + (G - \chi)(\Gamma)_\Sigma - G - AG\frac{(\mathbf{x} \cdot \mathbf{x})_\Sigma}{2d} \qquad \text{on} \quad \Sigma, \tag{3.20}$$

$$\Gamma = 1 \qquad \text{on} \quad \partial\Omega, \tag{3.21}$$

where $\mathcal{D} = \mathbf{1}_{\Omega_T} + D\mathbf{1}_{\Omega_H}$.

Finally, the tumor surface evolves at the normal velocity

$$V = -\mathbf{n} \cdot (\nabla p)_\Sigma + G\mathbf{n} \cdot (\nabla\Gamma)_\Sigma - AG\frac{\mathbf{n} \cdot (\mathbf{x})_\Sigma}{d}. \tag{3.22}$$

It should be noted that the models proposed in [149, 416] may be recovered as limiting cases of vanishing chemotaxis due to cell substrates and of a large healthy-tissue diffusion constant, i.e., as $\chi \to 0$ and $D \to \infty$, the governing equations (3.16)–(3.18) and boundary conditions (3.19)–(3.21) reduce to

$$\nabla^2\Gamma - \Gamma = 0 \qquad \text{in} \quad \Omega_T, \quad (\Gamma)_\Sigma = 1, \tag{3.23}$$

$$\nabla^2 p = 0 \qquad \text{in} \quad \Omega_T, \quad (p)_\Sigma = \kappa - AG\frac{(\mathbf{x} \cdot \mathbf{x})}{2d}, \tag{3.24}$$

the velocity at the tumor interface being given by Eq. (3.22).

To investigate the effect of nonlinearity, efficient numerical algorithms have been developed [149, 416] to solve Eqs. (3.16)–(3.22). The partial differential equations (3.23), (3.24) for the whole domain are reformulated into boundary integral equations that hold only on the tumor–host interface, using potential theory.

In Chapter 4 we present numerical simulations to analyze the nonlinear behavior of the system. We characterize the regimes of growth described by the model (3.23), (3.24) and present a linear analysis of the stability of (nonspherical) shape perturbations, including the necessary conditions for morphological stability.

3.2.2   Growth in heterogeneous tissue

Microenvironmental inhomogeneities play a significant role in the growth of a tumor [201, 315, 528, 586]. For example, hypoxic microenvironments induce both tumor and endothelial cells to upregulate the expression of HIF-1 genes, leading to the secretion of factors that promote angiogenesis and decreased cell–cell and cell–matrix adhesion [206, 362, 540]. In addition, hypoxic microenvironments affect the metabolism of tumor cells, leading to activation of the glycolytic pathway and acidosis in the microenvironment [264, 266, 268]. The presence of different tumor and host tissues and variations in the extracellular matrix may also influence tumor progression, through biological and physical interactions [705]. In addition, a more inhomogeneous microenvironment may induce tumor clonal diversity [366].

Gradients of oxygen in the presence of hypoxia are a key contributor to heterogeneity in the tumor microenvironment. These gradients may arise from inadequate vascularization exacerbated by disordered tumor-induced angiogenesis (see Section 3.3) and may lead to necrosis in the tumor interior. The basic model in Section 3.2 was extended by Byrne and Chaplain [99] to include necrosis. See the review papers listed in Chapter 1 for studies of the effects of necrosis on tumor growth. For example, Garner *et al.* [262] incorporated necrosis into a model of spherical tumor growth and used the conservation of energy to obtain scaling laws for the growth of the tumor and the necrotic core. In [435, 437, 438] a further extension was introduced to study the effects of oxygen variation in the tumor microenvironment. In this model, an avascular tumor is modeled as occupying volume $\Omega_T(t)$ with boundary $\partial\Omega$, denoted by Σ, viable region Ω_V, and necrotic region Ω_N where tumor cells die owing to low cell-substrate levels. The viable region is divided into a proliferating region Ω_P, where the substrate levels are high enough to permit cell proliferation, and a quiescent or hypoxic region Ω_Q where substrate levels are insufficient to sustain proliferation. The growing tumor also interacts with the surrounding microenvironment in the host tissue; this region is denoted by Ω_H. In addition, Eq. (3.2) governing the distribution of oxygen (or any other vital cell substrate) is generalized to allow for nonconstant diffusivity, variable uptake, and natural decay. Using the same nondimensional scaling as before, this gives

$$0 = \nabla \cdot (\mathcal{D}\nabla n) - \lambda^n(\mathbf{x}, n)n + \lambda^n_{\text{bulk}}(1 - n)\,B(\mathbf{x})\mathbf{1}_H, \qquad (3.25)$$

where λ^n is the uptake or decay rate of substrate n. It is assumed that oxygen is uptaken at different rates in the different domains of the tumor and of the host tissue region Ω_H. For example, the oxygen uptake may be different in domains where the cells are proliferating Ω_P or quiescent Ω_Q. In addition, oxygen may be degraded in the necrotic domain Ω_N owing to the presence of oxidizing agents [439]. These domains may actually be defined in terms of the oxygen concentration, as in Eq. (3.26) below. In Eq. (3.25), λ^n_{bulk} is the cell substrate delivery rate from the pre-existing vasculature, $B(\mathbf{x})$ is the pre-existing vascular density and $\mathbf{1}_H$ is the characteristic function of the host domain Ω_H. Normalized by the cell-substrate uptake in the proliferating tumor region, the uptake and decay functions and the definitions of the domains are taken to be as follows:

$$\lambda^n(\mathbf{x}, n) = \begin{cases} 0 & \text{if} \quad \mathbf{x} \in \Omega_H, \\ 1 & \text{if} \quad \mathbf{x} \in \Omega_P = \{\mathbf{x}|\ 1 \geq n > n_p\}, \\ \lambda_Q^n & \text{if} \quad \mathbf{x} \in \Omega_Q = \{\mathbf{x}|\ n_p \geq n > n_N\}, \\ \lambda_N^n & \text{if} \quad \mathbf{x} \in \Omega_N = \{\mathbf{x}|\ n_N \geq n\}, \end{cases} \tag{3.26}$$

where n_p and n_N are the cell-substrate thresholds for proliferation and necrosis, respectively. This function reflects the fact that substrates are uptaken much faster in the tumor than in the host tissue (hence the relative uptake in Ω_H is taken as zero) and the fact that when cells necrose they release their cellular contents, which are oxygen reactive [251, 383], and so this effect on the cell substrate concentration can be modeled through the decay rate λ_N^n. Note that with this choice of uptake and decay function, the cell-substrate equation is nonlinear.

Cells and ECMs in the host tissue Ω_H and in the viable tumor region Ω_V are affected by a variety of forces, each of which contributes to the cellular velocity field $\mathbf{u}$. Proliferating tumor cells in Ω_V generate an internal (oncotic) mechanical pressure (a hydrostatic stress) that also exerts a force on the surrounding noncancerous tissue in Ω_H. Tumor, noncancerous cells, and ECM can respond to pressure variations by overcoming cell–cell and cell–ECM adhesion and moving within the ECM scaffolding that provides the structure for the host tissue.

Using Darcy's law as a constitutive relation for the cell velocity in the host and tumor domains, and a domain-dependent net cell-proliferation rate λ_p, the mechanical pressure is given by

$$\nabla \cdot (\mu \nabla P) = \lambda_p = \begin{cases} 0 & \text{if} \quad \mathbf{x} \in \Omega_H, \\ n - A & \text{if} \quad \mathbf{x} \in \Omega_P, \\ 0 & \text{if} \quad \mathbf{x} \in \Omega_Q, \\ -G_N & \text{if} \quad \mathbf{x} \in \Omega_N. \end{cases} \tag{3.27}$$

Note that the cellular mobility μ also measures the permeability of tissue to tumor cells. See [22] and [103] for further motivation of this approach from the perspective of mixture modeling. In the host domain, the nondimensional net proliferation rate is assumed negligible since tumor cells proliferate more rapidly or die at a lower rate than host cells. In the proliferating domain, the proliferation rate is assumed to be linear in the oxygen concentration and apoptosis is assumed to result in volume loss as in Section 3.2.1. The parameter G_N is the nondimensional rate of volume loss due to necrosis [435, 722]. Following [149] and others, cell–cell adhesion forces are modeled in the tumor by generalizing the condition (3.8) by introducing a Laplace–Young jump condition:

$$[P] = G^{-1}\kappa, \qquad \mathbf{x} \in \Sigma, \tag{3.28}$$

where κ is the total curvature. It is also supposed that the pressure jump across the necrotic interface is zero, reflecting the low cell–cell adhesion in the perinecrotic region and the increased cellular mobility observed in hypoxic cells [82, 108, 317, 539, 540, 570]. The cell velocity is also assumed to be continuous across the tumor–host interface and across the boundary of the necrotic tumor domain. Cellular proliferation and apoptosis are in balance in the far field, i.e.,

$$P \equiv P_\infty, \qquad \mathbf{x} \in \partial(\Omega_T \cup \Omega_H). \tag{3.29}$$

3.2.3 Effect of stress

The mathematical modeling of residual stress development in growing tumors was recently reviewed by Araujo and McElwain [34], Roose et al. [573], Tracqui [661], and Lowengrub et al. [425]. The study of Shannon and Rubinsky in 1992 [604] showed that residual stresses are induced by any spatial variation in the growth process in a linear-elastic description of a growing tissue with spherical geometry, such as a tumor spheroid in vitro. This framework was extended by Jones et al. [352] by accommodating the continuous nature of the growth process instead of using a fixed growth-strain distribution. The dimensionless model equations from Jones et al. are:

$$0 = \nabla^2 n - n, \tag{3.30}$$

$$\nabla \cdot \mathbf{u} = n - A, \tag{3.31}$$

$$\nabla \cdot \sigma = 0, \tag{3.32}$$

$$\frac{1}{2}(\nabla \mathbf{u} + \nabla \mathbf{u}^T) = \frac{1}{3}(\nabla \cdot \mathbf{u})\mathbf{I} + \frac{1}{2}\left(\frac{D}{Dt}(3\sigma - \mathrm{Tr}(\sigma)\,\mathbf{I}) + 3(\omega \cdot \sigma - \sigma \cdot \omega)\right), \tag{3.33}$$

where the first equation describes cell substrate diffusion and uptake, the second describes the rate of volume growth, and the third describes mechanical equilibrium; here $\sigma(\mathbf{x}, t)$ is the stress tensor. The last equation is a differentiated version (using the material derivative) of the stress–strain relation, with $\omega = -\frac{1}{2}(\nabla \mathbf{u} - \nabla \mathbf{u}^T)$ and identity tensor $\mathbf{I}$. As before, $n(\mathbf{x}, t)$ is the cell substrate concentration and $\mathbf{u}(\mathbf{x}, t)$ is the cell velocity; the tumor–host interface Σ is advected with velocity $\mathbf{u}$. Traction-free boundary conditions are applied at the tumor–host interface, although a normal stress on the boundary could also be applied to mimic the effect of a gel containing the tumor.

This model does not accurately capture stress relaxation. In fact, the model predicts a linear increase over time of the compressive circumferential stresses in the peripheral region of a cell-substrate-regulated tumor spheroid at equilibrium. Araujo and McElwain [35] addressed this issue by incorporating anisotropy as a stress relaxation mechanism. This enables the stress to evolve to a steady-state distribution. In the model, the tumors grow preferentially in the direction of least stress (which is consistent with experimental observations [310]). Only if the radial and circumferential stresses are equal will the proliferation be uniform in all directions. In further work, Araujo and McElwain investigated the induced stresses for different distributions of growth strain in spherical geometries, considering the crucial role of spatial nonuniformity during the growth process in inducing residual stresses [39]; they found that a distribution of growth strains that decreases monotonically with radius induces stresses that become progressively less compressive with radius, with the circumferential component always less compressive than the radial component. This model also provides insight into the collapse of vasculature during growth [36] and highlights the role of anisotropic growth in relieving growth-induced stresses.

Another approach was used by Roose et al. [574] to correct the quantitative estimate of the solid stress inside tumor spheroids. These authors considered a linear poroelastic model where a pressure is introduced and the stress–strain relation is not inverted, as it is in the models described above. The tissue stress–strain relation is:

$$\sigma_{ij} = 2G\epsilon_{ij} + \left(\lambda - \frac{2}{3}G\right)\epsilon_{ij}\delta_{ij} - p\delta_{ij} - \lambda\eta\delta_{ij}, \tag{3.34}$$

$$\epsilon_{ij} = \frac{1}{2}\left(\frac{\partial u_i}{\partial x_j} + \frac{\partial u_j}{\partial x_i}\right), \tag{3.35}$$

where σ_{ij} is the effective stress in the tissue, G is the shear modulus of the tissue, λ is the drained bulk modulus of the tissue, ϵ_{ij} is the strain tensor of the tumor tissue, δ_{ij} is the Kronecker delta, p is the fluid pressure in the tissue, η is the volume of new tissue created per unit volume of original tissue (given by $\partial\eta/\partial t = S_C$, where S_C is the rate of production of solid-phase tumor tissue), and the u_i are the displacement components. The stress that builds up is proportional to the bulk modulus and the volume of material created per unit volume if there is no displacement of the tissue. In Sarntinoranont et al. [589], the poroelastic continuum model was extended to account for the effects of leaky vessels and of the lack of lymphatics.

As Ambrosi and Preziosi [23] have pointed out, the strain is not frame invariant [450] and hence the Jaumann derivative (the material derivative for tensors) is inappropriate. Ambrosi and Preziosi instead demonstrate that the use of accretive forces (e.g., [19, 62, 261, 328, 329, 390]) is required to derive the equations for growth in a consistent manner. In this approach, the deformation is decomposed multiplicatively to account for the contributions of pure growth, cell reorganization, and elastic deformation (see Chapter 5). In essence, the models reviewed here reduce to the one presented by Ambrosi and Preziosi in the limit of small deformations and when convection terms are dropped. As described in [23], Ambrosi and Preziosi developed a mixture model that accounts for both stress relaxation and cell adhesion, and they demonstrated how tumor growth may lead to cell reorganization to relieve stress. Previously published models, such as fluid-like or linearly elastic models, may be obtained as limiting cases of this more general model. Further details are provided in Chapter 5.

In another approach to address stress relaxation, MacArthur and Please [429] developed a viscoelastic model of residual stresses in a tumor spheroid, more closely representing the conditions in living tissues (e.g., [249]). In the model, necrotic regions developed based on adverse mechanical stress instead of cell-substrate limitations.

Nonlinear elastic models were employed by Chaplain and Sleeman [125] to classify solid tumors and by Chen et al. [130] to investigate the influence of growth-induced stress in the medium surrounding a tumor spheroid on the growth of the tumor. The growth rate and equilibrium size of the tumor were found to decrease as the stiffness of the surrounding medium increased, in consistency with experiments (e.g. [310]). Ambrosi and Mollica [20, 21] also employed a nonlinear approach to analyze the role of mechanical stress on the growth of a tumor spheroid. In [20] Ambrosi and Mollica considered a mechanical description where volumetric growth and mechanical responses were divided into two separate contributions. One was growth measured as an increase in the mass of the particles of the body instead of an increase in their number. The description included the diffusion of cell substrates through the growing material. The model was applied to describe homogeneous growth inside a rigid cylinder, modeling ductal carcinoma, and to the growth of a tumor spheroid with nonhomogeneous diffusion of cell substrates, which generates residual stresses because the nonuniform distribution of substrates leads to inhomogeneous growth. In a later work [18], Ambrosi and Guana provided a qualitative analysis of the stress-modulated growth of a continuum body as predicted by equations that satisfy an *a priori* dissipation principle. Accretive forces [177] were reinterpreted as a homeostatic value of the Eshelby stress, coinciding with the classical biomechanical concept of homeostatic stress in the case of infinitesimal strain.

The models reviewed above were all spherically symmetric and hence can be reduced to one-dimensional problems. Ben Amar and Goriely [62] performed a general stability analysis of nonlinear elastic models under asymmetric perturbations in higher dimensions. The crucial role of residual stress was established by showing that a spherical shell without any external loading could become spontaneously unstable under large anisotropic growth.

Recently, Lloyd *et al.* [422] developed a finite-element-based method to simulate the elastic tissue response during three-dimensional tumor growth. Although deformations are induced by a prescribed strain determined from cell proliferation, the accumulation of stress is neglected and it is assumed that residual stress dissipates completely, which would be consistent with an elasto-plastic growth law.

Several other models of stress effects in tumors have been developed in the context of multiphase mixture models (Chapter 5).

3.3 Vascularized tumor growth

3.3.1 Background

To transition from the avascular to the vascular phase of growth, a tumor must induce new blood vessels to sprout from the existing vascular network and grow towards the tumor, eventually penetrating it. This process, known as tumor-induced angiogenesis, is a critical milestone in the development of invasive and malignant cancer [300]. The process is thought to start when a small avascular tumor exceeds a critical size greater than can be sustained by the normal tissue vasculature [112]. Accordingly, tumor cells become hypoxic and secrete diffusible chemical signals, collectively called angiogenic factors, such as the vascular endothelial cell growth factor (VEGF). These molecules diffuse into the host microenvironment and bind to specific cell membrane receptors on the (vascular) endothelial cells that line existing blood vessels. This process activates the endothelial cells, which respond by degrading the basement membrane surrounding the existing vessel to form new vessel sprouts. The endothelial cells then proliferate as they

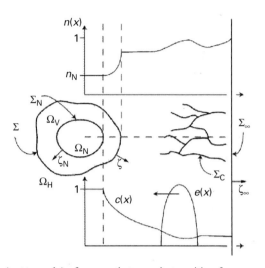

Figure 3.1 Schematic (not to scale) of a necrotic tumor in transition from avascular to vascular growth. The disjoint regions Ω_H, Ω_V, and Ω_N represent the host tissue, the viable tumor tissue, and the necrotic core domains, respectively. The tumor region is $\Omega_T = \Omega_V \cup \Omega_N$, Σ_∞ is the far-field boundary, Σ is the tumor interface, and Σ_N is the necrotic rim. Capillaries are defined on Σ_C. The cell substrate concentration $n(x)$, the tumor angiogenic factor concentration $c(x)$, and the endothelial cell density $e(x)$ along the horizontal broken line are plotted (n_N denotes the substrate threshold for cell viability). With kind permission from Springer Science and Business Media: *Bull. Math. Biol.*, Zheng *et al.*, vol. 67, p. 215, © 2005 Springer.

form new vessels in the direction of the tumor. Migration is mediated by the chemotactic response to VEGF and other tumor-angiogenic factors, by proteolytic enzymes that degrade the ECM, providing space for the cells to move, and by a haptotactic response to variable cell–matrix adhesion. As the endothelial cells proliferate through the ECM, they form tubular structures that fuse (anastomose) to form loops. Eventually blood flows through the neovascular network, providing the tumor and host microenvironments, with an additional, although inefficient, supply of cell substrates. This process is illustrated in Figure 3.1. See also Section 2.2.4.

Tumor growth and angiogenesis are coupled, in that hypoxic tumor cells release a net balance of tumor angiogenic factors that attract the endothelial cells and incite the neovascular network to approach the tumor. This creates additional sources of cell substrates in the microenvironment. The tumor responds by upregulating cell proliferation in the regions where cell substrates are increased. The additional substrates affect the hypoxic tumor regions, which in turn downregulate the net release of tumor angiogenic factors, and this affects the formation of the neovascular network.

In addition to heterogeneous blood flow, a neovascular network contends with pressure variation introduced by increased tumor cell proliferation and migration. Normally, the network responds by remodeling itself. Compared with the vessel networks formed during normal biological processes such as embryonic development and wound healing; however, tumor-induced neovascular networks may become leaky and inefficient, producing immature and tortuous vessels [305] that lead to increased flow resistance and ultimately to the heterogeneous supply of oxygen and nutrients in the tumor microenvironment [247].

3.3.2 Modeling of angiogenesis

Mathematical models of tumor-induced angiogenesis date from the work of Balding and McElwain [47]. Continuum, fully discrete, and hybrid mathematical models have been developed (see the recent reviews [14, 124, 353, 409, 411, 452, 523, 533, 543, 544]). In most models, the coupling between tumor growth and angiogenesis is simplified in that one of the two processes is taken to be static while the other is dynamic. There are a few models in which the two processes are coupled dynamically.

In models that focus on the angiogenic response, two approaches have been taken. One approach focuses on blood vessel densities rather than vessel morphology. Continuum conservation laws are introduced to describe the dynamics of the vessel densities and angiogenic factors. See for example Byrne and Chaplain [98], Orme and Chaplain [505], Sansone *et al.* [587], Levine *et al.* [410, 412, 414], Hogea *et al.* [318], and Peterson *et al.* [531]. Recently, Addison-Smith *et al.* [7] developed a simple mathematical model of the siting of capillary sprouts on an existing blood vessel during the initiation of tumor-induced angiogenesis. These models do not provide morphological or blood-flow information about the vasculature. In the context of vasculogenesis *in vitro*, biomechanical models have been developed that account for cell–ECM interactions and chemotactic response and are capable of describing the morphology of the vasculature. In these models, the network morphology emerges, roughly speaking, from a homogeneous distribution through a type of phase transition. See for example [16, 144, 256, 322, 451, 485, 496, 600, 660].

The other approach represents vessels as line segments, continuous curves, or interconnected lattice patterns, with a static tumor. The mechanisms modeled include vessel sprout branching and anastomosis; vascular endothelial cell activation, proliferation, and migration via chemotaxis up gradients of tumor angiogenic factors (e.g., VEGF); haptotaxis up gradients of ECM-bound chemokines (e.g., fibronectin); and proteolysis of the ECM. See, for example, Stokes and Lauffenberger [637], Anderson and Chaplain [26], Tong and Yuan [658], Plank and Sleeman [534, 535], Sun *et al.* [642, 643], Kevrekidis *et al.* [371], Bauer *et al.* [53], Milde *et al.* [470], and Capasso and Morale [110]. Blood flow and network remodeling have also been simulated (e.g., Pries *et al.* [550], McDougall *et al.* [458, 459], Stephanou *et al.*, [634, 635], Wu *et al.* [702], Zhao *et al.* [721], Sun and Munn [641], and Pries and Secomb [549]). In addition, models of tumor growth in static network topologies have been formulated (e.g., Alarcón *et al.* [9], Betteridge *et al.* [67]).

The first model in which tumor growth was fully coupled with tumor-induced angiogenesis for an arbitrary network topology was presented in Zheng *et al.* [722]. The vasculature was modeled in two dimensions using the continuum–discrete model of Anderson and Chaplain [26] coupled with a version of the nonlinear (continuum) tumor growth model of Cristini *et al.* [149]. This approach is discussed in further detail below. Later, in the context of discrete cell-based systems, models of angiogenesis and vascular tumor growth were also implemented; see for example Gevertz and Torquato [275], Bartha and Rieger [51], Bauer *et al.* [53], and Wcislo and Dzwinel [683]. The effects of the blood flow through the neovascular network on tumor growth were also recently considered in Alarcón *et al.* [10], Bartha and Rieger [51], Lee *et al.* [404], Welter *et al.* [690], and Owen *et al.* [511] using cellular automaton models for tumor growth combined with network models for the vasculature. The effects of an arterio-venous network were considered in Welter *et al.* [691]. Very recently, Macklin *et al.* [439] extended the model of Zheng *et al.* and incorporated a version of the dynamic model of tumor-induced angiogenesis developed by McDougall *et al.* [458] to include explicitly blood flow and vessel remodeling. In other recent work, Lloyd *et al.* [421] simulated the vascular growth of a three-dimensional tumor by coupling models for angiogenesis, for flow through the developing neovascular network, and for network remodeling with an elastic tumor growth model which they had developed earlier (Lloyd *et al.* [422]). In three dimensions, vascular tumor growth has been studied, using a mixture model and a lattice-free description of tumor-induced angiogenesis, by Frieboes *et al.* [229] and Bearer *et al.* [57].

3.3.3   Basic model

A basic angiogenesis model can be constructed [722] on the basis of that of Anderson and Chaplain [26], with some additions taken from Chaplain and Stuart [126], Paweletz and Knierim [527], and Paku [517]. The model of Anderson and Chaplain uses a hybrid continuum–discrete approach that has the ability to follow the motion of individual endothelial cells at the capillary tips and control important processes such as migration, proliferation, branching, and anastomosis using a discrete random-walk algorithm. The ECM and cell substrates such as tumor angiogenic factors were described using continuum fields. The objective was to replicate angiogenesis as observed in the experimental "rabbit-eye model" [283], where a tumor is implanted in the cornea, thus inducing angiogenesis that can be readily observed (since the cornea is normally avascular).

The following field variables were introduced [26]:

- the concentration of the tumor angiogenic factor (e.g. the vascular endothelial growth factor, VEGF), c
- the endothelial cell density (ECD), e
- the density of the ECM (e.g. of the matrix macromolecule fibronectin), f

Once a tumor cell senses that the cell substrate level has dropped below the minimum for viability, the cell releases diffusible tumor angiogenic factors. This release may be described through a reaction–diffusion equation with a point or line boundary condition at the necrotic–viable tumor-cell interface. The tumor angiogenic factor molecules are much smaller than cells and diffuse quickly through the extracellular spaces. As a result, a quasi-steady reaction–diffusion equation can be assumed for the nondimensional concentration c of the angiogenic factors:

$$0 = \nabla \cdot (D_c \nabla c) - \bar{\lambda}^c_{\text{decay}} c - \bar{\lambda}^c_{\text{binding}} c \mathbf{1}_{\text{sprout-tips}} + \lambda^c_{\text{prod}}, \qquad (3.36)$$

where D_c is the diffusion coefficient, $\mathbf{1}_{\text{sprout-tips}}$ is the characteristic function of the sprout tips, $\bar{\lambda}^c_{\text{decay}}$ and $\bar{\lambda}^c_{\text{binding}}$ denote the nondimensional natural decay and binding rates of the angiogenic factors, and λ^c_{prod} is the production rate of TAF by the tissue (see Eq. (3.42) below). In the far field at the boundary of the computational domain, zero Neumann boundary conditions, $\partial c / \partial n = 0$, may be taken.

A primary component of the extracellular matrix is fibronectin, a long, nondiffusible, binding molecule. Endothelial cells produce, degrade, and attach to these molecules during their migration toward the tumor. The concentration of fibronectin, $f(\mathbf{x}, t)$, satisfies [26]

$$\frac{\partial f}{\partial t} = \eta_p e - \eta_U f e - \eta_N \mathbf{1}_{\Omega_N} f, \tag{3.37}$$

where η_p is the rate of production of fibronectin by endothelial cells and η_U and η_N are the rates of degradation of fibronectin by endothelial cells and proteolytic enzymes in the necrotic region, respectively [722] ($\mathbf{1}_{\Omega_N}$ is the characteristic function of the necrotic region).

While endothelial cells are comparable in size to host cells and tumor cells, it may be assumed [722] that there are not enough endothelial cells to modify the cell velocity $\mathbf{u}$. The ratio of endothelial to tissue cells is on the order of $1/50$ or $1/100$ [89].

At the continuum level, the density of endothelial cells obeys a reaction–diffusion–convection equation. The problem is convection-dominated, with a primary source of convection driven by the chemotaxis and haptotaxis of endothelial cells in response to gradients of tumor angiogenic factors and ECM (fibronectin), respectively. In addition, the endothelial cells may be affected by the cell velocity. At the continuum level, the density $e(\mathbf{x}, t)$ of endothelial cells, which is related to the probability of finding the tip of a capillary at location $\mathbf{x}$ and time t, obeys a convection–reaction–diffusion equation in Ω_H and Ω_V:

$$\frac{\partial e}{\partial t} = D_e \nabla^2 e - \nabla \cdot \left[\left(\frac{\chi_c}{1 + \alpha c} \nabla c + \chi_f \nabla f + \chi_\mathbf{u} \mathbf{u} \right) e \right]$$
$$- \rho_D e + \rho_p e (1 - e)(c - c^*)_+ - \rho_N \mathbf{1}_{\Omega_N} e, \tag{3.38}$$

where D_e is the EC diffusion constant, χ_c and χ_f are chemotaxis and haptotaxis coefficients, respectively, and $\chi_\mathbf{u}$ and α are dimensionless constants. The parameter $\chi_\mathbf{u}$ measures the degree to which the capillaries are influenced by the cell velocity, c^* is the concentration of angiogenic factors above which proliferation occurs, and ρ_D, ρ_p, and ρ_N are the rates of natural degradation, proliferation, and death of endothelial cells, respectively.

In the original work of Anderson and Chaplain [26], the model for the motion of capillary sprout-tips comprised continuum and discrete components. Equations (3.36)–(3.38) constitute the continuum component. The discrete component was derived from Eq. (3.38) under the assumption that the growth of the capillary is determined by the biased random migration (random walk) of a single endothelial cell at the sprout-tip. In particular, it was assumed that there is a trail of endothelial cells that follow the sprout-tip. In later work, McDougall $et\ al.$ [458] omitted the continuum equation (3.38) and simply used the discrete approach. In the discrete algorithm, probabilities are generated from a finite difference approximation that describes the tendency of the tip endothelial cell to migrate and proliferate on a Cartesian lattice. In addition, capillary branching and anastomosis are incorporated. The entire capillary vessel may be convected by the external cell velocity using the kinematic condition [722]

$$\frac{d\mathbf{x}}{dt} = \mu_C \mathbf{u},\qquad(3.39)$$

where $\mathbf{x}$ is the position on the capillary and μ_C is the capillary mobility [722]. As the vessel network becomes more established in time, the capillaries mature into larger vessels and become more rigid; this can be accounted for by decreasing μ_C.

The flow in the neovascular network and its effects on network remodeling have been explicitly simulated using network models by Pries *et al.* [546, 547]. Following this work, McDougall *et al.* [459] extended the basic angiogenesis model by incorporating blood flow and treating the network as a series of straight rigid cylindrical capillaries that join adjacent nodes. The blood flow was modeled by considering the elemental flow-rate in each segment using Poiseuille's law, which describes the flow-rate as a function of the capillary lumen, fluid viscosity, capillary length, and pressure drop. Stephanou *et al.* [634] extended this model to three dimensions and found that the highly interconnected nature of irregular vasculature produced by tumor-induced angiogenesis could cause low rates of blood flow to the tumor, with the potential for blood-borne drugs to bypass the entire mass, depending on the tumor shape. They also examined the effect of vessel pruning on the flow through the vascular network (e.g., this may occur during anti-angiogenic therapy). In order to investigate how adaptive remodeling affects the oxygen and drug supply to tumors, these authors later included vascular adaptation effects [635] (due to shear and circumferential stresses generated by the flowing blood [547, 548]). This model was further updated by McDougall *et al.* [458] to incorporate dynamic vessel-radius adaptation, thus coupling vessel growth with blood flow in contrast with earlier flow models where the effects of blood flow were evaluated after the generation of a hollow network (e.g., as in Stephanou *et al.* [635] and Alarcón *et al.* [9]). The nonlinear coupling of the vessel growth and adaptation with the flow has a nontrivial effect on the vascular morphology and development.

3.3.4   Coupling of tumor growth with angiogenesis

The transfer of cell substrates from the vascular system to the host and tumor tissues needs to be modeled in order to couple the tumor growth and the angiogenesis process. In Zheng *et al.* [722] and Macklin *et al.* [439], the tumor growth model presented in Section 3.2 was coupled with the angiogenesis described above. Equation (3.25) was updated by a source term that accounts for cell substrates released by the neovascular network:

$$0 = \nabla \cdot (D\nabla n) - \lambda^n(\mathbf{x}, n)n + \lambda_{\text{bulk}}^n(1 - n)B(\mathbf{x})\mathbf{1}_H + \lambda_{\text{neo}}^n.\qquad(3.40)$$

Following Macklin *et al.* [439], the source term was defined as:

$$\lambda_{neo}^{n} = \overline{\lambda}_{neo}^{n} B_{neo}\left(\mathbf{x}, t\right) \left(\frac{h}{\overline{H}_{D}} - \overline{h}_{\min}\right)^{+} [1 - \mathcal{C}\left(P_{vessel}, P\right)] \left(1 - n\right), \qquad (3.41)$$

where $\overline{\lambda}_{neo}^{n}$ is the transfer rate from the neovascular vessels. The function $B_{neo}\left(\mathbf{x}, t\right) = \mathbf{1}_{neo}$ is the characteristic or indicator function of the neovasculature. Further, P is the oncotic (solid or mechanical or hydrostatic) pressure and P_{vessel} and h are the dimensional pressure and the hematocrit in the neovascular network, respectively. The constant $\overline{H}_{D}$ reflects the normal value of hematocrit in the blood (generally about 0.45), while $\overline{h}_{\min}$ is the minimum hematocrit needed to extravasate oxygen. The hematocrit may be modeled via the blood flow in the vascular network and is determined from the angiogenesis model [458]. This provides one aspect of the coupling between the tumor growth model and the angiogenesis model. A second type of coupling between the two models occurs through the cutoff function $\mathcal{C}\left(P_{vessel}, P\right)$, such that a large oncotic pressure P relative to the vessel pressure P_{vessel} may prevent the extravasation and transfer of oxygen from the vessels into the tissue.

The oxygen source term in Eq. (3.41) is designed in such a way that, for a sufficiently large transfer rate $\overline{\lambda}_{neo}^{n}$, the oxygen concentration $n \approx 1$ at the spatial locations of the neovessels. Note that oxygen flux conditions across the neovasculature could be imposed instead, see for example Alarcón *et al.* [10].

Another form of coupling between the growing tumor and the developing vascular network occurs through the angiogenic factors released in the tumor microenvironment. In Zheng *et al.* [722] and Macklin *et al.* [439], hypoxic or quiescent tumor cells are assumed to secrete tumor angiogenic factors, which diffuse into the surrounding tissue and attract endothelial cells. These cells respond by binding with the factors, proliferating, and chemotaxing up the gradient of these factors. Following Macklin *et al.* [439], the production rate of angiogenic factors λ_{prod}^{c} in Eq. (3.36), which describes the source of factors due to secretion by hypoxic cells in the perinecrotic region, can be defined as

$$\lambda_{prod}^{c} = \overline{\lambda}_{prod}^{c} \left(1 - c\right) \mathbf{1}_{\Omega_{Q}}, \qquad (3.42)$$

where $\overline{\lambda}_{prod}^{c}$ is the nondimensional production rate of TAF and $\mathbf{1}_{\Omega_{Q}}$ is the characteristic function of the region of quiescent (including hypoxic) cells (defined in the tissue domain Ω_{Q}).

The reader may refer to Chapter 4 for modeling results of the coupling of tumor growth and angiogenesis.

3.4 Conclusion

The tumor model described so far treats the tumor mass as a single component material that locally expands and contracts in response to variable rates of cell adhesion, mitosis, and apoptosis, which are modulated by their access to cell substrates delivered from the culture medium (*in vitro*) or from the vasculature (*in vivo*). In fact, tumors consist of multiple "phases" that include a variety of different cell genotypes and phenotypes, as well as extracellular matrix (ECM) and water. Being able to describe the dynamics of multiple cell species is crucial because the tumor microenvironment and impaired cell genetic mechanisms can lead to multiple cell genotypes and phenotypes that select for cell survival under abnormal conditions [289], with profound consequences for overall tumor growth, invasion, and response to treatment. Furthermore, the microenvironment of invasive tumors may be characterized by nonsharp boundaries between the tumor and host tissues and between multiple species within the tumor [392, 398, 418]. Thus, a multiphase approach represents a more general, and natural modeling framework for studying solid tumor growth and able to provide a more detailed account of the biophysical process than single-phase models. This is an important direction for future research, which we will explore in Chapter 5. In Chapter 4, however, we will first look at the theory of stability for the basic single-phase model.

4 Analysis and calibration of single-phase continuum tumor models

With X. Li

In this chapter we present analyses and calibrations of some single-phase tumor models discussed in Chapter 3. We begin with spherically symmetric solutions, to identify fundamental regimes of growth, and then present a linear stability analysis to characterize the growth of perturbations [149, 416]. After that we calibrate the continuum model with data from tumor spheroid experiments [230] and consider the effect of nonlinearity using numerical simulations. The linear stability analyses are further extended to take into account chemotaxis and necrosis [416]. We also present simulation results for tumors growing into heterogeneous tissue [438] and the coupling of their growth with angiogenesis [439].

4.1   Regimes of growth

The study of spherically symmetric tumor growth provides insight into the regimes of growth described by the basic tumor growth model in Chapter 3 (e.g., [97, 149, 416]). In this case, the PDEs reduce to ODEs in the radial polar coordinate r, $0 \leq r \leq R$, R being the tumor radius. Equations (3.23), (3.24) have the nonsingular solutions for a d-dimensional tumor

$$\Gamma(r, t) = \begin{cases} \dfrac{I_0(r)}{I_0(R)}, & d = 2, \\[2ex] \left(\dfrac{\sinh(R)}{R}\right)^{-1} \dfrac{\sinh(r)}{r}, & d = 3, \end{cases} \tag{4.1}$$

where $I_0(R)$ is the modified Bessel function of the first kind of zeroth order, and

$$p(r, t) = \frac{d - 1}{R} - \frac{AGR^2}{2d}; \tag{4.2}$$

G and A are the adhesion and apoptosis parameters, respectively (Eqs. (3.14), (3.15)). Note that $p(r, t) \equiv p(R, t)$, i.e., p is uniform across the tumor volume.

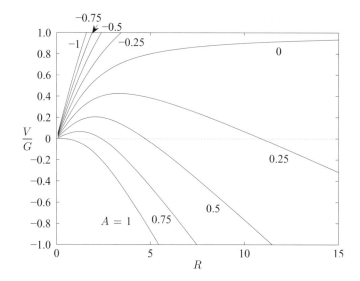

Figure 4.1 Rescaled rate of growth V/G as a function of the rescaled tumor radius R for radially symmetric tumor growth, for $d = 3$; plots are shown for various values of the apoptosis parameter A.

The evolution equation for the tumor radius R is given by

$$\frac{dR}{dt} = V = -AG\frac{R}{d} + G\begin{cases} \dfrac{I_1(R)}{I_0(R)}, & d = 2, \\[2ex] \dfrac{I_{3/2}(R)}{I_{1/2}(R)}, & d = 3. \end{cases} \tag{4.3}$$

where $I_\ell(x)$ is the ℓth-order modified Bessel function of the first kind. We have also used the fact that $1/\tanh(R) - 1/R = I_{3/2}(R)/I_{1/2}(R)$.

In both two and three dimensions, growth is unbounded ($R \to \infty$) if and only if $AG \le 0$. Figure 4.1 shows the rescaled growth velocity for $d = 3$. Note that for $d = 2$ the results are qualitatively similar to those shown.

For a given value of A, evolution from the initial condition $R(0) = R_0$ occurs along the corresponding curve. Three regimes can be identified using the apoptosis parameter A and the vascular parameter B (see the text before Eq. (3.11)); this behavior is qualitatively unaffected by the number of spatial dimensions d.

1. *Low vascularization*: $G \ge 0$ and $A > 0$ (i.e., $B < \lambda_A/\lambda_M$, where λ_A is the rate of apoptosis and λ_B is the blood–tissue transfer rate (see Section 3.2.1)). Note that the special case of avascular growth ($B = 0$) belongs to this regime. The evolution is monotonic and always leads to a stationary state R_∞, which corresponds to the intersection of a curve in Figure 4.1 with the dotted line $V = 0$. This behavior is in agreement with experimental observations of the *in vitro* diffusional growth [293] of avascular spheroids to a dormant steady state [479, 647]. In experiments, however, tumors always develop a necrotic core that further stabilizes their growth [99].

2. *Moderate vascularization*: $G \geq 0$ and $A \leq 0$ (i.e., $1 > B \geq \lambda_A/\lambda_M$). Unbounded growth occurs from any initial radius $R_0 > 0$. The growth tends to exponential for $A < 0$, with velocity $V \to -AGR/d$ as $R \to \infty$, and to linear for $A = 0$ with velocity $V \to G$ as $R \to \infty$.

3. *High vascularization*: $G < 0$ (i.e., $B > 1$). For $A > 0$, growth (i.e., $V > 0$) may occur, depending on the initial radius, and is always unbounded; for $A < 0$ (for which cell apoptosis is dominant, so that $\lambda_A/\lambda_M > B$), evolution is always to the only stationary solution, $R_\infty = 0$. This stationary solution may also be achieved for $A > 0$.

In Figure 4.2, top, the evolution of tumor radii as a function of time is shown for $A = 0.5$, $G = 20$ in the low-vascularization regime. All the initial data converge to the steady-state radius, indicating that the evolution is stable with respect to radially symmetric perturbations [241]. Figure 4.2, bottom, shows the evolution of the cell substrate concentration inside the expanding tumor for $d = 3$ and initial radius $R = 3$.

The stationary radius R_∞ is independent of G, and is the solution for $V = 0$ in Eq. (4.3). Figure 4.3 shows the relation between A and the stationary radius R_∞.

Note that the stationary radius has the following limiting behaviors for $d = 2$ and 3:

$$
\begin{aligned}
R_\infty &\to dA^{-1}, & A &\to 0, \\
R_\infty &\to d^{1/2}(d+2)^{1/2}(1-A)^{1/2}, & A &\to 1,
\end{aligned}
\tag{4.4}
$$

where R_∞ vanishes in the second case. Note that the limit $A \to 1$ corresponds to $\lambda_A \to \lambda_M$.

The pressure P_C at the center of the tumor ($r = 0$) is obtained from Eq. (3.12):

$$
\frac{P_C}{\gamma/L_D} = \frac{d-1}{R+G} - \frac{AGR^2}{2d} - G
\begin{cases}
\dfrac{1}{I_0(R)}, & d = 2, \\[2mm]
\dfrac{R}{\sinh(R)}, & d = 3,
\end{cases}
\tag{4.5}
$$

which has the asymptotic behavior $P_C(\gamma/L_D)^{-1} \to -AGR^2/(2d)$ as $R \to \infty$, indicating that if a tumors grows unboundedly ($AG \leq 0$) the pressure at the center also does (unless $A = 0$). This is a direct consequence of the absence of a necrotic core in this model. In reality, the increasing pressure may itself contribute to necrosis [494, 536]. It is known [119] that tumor cells continuously replace the loss of cell volume in the tumor because of necrosis, thus maintaining a finite pressure.

4.2   Linear analysis

We now derive equations that describe how a small perturbation from radial symmetry evolves in time. Consider a perturbation of the radially symmetric tumor interface Σ:

$$
r_\Sigma = R(t) + \delta(t)
\begin{cases}
\cos(\ell\theta), & d = 2, \\
Y_{\ell,m}(\theta, \varphi), & d = 3,
\end{cases}
\tag{4.6}
$$

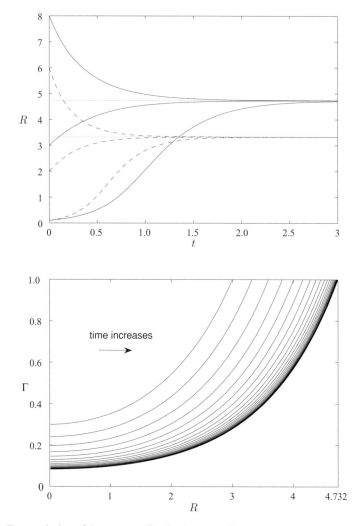

Figure 4.2 Top: evolution of the tumor radius leads to a stationary state for $A = 0.5$, $G = 20$. For $d = 2$ (broken lines) the initial radii are 0.1, 2, and 6. For $d = 3$ (solid lines) the initial radii are 0.1, 3, and 8. Bottom: evolution of the cell substrate concentration inside the expanding tumor with initial radius 3 and $d = 3$; profiles are plotted for 0.1 time increments.

where r_Σ is the radius of the perturbed tumor–host interface, δ is the dimensionless perturbation size, and $Y_{\ell,m}$ is a spherical harmonic, where ℓ and θ are the polar wavenumber and angle and m and φ are the azimuthal wavenumber and angle, respectively. By solving the system of Eqs. (3.23), (3.24) in the presence of a perturbed interface and matching the linear terms, a linear evolution equation is obtained for the shape factor δ/R in terms of Bessel functions (see Section 4.1). For $d = 2$ we have

$$\left(\frac{\delta}{R}\right)^{-1}\frac{d\,(\delta/R)}{dt} = \frac{-\ell(\ell^2 - 1)}{R^3} + \frac{\ell A G}{2} + G - G\left(\frac{\ell + 2}{R} + \frac{I_{\ell+1}(R)}{I_\ell(R)}\right)\frac{I_1(R)}{I_0(R)},$$

$$(4.7a)$$

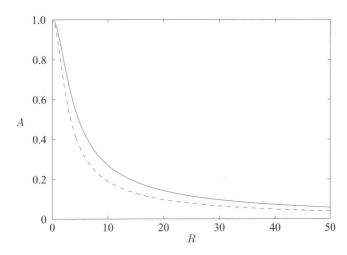

Figure 4.3 The apoptosis parameter A as a function of the stationary radius R_∞, obtained when $V = 0$ in Eq. (4.3); $d = 2$ (broken lines) and $d = 3$ (solid line).

and for $d = 3$ we have

$$\left(\frac{\delta}{R}\right)^{-1} \frac{d(\delta/R)}{dt} = \frac{-\ell(\ell+2)(\ell-1)}{R^3} + \frac{\ell A G}{3}$$
$$+ G - G \left(\frac{\ell+3}{R} + \frac{I_{\ell+3/2}(R)}{I_{\ell+1/2}(R)}\right) \frac{I_{3/2}(R)}{I_{1/2}(R)}. \qquad (4.7b)$$

Note that δ/R is the appropriate way to measure the perturbation since the underlying radius of the symmetric tumor is time dependent. Also observe that the linear term in the evolution of the perturbation depends on ℓ but is independent of the azimuthal wavenumber m and that there is a critical mode ℓ_c such that perturbations grow for $\ell < \ell_c$ and decay for $\ell > \ell_c$. The critical mode depends on the parameters A, G and on the evolving radius R. This extends prior linear analyses [97, 101, 293], in which only the special case in which the unperturbed configuration is stationary (i.e. R is constant in time) was considered.

The condition $(d/dt)\,(\delta/R) = 0$ defines the boundary between the regimes of stable $(\delta/R \to 0)$ and unstable evolution $(|\delta/R| \to \infty)$. Let us take G to be constant and take $A = A(\ell, G, R)$ such that $(d/dt)\,(\delta/R) = 0$. This gives for $d = 2$

$$A(\ell, G, R) = \frac{2(\ell^2-1)}{GR^3} + \left(\frac{2(1+2/\ell)}{R} + \frac{2}{\ell}\frac{I_{\ell+1}(R)}{I_\ell(R)}\right) \frac{I_1(R)}{I_0(R)} - \frac{2}{\ell}, \qquad (4.8a)$$

and, for $d = 3$,

$$A(\ell, G, R) = \frac{3(\ell-1)(\ell+2)}{GR^3} + \left(\frac{3(1+3/\ell)}{R} + \frac{3}{\ell}\frac{I_{\ell+3/2}(R)}{I_{\ell+1/2}(R)}\right) \frac{I_{3/2}(R)}{I_{1/2}(R)} - \frac{3}{\ell}.$$
$$(4.8b)$$

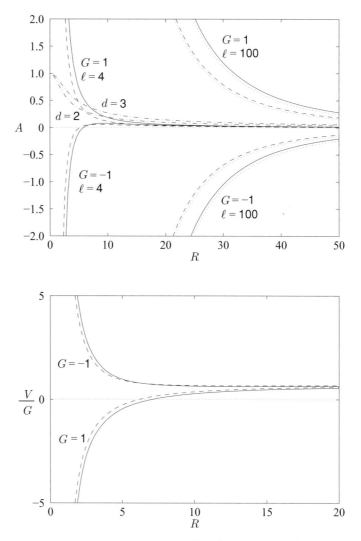

Figure 4.4 Top: the apoptosis parameter A as a function of the unperturbed radius R from condition (4.8b); G and ℓ label the groups of curves. Broken lines, $d = 2$; solid lines, $d = 3$; dotted lines, asymptotic behaviors from condition (4.9). The two broken-and-dotted curves for $d = 2, 3$ correspond to stationary radii. Bottom: the corresponding growth velocities $G^{-1}V$ for $\ell = 4$.

Note as $\ell \to \infty$ that $A(\ell, G, R)$ has the following limiting behavior:

$$\begin{cases} A(\ell, G, R) \to \dfrac{2(\ell^2 - 1)}{GR^3}, & d = 2, \\[3mm] A(\ell, G, R) \to \dfrac{3(\ell - 1)(\ell + 2)}{GR^3}, & d = 3. \end{cases} \tag{4.9}$$

In Figure 4.4, top, the apoptosis parameter $A(\ell, G, R)$ is shown for $d = 2$ (broken lines) and $d = 3$ (solid lines). The corresponding growth velocity V, obtained from (4.3) with $A = A(\ell, G, R)$, is plotted in Figure 4.4, bottom. The A-curves separate the parameter space into regions of stable and unstable growth for a given mode ℓ. The stable regions are $A < A(\ell, G, R)$, i.e., below the curve, for $G > 0$ and $A > A(\ell, G, R)$ for $G < 0$. The unstable regions are $A > A(\ell, G, R)$ for $G > 0$ and $A < A(\ell, G, R)$ for $G < 0$.

The linear analysis reveals that with constant parameters A and G the evolution may be either stable or unstable. Moreover, unstable growth is possible *only* in the low-vascularization regime (see [149]). In particular, in the low-vascularization regime and in the high-vascularization regime with $A < 0$, either bounded growth or shrinkage may occur. In these regimes the perturbation may either grow or decay and thus complicated tumor morphologies can develop, as will be shown later. In the moderate-vascularization regime and in the high-vascularization regime for $A > 0$, unbounded growth may occur and is always stable. The limiting behaviors as $R \to \infty$ are $A \approx d/(\ell R)$ for $d = 2$ and 3 and are independent of G. For $G > 0$ and $A < 0$ (the moderate-vascularization regime), growth is always stable for both $d = 2$ and 3 since $A < A(\ell, G, R)$ ($A(\ell, G, R) > 0$); for $G < 0$ and $A > 0$ (the high-vascularization regime), the stability condition is in this case $A > A(\ell, G, R)$. Note that growth occurs for A values above the curves describing the stationary radii, and thus growth is always stable for both $d = 2$ and 3 since the curves describing the stationary radii are always above the curves of $A(\ell, G, R)$. In these regimes the behaviors of the shape factor δ/R as $R \to \infty$ (for both $d = 2$ and 3) are

$$\frac{\delta}{R} \sim \begin{cases} R^{-\ell}, & A \neq 0, \\ R^{-1}, & A = 0. \end{cases} \tag{4.10}$$

Note that in this case $V \to G$ for both $d = 2$ and $d = 3$. Figure 4.5 shows the prediction from linear analysis of the evolution of a perturbation during unbounded tumor growth ($R \to \infty$).

Linear stationary states
In the low-vascularization regime, the existence of nonsymmetric steady-state tumor shapes is predicted by linear theory. This is seen as follows. The stationary radius $R_\infty = R_\infty(A)$ is the solution of Eq. (4.3) for $V = 0$ and is a function of $0 < A < 1$. There exists a non-negative critical $G_\ell(R_\infty, A)$, given by

$$G_\ell(R_\infty, A) = \begin{cases} \dfrac{2\ell(\ell^2 - 1)}{R_\infty^3 \left\{ 2 - A\left[2 + R_\infty I_{\ell+1}(R_\infty)/I_\ell(R_\infty) \right] \right\}}, & d = 2, \\[4ex] \dfrac{3\ell(\ell - 1)(\ell + 2)}{R_\infty^3 \left\{ 3 - A\left[3 + R_\infty I_{\ell+3/2}(R_\infty)/I_{\ell+1/2}(R_\infty) \right] \right\}}, & d = 3, \end{cases}$$

$$\tag{4.11}$$

such that for $G = G_\ell(R_\infty, A)$ the perturbation remains stationary. It can be shown that, for both $d = 2$ and 3, $G_\ell(R_\infty, A) > 0$ (Figure 4.6) and a perturbed stationary shape always exists. The perturbation δ/R_∞ grows unbounded for $G > G_\ell(R_\infty, A)$ and decays to zero for $G < G_\ell(R_\infty, A)$. At large radii R_∞, $G_\ell(R_\infty, A) \to 0$; thus in this limit perturbations of stationary states always grow unboundedly owing to the reduced cell–cell adhesion.

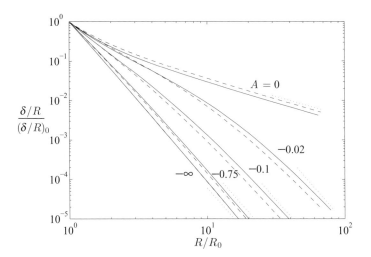

Figure 4.5 The rescaled growth ratio as a function of the rescaled radius R/R_0 during unbounded growth in the moderate-vascularization regime for $d = 2$ (broken lines) and $d = 3$ (solid lines); $G = 1$, $\ell = 4$, and A labels the groups of curves. The initial radius $R_0 = 10$. The asymptotic behaviors (dotted lines) are from Eq. (4.10). For $A = -\infty$ the broken and solid curves are indistinguishable from each other.

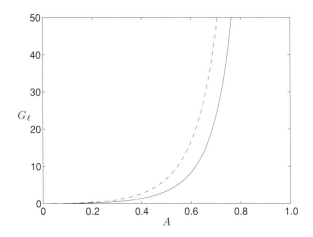

Figure 4.6 The parameter G_ℓ from the condition (4.11) for shape invariance with $R_\infty(A)$ vs. A $(0 < A < 1)$, for $d = 2$ (broken line) and $d = 3$ (solid line). Here $\ell = 4$.

Linear self-similar evolution

Here we study the self-similar evolution of the tumor (i.e., evolution where the overall scale changes but the shape does not), using linear analysis, and explore the possibility of controlling the shape during evolution to take advantage of the self-similar conditions and thus prevent the formation of shape instabilities.

Note that when $(d/dt) \, (\delta/R) = 0$ the tumor shape does not change in time and the evolution is linearly self-similar. Equation (4.8b) shows that self-similar evolution is possible, for both $d = 2$ and 3, if G is kept constant and A is varied as a function of the unperturbed radius R. In particular, Figure 4.4, top, indicates that in the low-vascularization regime self-similar evolution towards a stationary state is not possible for G constant (the stationary states R_∞ correspond to the intersection of the curves in Figure 4.4, top, with the curves describing stationary radii). Instead, one obtains self-similarly growing and shrinking tumors. For instance, the growth velocity V is less than zero for initial radius $R_0 < R_\infty$, and thus self-similar shrinkage of the tumor to zero occurs. However, for $R_0 > R_\infty$, V is greater than zero and thus self-similar growth away from the stationary radius occurs. In the high-vascularization regime, during self-similar evolution, V is less than zero. Thus self-similar shrinkage of a tumor from an arbitrary initial condition to a point occurs, and self-similar unbounded growth is not possible. Similarly, self-similar unbounded growth does not occur in the moderate-vascularization regime. Therefore, self-similar growth is possible only in the low-vascularization regime.

These considerations show that if one is able to control the apoptosis parameter in a subtle way, using Eq. (4.8b) to devise appropriate therapeutic treatments, it may be possible to prevent a tumor from becoming unstable and invasive. This has important implications for the angiogenic response of the host – a smaller surface area to volume ratio means a lower flux of angiogenic factors – as well as for the resectability of the tumor since a compact shape is easier to remove surgically.

Diffusional instability

The linear stability analysis shows that, during growth, perturbations can increase only in the low-vascularization regime. In the moderate- and high-vascularization regimes, perturbations always decay during growth. For example, taking A to be a non-negative constant, observe from Figure 4.4 that the evolution will cross the $A(\ell, G, R)$ curves and hence become unstable only for $G > 0$. Instability arises because growth is limited by the diffusion of cell substrates (i.e. diffusional instability). This is analogous to the Mullins–Sekerka diffusional instability that occurs in crystal growth [483]. In the high-vascularization regime, shrinkage may be unstable.

The biological significance of this diffusional instability [149] is that the instability allows the tumor to increase its surface area relative to its volume, thereby allowing the cells in the tumor bulk greater access to cell substrates. This in turn allows the tumor to overcome the diffusional limitations on growth and to grow to larger sizes than would be possible if it were spherical. Thus, diffusional instability provides an additional pathway for tumor invasion that does not require an additional cell substrate source such as would be provided from a newly developing vasculature through angiogenesis. This is discussed further in the next section and in Chapter 9.

4.3   Nonlinear results

To investigate the effect of nonlinearity, efficient numerical algorithms have been developed [149, 416] to solve Eqs. (3.23), (3.24). These partial differential equations over the whole domain are reformulated into boundary integral equations that hold only on the tumor–host interface, using potential theory. The boundary integral equations are then discretized and solved iteratively using a generalized minimal residual (GMRES) algorithm [578] and a discretized version of the Dirichlet–Neumann map is used to determine the normal velocity of the tumor interface. In addition, to gain enhanced accuracy in three dimensions a spatial rescaling scheme is used to scale out the overall growth of the evolving tumor in such a way that the volume of the tumor remains unchanged. It is important to note that, particularly in three dimensions, it is nontrivial to maintain stability and accuracy, because of singularities in the kernels, the high-order effects of cell–cell adhesion, and the strong nonlinear, nonlocal, character of the free-boundary system.

Nonlinear stationary states

Friedman and Reitich [242, 243] proved that there exist nonsymmetric steady tumor shapes that are solutions of the fully nonlinear equations. Their proof was not constructive, however. In [149], the numerical scheme described above was used to obtain approximations of these solutions in two dimensions. In the nonlinear case, nonsymmetric steady tumor shapes may be found by taking $G = G_\ell^{\mathrm{NL}} < G_\ell$, where $G_\ell^{\mathrm{NL}} - G_\ell = O(\delta^2)$. Thus, nonlinearity is destabilizing for stationary shapes.

Nonlinear self-similar evolution

We now investigate the effect of nonlinearity on the self-similar evolution for $d = 2$ predicted by the linear analysis. As discussed above, self-similar evolution requires a time-dependent apoptosis parameter $A = A(\ell, G, R)$, as plotted in Figure 4.4, top. The radius R used in the nonlinear simulation is determined from the area of an equivalent circle: $R = \sqrt{\mathrm{area}/\pi}$.

    An example of nonlinear self-similar evolution in the low-vascularization regime with $\ell = 4$ and $G = 1$ is shown in Figure 4.7. In the simulation, the initial radius and perturbation are $R_0 = 7$ and $\delta_0 = 0.3$. Since the velocity $V > 0$, from Figure 4.4, bottom, the tumor grows and correspondingly A decreases. The solid curves in Figure 4.7 correspond to the nonlinear solution, at time $t = 0$ and at a later time, and the broken curves correspond to the linear solution. The two are virtually indistinguishable, revealing that self-similar evolution is robust to nonlinearity for small perturbations.

    In Figure 4.8 the linear (broken-line) and nonlinear (solid-line) solutions are compared in the low-vascularization regime, for $\ell = 5$, $G = 1$, $A = A(\ell, G, R)$, and $R_0 = 4$. Since $V < 0$, the tumors are shrinking and A is increasing. In the left-hand frame $\delta_0 = 0.2$ and in the right-hand frame $\delta_0 = 0.4$. The results reveal that large perturbations are nonlinearly unstable and grow, leading to tumor fragmentation. This can have significant implications for therapy. For example, one can imagine an experiment in which a tumor is made to shrink by therapy in such a way that as a result A is increased by increasing the apoptosis rate λ_A. This example shows that a rapid decrease in size can result in shape instability leading to tumor breakup and the formation of microscopic tumor fragments that can enter the blood stream through leaky blood vessels, thus leading to metastases.

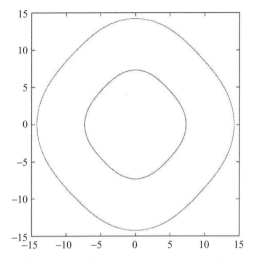

Figure 4.7 Self-similar nonlinear growth for $R_0 = 7$ and $\delta_0 = 0.3$ at times $t = 0$ and $t = 20$. The solid curves correspond to the nonlinear solution and the broken curves to the linear solution. The solid and dashed curves are virtually indistinguishable, revealing that the self-similar evolution is robust to nonlinearity for small perturbations. The evolution is in the low-vascularization regime with $d = 2$, $G = 1$, $\ell = 4$, and the time-dependent $A = A(\ell, G, R)$ given in Eq. (4.8a). With kind permission of Springer Science and Business Media: *J. Math. Biol.*, Cristini *et al.* vol. 46, p. 214, © 2003 Springer.

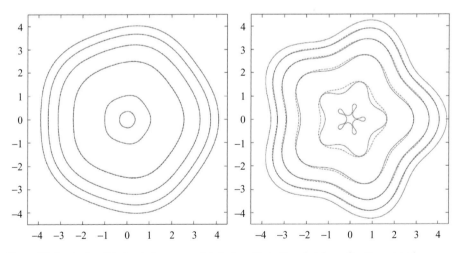

Figure 4.8 Analysis of the basic tumor model. Left: self-similar shrinkage for $R_0 = 4$ and $\delta_0 = 0.2$ ($t = 0$ to 0.96 shown). Right: Unstable shrinkage for $R_0 = 4$ and $\delta_0 = 0.4$ ($t = 0$ to 0.99). The solid curves correspond to the nonlinear solution and the broken curves to the linear solution (note that most solid and broken lines overlap). In both cases $d = 2$, $G = 1$, $\ell = 5$, and the evolution is in the low-vascularization regime; $A = A(\ell, G, R)$ is given in Eq. (4.8a). With kind permission of Springer Science and Business Media: *J. Math. Biol.*, Cristini *et al.*, vol. 46, p. 215, © 2003 Springer.

Nonlinear evolution in the high-vascularization regime

In the high-vascularization regime, both the shrinkage and growth of tumors may occur. Shrinkage may be stable, self-similar, or unstable. In contrast, unbounded growth is always stable for both $d = 2$ and 3. In the nonlinear regime, self-similar shrinkage and unstable shrinkage are qualitatively very similar to that presented in Figure 4.8. Examples of stable shrinkage and unbounded stable growth can be found in [149]. In fact, all the nonlinear simulations of growth in the high-vascularization regime lead to stable evolution, in agreement with the linear analysis, i.e., well-vascularized tumors tend to grow in compact, nearly spherical, shapes showing little or no signs of invasiveness. This prediction suggests that tumors could maintain stable morphology under more normal microenvironmental conditions, as has been observed in experiments [58, 392, 398, 493, 576, 629].

In some cases, it has been observed experimentally that highly vascularized cancers evolve invasively by extending branches into regions of the external tissue where the mechanical resistance is lowest (e.g. [119]). These results suggest that the formation of invasive tumors is due to anisotropies rather than to vascularization alone. Anisotropies (e.g., in the distribution of the resistance of the external tissue to tumor growth or in the distribution of blood vessels) are neglected in the model here but will be included in the next section. This consequence, which had not been recognized before, is supported by recent experiments [499] of *in vivo* angiogenesis and tumor growth.

Nonlinear unstable growth

In Figure 4.9 the evolution of a two-dimensional tumor surface from a nonlinear boundary integral simulation (solid lines) is compared with the result of the linear analysis (dotted lines), using $A = 0.5$, $G = 20$ [149]. According to the linear theory (Eq. (4.3)), the tumor grows. The radially symmetric equilibrium radius $R_\infty \approx 3.32$. The $\ell = 2$ mode is linearly stable initially but becomes unstable at $R \approx 2.29$. The linear and nonlinear results in Figure 4.9 are indistinguishable up to $t = 1$ and gradually deviate thereafter. A shape instability develops and forms a neck. At $t \approx 1.9$ the linear solution collapses, suggesting a pinch-off. However, the nonlinear solution is stabilized by the cell-to-cell adhesive forces (due to surface tension) that resist the development of high negative curvatures in the neck. This is not captured by the linear analysis. Instead of pinching off, as is predicted by a linear evolution, the nonlinear tumor continues to grow and develops large bulbs that eventually reconnect, thus trapping healthy tissue (the shaded regions in the last frame in Figure 4.9) within the tumor. The frame at $t = 2.531$ describes the onset of reconnection of the bulbs. It is expected that reconnection is affected by the diffusion of cell substrates outside the tumor, which is not included in the model used here but will be considered in the next section.

An analogous evolution is observed in three dimensions. See Figure 4.10, where two three-dimensional views of the morphology are shown. The tumor appears not to change volume in the simulation, because of the spatial rescaling. At early times the perturbation decreases and the tumor becomes sphere-like. As the tumor continues to grow, the perturbation starts to increase around time $t \approx 0.4$. The tumor begins to take on a flattened ellipse-like shape. Around time $t \approx 2.2$ the perturbation growth accelerates

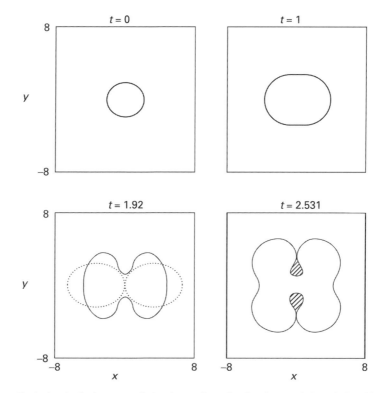

Figure 4.9 Evolution in the low-vascularization regime, for $d = 2$, $A = 0.5$, and $G = 20$, of the initial tumor surface shown. Dotted lines, the solution from linear analysis; solid lines, the solution from a nonlinear calculation with time step $\triangle t = 10^{-3}$ and a number of marker points $N = 1024$, reset, after time $t = 2.51$, to $\triangle t = 10^{-4}$ and $N = 2048$. With kind permission of Springer Science and Business Media: *J. Math. Biol.*, Cristini *et al.* vol. 46, p. 202, © 2003 Springer.

dramatically, and dimples form around time $t \approx 2.42$. The dimples deepen, and the tumor surface buckles inwards. The mesh adapts accordingly.

Here the diffusional instability leads to repeated sub-spheroid development in the nonlinear regime. This provides an even larger source of cell substrates to interior cells than is predicted by linear theory. Furthermore, growth by the successive formation of sub-spheroids or budding is observed in experiments [230].

4.4 Model calibration from experimental data

The parameters G and A in the basic tumor growth model, which measure the strength of cell adhesion and apoptosis, can be estimated by comparison with experimental tumor spheroid results. This work was performed by Frieboes *et al.* [230] using ACBT (grade IV human glioma multiforme) tumor spheroids, with experiments performed *in vitro*; the results were analyzed using the mathematical model. Cell adhesion and proliferation rates were modulated by varying the concentrations of glucose and of fetal bovine

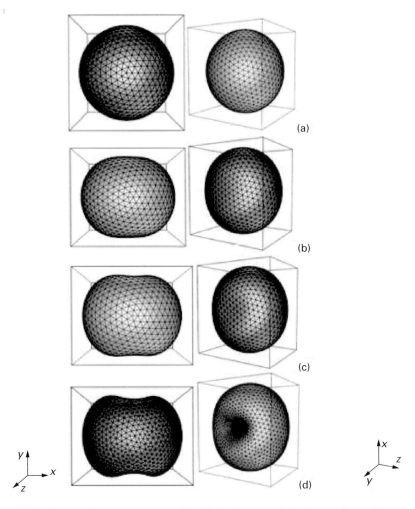

Figure 4.10 Evolution of the basic tumor surface in the low-vascularization regime, with $A = 0.5$, $G = 20$, and the initial tumor surface as defined in [149]: (a) $t = 0$, $\bar{\delta} = 0.0137$, $S = 3.0$; (b) $t = 2.21$, $\bar{\delta} = 0.12$, $S = 4.732$; (c) $t = 2.42$, $\bar{\delta} = 0.2$, $S = 4.745$; (d) $t = 2.668$, $\bar{\delta} = 0.496$, $S = 4.781$. Reprinted from *Disc. Cont. Dyn. Syst. – Series B*, Li *et al.*, vol. 7, p. 599, © 2007, with permission from the American Institute of Mathematical Sciences.

serum (FBS), respectively. As expected, the rates of proliferation increased with the serum concentration. Cases with 1% FBS had the slowest tumor mass growth, whereas 10% FBS cases had the fastest growth. Serum may also decrease cell adhesion while increasing proliferation. Furthermore, cell mobility (adhesion) was found to increase (decrease) with glucose concentration, in agreement with previous observations that higher levels of glucose reduce oxygenation [481] and that hypoxia can increase tumor cell motility by increasing the production of autocrine motility factor and by upregulation of hepatocyte growth factor (HGF) [108, 539, 710, 711]. Thus, an important effect of higher glucose was to decrease cell adhesion forces and thereby contribute to tumor morphologic instability. In the model, this corresponds to larger values of the parameter G. The most stable, compact, morphologies were observed with low or medium levels of both glucose and FBS; in these morphologies tumors shed the fewest cells and attained the smallest overall sizes. In contrast, the combination of high glucose and any serum concentration exhibited very unstable morphologies because the cells were very motile. Similarly, the combination of any glucose and 10% FBS developed unstable morphologies apparently because cells proliferated faster than the time needed for them to connect into a stable structure.

By estimating the size of the viable rim of cells on the tumor periphery, the diffusion length $L_D \approx 100$ μm is obtained. By fitting Eq. (4.3) to the spheroid growth curves at small times (when the growth is nearly exponential), the proliferation rate λ_M is estimated to be ≈ 1 day^{-1}. The calibrated model is consistent with other measurements. For instance, Eq. (3.2) gives an oxygen-penetration length scale $L_{oxy} = (D_{oxy}/\lambda_{oxy,uptake})^{1/2}$. By measuring the distance between the necrotic core and the basement membrane, this length can be estimated to be 100–140 μm. Using previously published values, $\lambda_{oxy,uptake} = 9.41 \times 10^{-2}$ s^{-1} [115] and $D_{oxy} = 1.45 \times 10^{-5}$ cm^2 s^{-1} [497], gives $L_{oxy} = 124$ μm, in agreement with these results. Similar calculations were consistent for calculating the glucose penetration length and uptake rate [267, 295, 358], confirming that hypoxia is the limiting condition for tumor cell viability.

An analytical relation between A and R can be established for a steady-state solution by taking $V = 0$ in Eq. (4.3). The corresponding steady-state relation $A = A_s(R)$ can be used to estimate the value of A in the experiment by taking R to be the average experimental tumor radius, nondimensionalized by L_D, of morphologically stable spheroids (for low or medium levels of both glucose and FBS). The parameter A was thus estimated as $0.26 \leq A \leq 0.38$. This is an overestimate in the proliferating rim, however, because the mathematical model as described assumes that cell death occurs uniformly throughout the spheroid. In fact, cell death is spatially heterogeneous, the largest values occurring in the interior hypoxic region where cells are starved of oxygen and nutrients.

Frieboes *et al.* showed that the parameter G can be estimated by comparing the pressure in the proliferating rim with the pressure at the tumor boundary. In the proliferating rim at steady state the dimensional pressure is from Darcy's law approximately $P \approx L_D^2 \lambda_M/\mu$ where μ is the mobility, while at the tumor boundary $P \approx \tau/(L_D R)$, where R is the nondimensional (stable) tumor spheroid radius; this follows from Eq. (3.8). Equating the two and using the definition of G with $B = 0$, again with R as the average experimental tumor radius, nondimensionalized by L_D, the estimate $0.6 \leq G \leq 0.9$ is obtained for stable tumor spheroids. Spheroids with values of G above this range will be morphologically unstable owing to weak adhesive forces. Interestingly, as Frieboes *et al.* pointed out, this approach provides an indirect method for estimating G without directly measuring cell–cell adhesion.

Given A and G, Frieboes *et al.* used the linear stability analysis to predict the morphological stability of the tumor spheroids. In Figure 4.11 [230] a phase diagram is presented in which the marginal stability curves $A(\ell, G, R)$ from Eq. (4.8a) are plotted for $\ell = 4$ and $G = 0.9$ and 0.6. The parameter $\ell = 4$ was chosen because unstable morphologies

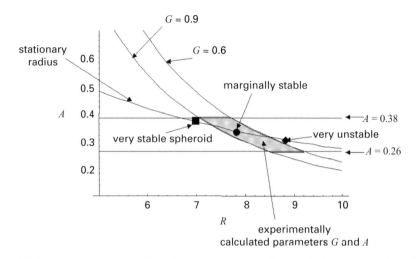

Figure 4.11 Evaluation of tumor stability. The diagram shows the apoptosis parameter A vs. the spheroid radius R (rescaled by the diffusion length L). The curves for the given values of G were obtained from [230]. The experimental conditions for morphologically stable spheroids (shaded area) are enclosed by these curves and by the horizontal lines delimiting the range of values of A, all estimated by fitting the mathematical model to the *in vitro* data. The curve labeled "stationary radius" was obtained by setting $dR/dt = 0$ in Eq. (4.3) in [230]. Three representative stationary spheroid radii are reported on this curve as sampled *in vitro*. Since the curve crosses and continues beyond the shaded region, most glioma spheroids under these *in vitro* conditions are marginally stable. Reprinted from Frieboes *et al.* [230], with permission from the American Association for Cancer Research.

in vitro seemed to exhibit mostly tumor surface perturbations characterized by low wave numbers (e.g., 3 or 4) at the onset of instability. The stationary relation $A_s(R)$ is also shown. The horizontal lines indicate the experimentally estimated values $A = 0.26$ and $A = 0.38$, obtained as described above. The experimental range of the parameters A and G is indicated by the shaded region. In the presence of cell-substrate gradients, the morphology can be unstable when the cell adhesion is weak (which occurs for large G), whereas for small G the tumor morphology is stabilized by cell adhesion [149]. The larger a tumor grows, the weaker the stabilizing effect of cell adhesion. Each $A(\ell, G, R)$ curve describes a tumor with specific cell characteristics and divides the parameter space into a stable region (on the left) and an unstable region (on the right). The lower the cell adhesion, the more shifted to the left is the G curve, thus reducing the range of tumor sizes that will be morphologically stable. As a tumor grows, this corresponds to moving from left to right and thus may lead to the eventual crossing of the G curve corresponding to that tumor's degree of morphological stability.

In Figure 4.11, the solid symbols denote experimental spheroids. The spheroid denoted by the solid square is very stable, being on the left of all the curves that are compatible with stable morphologies; the spheroid denoted by the solid diamond is very unstable; and the spheroid denoted by the solid circle is marginally stable (i.e., it may develop morphologic instability, depending on the value of G). In their experiments, Frieboes *et al.* were able to observe both stable and unstable spheroids, consistently with the theory, by varying the parameter G. This was done by using different glucose and FBS concentrations, since these cell substrates affect the parameter G by modulating the cell adhesion and proliferation as described above. This work shows that by training the model to estimate G from stable spheroid data, it can predict both stable and unstable spheroid morphologies.

4.5 Tumor growth with chemotaxis

In this section, the analysis is extended to take into account chemotaxis, using Eqs. (3.23), (3.24). The cell substrates are assumed to diffuse from the surface of a sphere of radius R_0, where $R_0 \gg R$, the tumor radius, and the cell substrate concentration at R_0 is assumed to be 1. The evolution equation for the tumor radius R is given by

$$\frac{dR}{dt} = V = -AG\frac{R}{d} + G \begin{cases} C_2, & d = 2, \\ C_3, & d = 3, \end{cases} \tag{4.12}$$

Where

$$C_2 = \frac{DI_1/I_0}{D + R[\log(R_0/R)]I_1/I_0},$$

$$C_3 = \frac{DI_{3/2}/I_{1/2}}{D + R[(R_0 - R)/R_0)]I_{3/2}/I_{1/2}},$$

and the Bessel functions $I_1, I_0, I_{3/2}, I_{1/2}$ are functions of R. Note that the radial velocity in Eq. (4.12) is independent of the chemotaxis coefficient χ, which is defined in Eq. (3.13).

The equation for the shape perturbation δ/R is given for $d = 2$ by

$$\left(\frac{\delta}{R}\right)^{-1} \frac{d(\delta/R)}{dt} = -\frac{\ell(\ell^2 - 1)}{R^3} + \ell\frac{AG}{2} - \frac{(\ell + 2D)G - \ell\chi}{DR}C_2$$
$$+ \frac{(1 + D)\ell G - \ell\chi}{(1 + D)\ell + RI_{\ell+1}/I_\ell}C_2\left[\frac{I_0}{I_1} + \frac{1 - D}{D}\left(\frac{\ell}{R} + \frac{I_{\ell+1}}{I_\ell}\right)\right], \tag{4.13a}$$

and for $d = 3$ by

$$\left(\frac{\delta}{R}\right)^{-1} \frac{d(\delta/R)}{dt} = -\frac{\ell(\ell + 2)(\ell - 1)}{R^3} + \ell\frac{AG}{3} - \frac{(\ell + 3D)G - \ell\chi}{DR}C_3$$
$$+ \frac{[(1 + D)\ell + D]G - \ell\chi}{(1 + D)\ell + D + RI_{\ell+3/2}/I_{\ell+1/2}}C_3$$
$$\times \left[\frac{I_{1/2}}{I_{3/2}} + \frac{1 - D}{D}\left(\frac{\ell}{R} + \frac{I_{\ell+3/2}}{I_{\ell+1/2}}\right)\right]. \tag{4.13b}$$

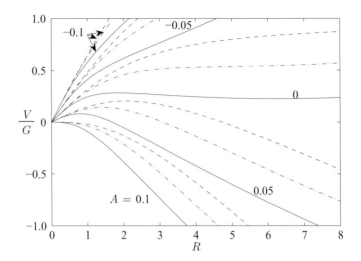

Figure 4.12 Rescaled rate of growth V/G as a function of the unperturbed tumor radius R, for $R_0 = 13$ and various values of A; $D = 1$ (solid lines), $D = 5$ (broken- and -dotted lines), $D = 1000$ (broken lines); $d = 3$.

Again, at the level of the linear theory, perturbations consisting of different spherical harmonics are superpositions of the above solutions. Also, observe that the shape perturbation depends on ℓ but not on m.

Figure 4.12 shows the rescaled rate of growth V/G as a function of the unperturbed tumor radius R for $d = 3$. For a given apoptosis parameter A and diffusion constant D, evolution from the initial condition $R(0)$ occurs along the corresponding curve. In Figure 4.13 the stability region is characterized by keeping G constant and varying A as a function of the unperturbed radius R in such a way that $d(\delta/R)/dt = 0$. We have for $d = 2$

$$
\begin{aligned}
A(\ell, G, R) = {} & \frac{2(\ell^2 - 1)}{GR^3} + 2\frac{1 + 2D/\ell - \chi/G}{DR}C_2 - 2\frac{1 + D - \chi/G}{(1 + D)\ell + RI_{\ell+1}/I_\ell}C_2 \\
& \times \left[\frac{I_0}{I_1} + \frac{1 - D}{D}\left(\frac{\ell}{R} + \frac{I_{\ell+1}}{I_\ell}\right)\right],
\end{aligned}
\tag{4.14a}
$$

and for $d = 3$

$$
\begin{aligned}
A(\ell, G, R) = {} & \frac{3(\ell + 2)(\ell - 1)}{GR^3} + 3\frac{1 + 3D/\ell - \chi/G}{DR}C_3 \\
& - 3\frac{1 + D + D/\ell - \chi/G}{(1 + D)\ell + D + RI_{\ell+3/2}/I_{\ell+1/2}}C_3 \\
& \times \left[\frac{I_{1/2}}{I_{3/2}} + \frac{1 - D}{D}\left(\frac{\ell}{R} + \frac{I_{\ell+3/2}}{I_{\ell+1/2}}\right)\right],
\end{aligned}
\tag{4.14b}
$$

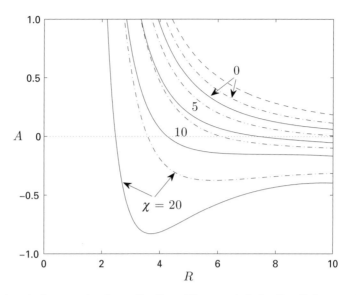

Figure 4.13 Apoptosis parameter A, as a function of the unperturbed radius R, from condition (4.14a) for various values of χ; $\ell = 4$, $G = 1$, $R_0 = 13$; $d = 3$, $D = 1$ (solid lines), $D = 5$ (broken- and -dotted lines). The broken curve at top right corresponds to condition (4.8b).

where $A(\ell, G, R)$ is the critical value dividing the plot into regions of stable growth and regions of unstable growth for a given mode ℓ. In Figure 4.13, $A(\ell, G, R)$ is plotted as a function of R, for various values of χ, $\ell = 4$, $G = 1$, $D = 1$ (solid lines), $D = 1000$ (broken- and -dotted lines), $d = 3$. The figure reveals that taxis destabilizes the tumor evolution.

4.6 Tumor growth with necrosis

The model is now updated to take into account the necrosis that may form in tumor regions depleted of oxygen and nutrients. Necrosis is a source of tumor volume loss in which cell mass is converted into water.

Tumors are assumed to consist of viable and necrotic cells, i.e., $\Omega_T = \Omega_V \cup \Omega_N$, where Ω_V denotes the viable tumor and Ω_N denotes the necrotic core. This necrotic core of the tumor contains tumor cells that are not viable owing to a lack of available cell substrates; it may be modeled as a region where the substrate level drops below n_N, where n_N represents the minimum cell substrate level for viability. For simplicity the parameter B denoting the extent of vascularization (see Section 3.2) is assumed to be 0, i.e., the growth is taken to be avascular.

The cell substrate concentration n obeys the following:

$$0 = D_T \nabla^2 n + \begin{cases} -\lambda n & \text{in} \quad \Omega_V, \\ 0 & \text{in} \quad \Omega_N. \end{cases} \tag{4.15}$$

Note that it is assumed that there is no transfer of substrates in the necrotic core.

The cell substrate concentration is fixed at the tumor interface Σ and is continuous at the interface Σ_N between the necrotic core Ω_N and viable tumor Ω_V:

$$n = n^\infty \qquad\qquad\qquad \text{on} \quad \Sigma, \qquad (4.16)$$

$$n = n_N, \quad [\mathbf{n} \cdot \nabla n] = 0 \qquad \text{on} \quad \Sigma_N, \qquad (4.17)$$

where it is assumed that the diffusion coefficient in the necrotic core is the same as that in the viable tumor. The local rate of volume change is given by

$$\nabla \cdot \mathbf{u} = \begin{cases} \lambda_p & \text{in} \quad \Omega_V, \\ -\lambda_N & \text{in} \quad \Omega_N, \end{cases} \qquad (4.18)$$

where λ_p is given by Eq. (3.5) and λ_N is the rate of volume loss in the necrotic core, modeling cell necrosis and disintegration therein. The velocity $\mathbf{u}$ obeys Darcy's law,

$$\mathbf{u} = -\mu \nabla P. \qquad (4.19)$$

The normal component of velocity is assumed to be continuous across the necrotic rim Σ_N and the pressure is assumed to extend continuously into the necrotic core:

$$[\mathbf{n} \cdot \mathbf{u}] = 0, \quad [P] = 0 \qquad \text{on} \quad \Sigma_N. \qquad (4.20)$$

The normal velocity of the tumor interface Σ is given by

$$V = -\mu \mathbf{n} \cdot (\nabla P)_\Sigma. \qquad (4.21)$$

The nondimensional equations are

$$\nabla^2 \Gamma = \begin{cases} \Gamma & \text{in} \quad \Omega_V, \\ 0 & \text{in} \quad \Omega_N, \end{cases} \qquad (4.22)$$

$$\nabla^2 p = G \begin{cases} \Gamma - A & \text{in} \quad \Omega_V, \\ -A_N & \text{in} \quad \Omega_N, \end{cases} \qquad (4.23)$$

where $A_N = \lambda_N / \lambda_M$ is the nondimensional rate of volume loss due to necrosis, $N = n_N / n^\infty$ is the nondimensional cell substrate limit for cell viability, $p = L_D P / \gamma$ is the modified pressure, and Γ, A, G are the same as in Chapter 3 (with $B = 0$). The nondimensionalized normal velocity of the tumor boundary can now be written as

$$V = -\mathbf{n} \cdot (\nabla p)_\Sigma. \qquad (4.24)$$

On Σ and Σ_N, Γ and p satisfy the following conditions:

$$\Gamma = 1, \quad p = \kappa \qquad\qquad\qquad \text{on} \quad \Sigma, \qquad (4.25)$$

$$\Gamma = N, \quad [\mathbf{n} \cdot \nabla \Gamma] = 0 \quad [\mathbf{n} \cdot \mu \nabla p] = 0 \qquad \text{on} \quad \Sigma_N. \qquad (4.26)$$

The evolution equation for the tumor radius R is (for $d = 3$)

$$\frac{dR}{dt} = V = -\frac{AG}{3}R + G\frac{A - A_{\mathrm{N}}}{3}\frac{R_{\mathrm{N}}^3}{R^2}$$
$$+ GN\frac{\sinh(R - R_{\mathrm{N}})}{R}\left(\frac{1}{\tanh(R - R_{\mathrm{N}})} - \frac{1}{R}\right)$$
$$+ GNR_{\mathrm{N}}\left(\frac{\sinh(R - R_{\mathrm{N}})}{R} - \frac{\cosh(R - R_{\mathrm{N}})}{R^2}\right). \qquad (4.27)$$

The necrotic core radius R_{N} satisfies

$$R = N\left(\sinh(R - R_{\mathrm{N}}) + R_{\mathrm{N}}\cosh(R - R_{\mathrm{N}})\right). \qquad (4.28)$$

The equation for the shape perturbation δ/R is given by, for $d = 3$,

$$\left(\frac{\delta}{R}\right)^{-1}\frac{d(\delta/R)}{dt}$$
$$= -\frac{\ell(\ell + 2)(\ell - 1)}{GR^3} + \frac{A}{3}\ell + \frac{A - A_{\mathrm{N}}}{3}(\ell - 3)\frac{R_{\mathrm{N}}^3}{R^3} + 1 - \left(\frac{\ell + 3}{R} + \frac{I_{\ell+3/2}}{I_{\ell+1/2}}\right)$$
$$\times N\left[\frac{\sinh(R - R_{\mathrm{N}})}{R}\left(\frac{1}{\tanh(R - R_{\mathrm{N}})} - \frac{1}{R}\right)\right.$$
$$+ R_{\mathrm{N}}\left.\left(\frac{\sinh(R - R_{\mathrm{N}})}{R} - \frac{\cosh(R - R_{\mathrm{N}})}{R^2}\right)\right]. \qquad (4.29)$$

Note that if $R_{\mathrm{N}} = 0$ then Eq. (4.27) reduces to Eq. (4.3) (with $B = 0$), and Eq. (4.29) reduces to Eq. (4.7b).

In Figure 4.14 the stability region is characterized for constant G: A is given as a function of the unperturbed radius R such that $d(\delta/R)/dt = 0$, thus the critical value of A, which divides the plot into regions of stable growth and regions of unstable growth for a given mode ℓ, is given by

$$A(\ell, G, R) = \frac{1}{\ell + (\ell - 3)R_{\mathrm{N}}^3/R^3}$$
$$\times \left(\frac{3\ell(\ell + 2)(\ell - 1)}{GR^3} + A_{\mathrm{N}}(\ell - 3)\frac{R_{\mathrm{N}}^3}{R^3} - 3 + 3\left(\frac{\ell + 3}{R} + \frac{I_{\ell+3/2}}{I_{\ell+1/2}}\right)\right.$$
$$\times N\left[\frac{\sinh(R - R_{\mathrm{N}})}{R}\left(\frac{1}{\tanh(R - R_{\mathrm{N}})} - \frac{1}{R}\right)\right.$$
$$\left.\left.+ R_{\mathrm{N}}\left(\frac{\sinh(R - R_{\mathrm{N}})}{R} - \frac{\cosh(R - R_{\mathrm{N}})}{R^2}\right)\right]\right). \qquad (4.30)$$

In this figure $\ell = 4$, $G = 1$, $A_{\mathrm{N}} = 0.1$, $d = 3$, and N takes the values shown. The figure reveals that necrosis destabilizes the tumor evolution.

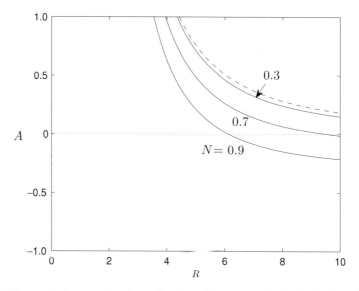

Figure 4.14 The apoptosis parameter A as a function of the unperturbed radius R from the condition (4.30), for various values of N; $\ell = 4$, $G = 1$, $A_N = 0.1$, $d = 3$. The broken curve corresponds to condition (4.8b).

4.7   Tumor growth in heterogeneous microenvironments

The approach described in Section 3.2.2 can be used to simulate tumor growth in a complex tissue. In Figure 4.15, tumor growth in a heterogeneous domain that mimics brain tissue is shown [438]. In the white region at the right-hand side of each frame, the pre-existing blood vessel density $B = 0$, the cell substrate diffusivity $D = 0.0001$, and the cell mobility $\mu = 0.0001$; this models a rigid material such as the skull. In the black regions $B = 0$, $D = 1$, and $\mu = 10$, which models the cerebrospinal fluid. The light-gray and dark-gray regions model the white and gray brain matter; in the light-gray regions $B = 1$, $D = 1$, and $\mu = 1.5$ and in the dark-gray regions $B = 1$, $D = 1$, and $\mu = 0.5$. The tumor is indicated by a thin white boundary in the center right of the frames. The proliferating, hypoxic, and necrotic regions in the tumor are shown in white, gray, and black, respectively.

The simulations are from time $t = 0$ to $t = 60$, i.e., approximately 45–90 days. The tumor initially grows rapidly until the cell substrate level drops below $n_P = 0.30$, at which time a large portion of the tumor becomes hypoxic and proliferation is down-graded. The tumor continues to grow at a slower rate until the interior of the tumor becomes necrotic ($t = 10.0$). This causes nonuniform volume loss within the tumor and contributes to morphological instability. Because the biomechanical responsiveness is continuous across the tumor boundary and the microenvironment has a moderate cell-substrate gradient, this simulation lies between the invasive, fingering, growth regime and the fragmenting growth regime (as described in [438]).

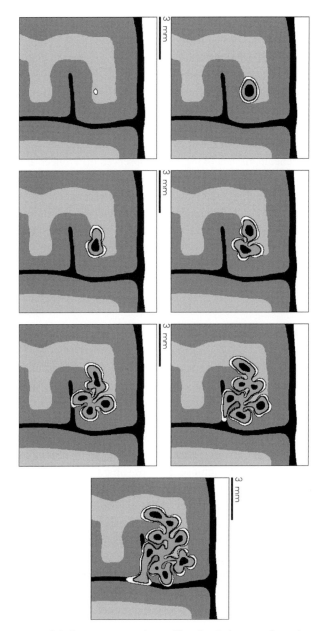

Figure 4.15 Tumor growth in heterogeneous tissue. The simulation runs from time $t = 0.0$ days (top left) to $t = 60.0$ days (bottom) in 10-day increments. The white band at the right of each frame models a rigid material such as the skull; the black regions represent an incompressible fluid (e.g., the cerebrospinal fluid); the light-gray and dark-gray regions represent tissues of differing biomechanical properties (e.g., white and gray brain matter). Tumor tissue is shown growing at center right, with viable outer layer (white), hypoxic middle layer (gray), and necrotic core (black). With kind permission from Springer Science plus Business Media: *J. Sci. Compat.*, Macklin and Lowengrub, vol. 35, pp. 293–4, ⓒ 2008 Springer.

As the tumor grows out of the biomechanically permissive tissue (light gray; $\mu = 1.5$) and into the biomechanically resistant tissue (dark gray; $\mu = 0.5$), its rate of invasion slows ($t = 20.0$). This results in preferential growth into the permissive (light-gray) material, a trend which can be clearly seen from $t = 30.0$ onward. When the tumor grows through the resistant tissue (dark gray) and reaches the fluid (black) ($t = 40.0$), it experiences a sudden drop in biomechanical resistance to growth. As a result, the tumor grows rapidly and preferentially in the 1/2 mm fluid structures that separate the tissue (see the last three frames). Such growth patterns are not observed in simulations of homogeneous tissues [438]. Other observed differences are due to the treatment of the hypoxic (quiescent) tumor cells. Regions that had previously been classified as necrotic (in [430, 435–437]) are now treated as quiescent. Tumor volume loss is thus reduced, and this may result in tumors with large hypoxic regions and small viable rims. Had the hypoxic regions been treated as necrotic, the invasive fingers would have been thinner and the tumor could have fragmented. Therefore, a separate treatment of the hypoxic regions can have a significant impact when simulating the details of invasive tumor morphologies.

4.8   Vascular tumor growth

The effect of solid or mechanical pressure-induced vascular response on tumor-induced angiogenesis and vascular growth is shown in Figure 4.16, which is taken from Macklin *et al.* [439]. With the parameter values used here, a solid pressure-induced vascular constriction occurs when the pressure $P \approx 0.8$. Angiogenesis is initiated from an avascular tumor configuration at $t = 45$ days (not shown), when 10 sprout tips are released from the parent vessel. At early times, the newly developing vessels migrate, proliferate, branch, and anastomose. It takes some time for flow to start, with significant flow developing only after about 10 days (in 55 days of total growth time). The blood flow in the neovasculature starts near the parent capillary and eventually reaches the tumor.

The solid pressure generated by tumor-cell proliferation prevents the delivery of oxygen internally to the tumor, and thus the delivery of oxygen is heterogeneous and significant oxygen gradients persist in the tumor interior. There is no functional microvasculature internal to the tumor. While the tumor responds by growing towards the oxygen-delivering neovasculature, the solid pressure constricts the neovasculature in the direction of growth (where the pressure is highest) and correspondingly inhibits the transfer of oxygen from those vessels. This situation slows the growth of the tumor.

The neovasculature in other areas of the host microenvironment thus provides a stronger source of oxygen. This triggers tumor-cell proliferation and growth in regions where previously proliferation had decreased. The heterogeneity of oxygen delivery and the associated oxygen gradients cause heterogeneous tumor cell proliferation. Proliferation is confined to regions close to the tumor–host interface. This results in morphological instability leading to the formation of invasive tumor clusters and complex tumor morphologies, which is consistent with the theory and the predictions presented earlier

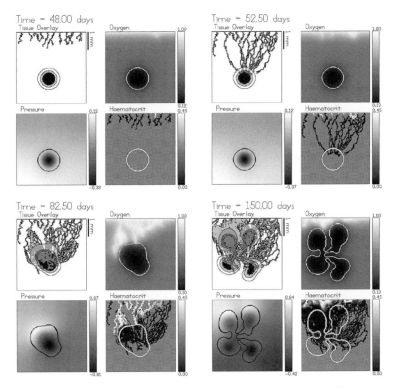

Figure 4.16 Tumor-induced angiogenesis and vascular tumor growth. The vessels respond to the mechanical pressure generated by the growing tumor. Accordingly, strong oxygen gradients are present that result in strongly heterogeneous tumor cell proliferation and shape instability. The tumor regions (outer, proliferating; middle, hypoxic or quiescent; inner, necrotic) and the oxygen, mechanical pressure, and ECM are shown. The times shown are $t = 48$ (three days after angiogenesis is initiated), 52.5, 82.5, and 150 days. Adapted from Macklin *et al.* [439].

[149, 416] that substrate inhomogeneities in the tumor microenvironment may cause morphological instability in growing tumors.

4.9 Conclusion

The analyses and nonlinear simulations presented in this chapter revealed that the two dimensionless parameters A and G uniquely subdivide tumor growth into three regimes, associated with increasing degrees of vascularization: low (diffusion dominated, e.g., *in vitro*), moderate, and high, which correspond to the regimes observed in *in vivo* experiments. This demonstrates, using nonlinear simulations, that critical conditions exist for which tumors evolve to nontrivial dormant states or grow self-similarly in the low-vascularization regime. The region of self-similar growth separates stable tumors, which grow while maintaining a compact shape, from unstable tumors, for which vascularization is favored and growth leads to invasive fingering into the healthy tissues. Nonlinear unstable growth may thus lead to topological transitions such as tumor breakup and reconnection, with the encapsulation of healthy tissue.

The results presented here correspond with the reality that tumor growth is limited by the supply of cell substrates provided by diffusion from blood vessels. Inhomogeneities in substrate availability, and in the biomechanical properties of the host tissue, can result in instability. These instabilities may enable the tumor to overcome the diffusional limitations on growth and to expand to larger sizes than would be possible if it had retained a more compact shape. Thus, diffusional instability may provide a physical mechanism to drive tumor invasion further.

In Chapter 9 we will evaluate the complex, nonlinear, coupling of tumor morphology and dynamics to molecular properties (e.g. phenotype) and phenomena in the environment (e.g. hypoxia).

5 Continuum tumor modeling: multiphase[1]

With H. B. Frieboes and F. Jin

5.1 Background

In this chapter we develop and analyze multiphase mixture models of solid tumor growth and include their coupling via angiogenesis to the tumor-induced vasculature. Mixture models incorporate more detailed biophysical processes than can be accounted for in single-phase models, and they provide a more general framework for modeling solid tumor growth. Phenotypic and genetic heterogeneity (e.g., mutations) in the cell population can be represented, and this is a critical improvement towards realistically simulating mutation-driven heterogeneity.

A multiphase approach describes a tumor as a saturated medium, comprising at least one solid phase and one liquid phase. The approach can be generalized to incorporate any number of additional phases describing multiple cell species and extracellular matrix components. At a given physical location, the mass or volume fractions describe the relative amounts of different cell clones, necrotic tissue and host tissue. The governing equations consist of mass and momentum balance for each phase, mass and momentum exchange between phases, and appropriate constitutive laws to close these equations. The mixture-model approach eliminates the need to enforce complicated boundary conditions across the tumor–host interface (and other species–species interfaces) that would have to be satisfied if the interfaces were assumed to be sharp. Also, this methodology does not require the interface to be tracked as is the case for a sharp interface.

Recently, multiphase mixture models have been developed to account for heterogeneities in cell type and in the mechanical response of the cellular and liquid tumor phases (e.g., [22, 37, 38, 80, 81, 103, 105, 122, 147, 220, 221, 228, 574, 575, 659, 661, 697]). In early work, Please *et al.* [536, 537] applied multiphase modeling to tumor growth by capturing both tumor cells and extracellular fluid as separate continuum phases. Ward and King [679, 680] and Breward *et al.* [80] modeled avascular cancer growth through a two-phase description comprising tumor and dead tissue (extracellular space), and incorporated a model of cell–cell adhesion. Ambrosi and Preziosi [22] developed a mixture model treating the tumor as a deformable porous material. Breward *et al.* [81] extended avascular modeling to include vascular tumor growth, thus incorporating a third phase to describe the spatial and temporal distribution of blood vessels. Byrne

---

[1] Sections 5.1–5.9 are partly based on a paper published in *Nonlinearity*, Lowengrub *et al.* [425], © 2010 Institute of Physics.

and Preziosi [105] extended the two-phase model of Ambrosi and Preziosi to investigate the effect of stress-dependent cell proliferation and external loads on spherical tumor growth and the equilibrium size attained. Roose *et al.* [574] studied the stress generated by solid tumor growth using a poroelastic model. Araujo and McElwain [37, 38] proposed an alternative multiphase model of tumor growth in an effort to capture residual stresses more accurately using an energetically derived thermodynamically consistent mixture model; this included a liquid phase and a solid phase representing the tumor cells and extracellular matrix. Isotropic and anisotropic growth models were considered, highlighting the need to incorporate stress relaxation in order to predict a stable evolution of stresses over a period of growth and equilibration to a steady avascular state. Preziosi and Tosin [545] performed two-dimensional simulations of a mixture model to study the development of tumor cords and the formation of fibrosis. Ambrosi and Preziosi [23] studied the elasto-viscoplastic response of the tumor and microenvironment during growth. Galle *et al.* [255] used a multiphase mixture model to study the effect of contact inhibition on the growth of cell colonies and compared the results with an agent-based discrete model of cell growth. Ambrosi and Preziosi [23] developed a mixture model that accounts for both stress relaxation and cell adhesion, and they demonstrated how tumor growth may lead to cell reorganization to relieve stress. The ECM is modeled as an elastic compressible material, and growth and rearrangement processes are represented using a multiplicative decomposition of the deformation gradient according to these processes. This corresponds to an evolving natural configuration. Previously published models, such as fluid-like or linearly elastic models, may be obtained as limiting cases of this more general model. Further references on multiphase modeling can be found in the recent reviews by Araujo and McElwain [34], Hatzikirou *et al.* [307], Quaranta *et al.* [553], Byrne *et al.* [96], Graziano and Preziosi [291], Astanin and Preziosi [42], and Roose *et al.* [573].

Owing to the complexity of multiphase models, most analyses and numerical simulations are one-dimensional or radially symmetric. Very recently, thermodynamically consistent mixture models in three dimensions have been developed for all phases of solid tumor growth, including neovascularization (Frieboes *et al.* [229], Cristini *et al.* [147], and Wise *et al.* [697]). Using a general approach based on energy variation, the nonlinear effects of cell-to-cell adhesion and taxis-inducing chemical and molecular species have been incorporated. This model enables a detailed description of tumor progression and the dependence of cell–cell and cell–matrix adhesion on cell phenotype and genotype as well as on the local microenvironmental conditions (e.g., oxygen levels). The system energy accounts for all the processes that are modeled. Adhesion is introduced through an interaction energy that leads to well-posed fourth-order equations. Sharp interfaces are replaced by narrow transition layers that arise owing to differential adhesive forces between the cell species. A related nonlocal continuum model of adhesion was recently developed by Armstrong *et al.* [40] and used later in the context of tumor growth by Gerisch and Chaplain [271]. The resulting diffuse-interface mixture equations are well-posed, unlike in some previous mixture models, and constitute a coupled system of equations. The system includes fourth-order nonlinear advection–reaction–diffusion equations of the Cahn–Hilliard type [107] for the cell-species volume fractions, coupled with reaction–diffusion equations for the substrate components. To account for angiogenesis, this mixture model has very recently been nonlinearly coupled to a composite continuum–discrete lattice-free model of tumor-induced angiogenesis originally developed by Plank and Sleeman [534, 535]. To solve the equations numerically, very efficient adaptive finite-difference nonlinear multigrid methods have been developed (see Chapter 8 and also [147, 695–697]. The model was employed to study the three-dimensional vascularized growth of malignant brain tumors, by Bearer *et al.* [57] and Frieboes *et al.* [229] (see also Sanga *et al.* [584]), as well as drug response (Frieboes *et al.* [227]).

5.2 General conservation equations

The primary dependent variables in an $(N + 1)$-species model can be specified as follows [697]:

- the volume fractions of the water, tumor, and host cell species $\varphi_0, \ldots, \varphi_N$;
- the densities of the components $\rho_0, \ldots, \rho_N$;
- the stresses $\sigma_0, \ldots, \sigma_N$;
- the component velocities $\mathbf{u}_0, \ldots, \mathbf{u}_N$.

It may be assumed that there are no voids (i.e., the mixture is saturated) and thus $\sum_{i=0}^{N} \varphi_i = 1$. For simplicity, it may be assumed that the densities are constant and equal to ρ, i.e., independent of temperature, pressure, and component type. We also assume that the system is isothermal. Without loss of generality, $i = 0$ can be identified as the water component. The volume fractions of the components are assumed to be continuous in a domain Ω, which contains both the tumor and host tissues.

The volume fractions obey the mass conservation (advection–reaction–diffusion) equations

$$\rho \left(\frac{\partial \varphi_i}{\partial t} + \nabla \cdot (\mathbf{u}_i \varphi_i) \right) = -\nabla \cdot \mathbf{J}_i + S_i, \tag{5.1}$$

where the $\mathbf{J}_i$ are fluxes that account for the mechanical interactions and diffusional fluxes among the cell species. The source terms S_i represent intercomponent mass exchange as well as gains due to cell proliferation and losses due to cell death and lysing.

The volume-averaged velocity of the mixture is defined as $\mathbf{u} = \sum_{i=0}^{N} \varphi_i \mathbf{u}_i$. Summing Eq. (5.1), the mass of the mixture is conserved only if [2]

$$\sum_{i=0}^{N} \mathbf{J}_i = \sum_{i=0}^{N} S_i = 0. \tag{5.2}$$

These conditions are posed as consistency constraints for the fluxes and sources. In the absence of inertial and external forces, the balance of linear momentum is

$$0 = \nabla \cdot \sigma_i + \pi_i, \tag{5.3}$$

---

[2] The sum $\sum_{i=0}^{N} \mathbf{J}_i$ can be taken to be constant. Without loss of generality, this constant is chosen as zero.

where the π_i are interaction forces among the species. Assuming that the mixture stress $\sigma = \sum_{i=0}^{N} \sigma_i$ satisfies $\nabla \cdot \sigma = 0$, this gives the constraint $\sum_{i=0}^{N} \pi_i = 0$.

Let u_i be the internal energy of the ith component and let the volume-averaged internal energy of the mixture be $u = \sum_{i=0}^{N} \varphi_i u_i$. Then, in its simplest form, the balance of energy equation is given by (e.g., [37, 147])

$$\rho \varphi_i \frac{D^i u_i}{Dt} = \sigma_i : \nabla \mathbf{v}_i + \rho_i \varphi_i r_i + \beta_i, \tag{5.4}$$

where the colon denotes a tensor product, D^i/Dt is the advective derivative with respect to the velocity $\mathbf{u}_i$, r_i is the heat added to each phase to keep the mixture isothermal, and the β_i are energy interaction terms that satisfy

$$\sum_{i=0}^{N} (\beta_i + S_i u_i - \pi_i \cdot \mathbf{u}_i) = 0. \tag{5.5}$$

Constitutive relations for the fluxes $\mathbf{J}_i$, the interaction forces π_i, and the interaction energies β_i may be posed consistently with the second law of thermodynamics. Briefly, the idea is as follows. Let η_i denote the entropy of each component. Then, the volume-averaged entropy of the mixture is $\eta = \sum_{i=0}^{N} \varphi_i \eta_i$. Defining the temperature to be θ (which is assumed to be constant), the second law of thermodynamics can be posed in terms of the Clausius–Duhem inequality [663],

$$\rho \left(\frac{\partial \eta}{\partial t} + \mathbf{u} \cdot \nabla \eta \right) + \nabla \cdot \mathcal{J} - \frac{\rho r}{\theta} \geq 0, \tag{5.6}$$

where $\mathcal{J}$ is an entropy flux and $r = \sum_{i=0}^{N} \varphi_i r_i$; the r_i are the volume-averaged rates of heat supply needed to keep each component isothermal. The entropy and internal energy u_i may be used to define the Helmholtz free energy of each component $\psi_i = u_i - \theta \eta_i$ and the free energy per unit volume $\Psi_i = \rho \varphi_i \psi_i$. The next step is to rewrite the energy equation (5.4) in terms of the Helmholtz free energy and to posit the dependence of the free energy upon independent variables for each phase [141].

5.3   A solid–liquid biphasic model

For binary mixtures of one solid ($i = 1$) and one liquid ($i = 0$) component, Araujo and McElwain [37] assumed that $\psi_i = \psi_i(\mathbf{F}_1, \mathbf{G}_1, \mathbf{u}_1 - \mathbf{u}_0)$. In this approach, $\mathbf{F}_1$ is the deformation gradient in the solid phase, and $\mathbf{G}_1 = \nabla \mathbf{F}_1$. If one assumes further that the volume fractions of the solid and liquid phases are constant, i.e., $\mathbf{J}_i = 0$, and that the liquid phase is inviscid then Araujo and McElwain showed that the stresses become

$$\sigma_0 = -\varphi_0 p \mathbf{I}, \qquad \sigma_1 = -\varphi_1 p \mathbf{I} + \varphi_1 \mathbf{F}_1 \left(\frac{\partial \psi_1}{\partial \mathbf{F}_1} \right)^T, \tag{5.7}$$

where p is a pressure that is introduced to maintain $\varphi_0 + \varphi_1 = 1$, $\mathbf{I}$ is the identity tensor, and $\partial \psi_1 / \partial \mathbf{F}_1 \equiv \nabla \psi_1 \cdot \mathbf{F}_1$ is the directional derivative. Assuming that displacements are small and that the solid component is incompressible, Araujo and McElwain derived the mixture system for conservation of mass:

$$\varphi_1 \rho \nabla \cdot \mathbf{u}_1 = S_1, \qquad \rho \frac{D^1 \varphi_1}{Dt} = 0, \qquad \nabla \cdot (\varphi_1 \mathbf{u}_1 + \varphi_0 \mathbf{u}_0) = 0. \qquad (5.8)$$

To account for growth, Araujo and McElwain [37] proposed a conservation equation for the strain involving the Jaumann derivative and utilized an anisotropic growth tensor. Related approaches were taken earlier by Jones et al. [352] in a single-phase model and Roose et al. [574] in the context of a poroelastic mixture model. However, as Ambrosi and Preziosi [23] pointed out, the strain is not frame invariant [450] and thus the Jaumann derivative is inappropriate. Ambrosi and Preziosi instead demonstrated that the use of accretive forces (e.g., [19, 62, 261, 328, 329, 390]) is required to derive the equations for growth in a consistent manner. In this approach the deformation is decomposed multiplicatively to account for the contributions of pure growth, cell reorganization, and elastic deformation, writing

$$\mathbf{F}_1 = \mathbf{F}_n \mathbf{F}_p \mathbf{G}_1, \qquad (5.9)$$

where $\mathbf{F}_n$ describes the deformation in the natural state (without growth), $\mathbf{F}_p$ describes the cell reorganization after growth, and $\mathbf{G}_1$ describes the deformation due to growth. Assuming that the tumor is incompressible, responds elastically (i.e., $\dot{\mathbf{F}}_p = 0$, where the overdot denotes the time derivative), and that the growth is isotropic (i.e., $\mathbf{G}_1 = g\mathbf{I}$), Ambrosi and Preziosi rewrote the stress as $\sigma_1 = -p_1 \mathbf{I} + \sigma_1'$, where p_1 is a Lagrange multiplier pressure introduced to maintain incompressibility, and $\sigma_1' = \varphi_1 \left(\mathbf{B}_n - \frac{1}{3} \operatorname{Tr}(\mathbf{B}_n) \mathbf{I} \right)$, where $\mathbf{B}_n = \mathbf{F}_n \mathbf{F}_n^T$. Taking the time derivative in the limit of small deformations and in the absence of cell rearrangement this gives [23]

$$\dot{\sigma}_1' = 2\varphi_1 \left(\mathbf{D}_1 - \frac{S_1}{3} \mathbf{I} \right), \qquad (5.10)$$

where $\mathbf{D}_1 = \left(\nabla \mathbf{u}_1 + \nabla \mathbf{u}_1^T \right) (\cdots)/2$. Ambrosi and Preziosi also considered a viscoplastic law in which there is a stress threshold above which cell reorganization occurs. This yields, in the same limit as above,

$$\eta_p \dot{\sigma}_1' + \left(1 - \frac{\varphi_1 \tau}{f(\sigma_1')} \right)_+ = 2\eta_p \varphi_1 \left(\mathbf{D}_1 - \frac{S_1}{3} \mathbf{I} \right), \qquad (5.11)$$

where η_p is a characteristic time for stress relaxation to the yield value, τ is a yield stress, and f is a frame invariant measure of the stress that is used to determine the yield threshold (e.g., $f(\mathbf{T}) = \sqrt{\mathbf{T} : \mathbf{T}}$ where as before the colon denotes a tensor product). For the more general case, we refer the reader to Ambrosi and Preziosi [23].

5.4   Liquid–liquid mixture model I

An alternative approach was taken by Cristini et al. [147]. In their work, the solid fraction was treated as a viscous fluid, cell–cell and cell–matrix adhesive interactions were incorporated, and thermodynamically and mechanically consistent equations were derived. In addition, taxis-inducing chemical and molecular species were included. Accordingly, the Helmholtz free energy was taken to be

$$\Psi_i = \tilde{\Psi}_i(\varphi_0, \ldots, \varphi_N) + \varphi_i \sum_{l=1}^{L} \chi_{il} c_l + \sum_{j=0}^{N} \frac{\epsilon_{ij}^2}{2} |\nabla \varphi_j|^2, \tag{5.12}$$

where $c_1, \ldots, c_L$ are the concentrations of the chemical and molecular species, the χ_{il} are taxis coefficients, and the ϵ_{ij} measure the strength of the component interactions (see Wise *et al.* [697]). The dependence of Ψ_i upon the volume fractions and the gradients of the volume fractions arises naturally through the expansion of a nonlocal interaction potential between the phases (e.g., see [107, 697]) that represents adhesive interactions between the species. The resulting system is:

$$\frac{\partial \varphi_i}{\partial t} + \nabla \cdot (\varphi_i \mathbf{u}_i) = \frac{1}{\rho} (S_i - \nabla \cdot \mathbf{J}_i), \qquad \sum_{i=0}^{N} \nabla \cdot (\varphi_i \mathbf{u}_i) = 0, \tag{5.13}$$

for the conservation of mass. In Eq. (5.13) the adhesion fluxes may be given for $i \neq 0$ as [147] $\mathbf{J}_i = -M \nabla \mu_i$, which is a generalized Fick's law, where μ_i is the chemical potential,

$$\mu_i = F_i(\varphi_0, \ldots, \varphi_N) + \sum_{l=1}^{L} (\chi_{il} - \chi_{0l}) c_l - \sum_{j=0}^{N} \left(\epsilon_{ji}^2 \Delta \varphi_i - \epsilon_{j0}^2 \Delta \varphi_0 \right), \tag{5.14}$$

and

$$F_i = \sum_{j=0}^{N} \left(\frac{\partial \tilde{\Psi}_j}{\partial \varphi_i} - \frac{\partial \tilde{\Psi}_0}{\partial \varphi_0} \right).$$

The flux in the liquid is $J_0 = -\sum_{i=1}^{N} \mathbf{J}_i$. In the liquid (assumed to be inviscid), the momentum balance equation is the multicomponent Darcy's law:

$$\varphi_0 \nabla p = \sum_{i=0}^{N} \alpha_i (\mathbf{u}_i - \mathbf{u}_0), \tag{5.15}$$

where the α_i are drag coefficients. For the solid components, the general viscous law is obtained (for $i > 0$):

$$\alpha_i (\mathbf{u}_i - \mathbf{u}_0) = -\varphi_i \nabla (p + \mu_i) + \nabla \cdot \left[\mathcal{L}_i \left(\nabla \mathbf{u}_i + \nabla \mathbf{u}_i^T \right) \right], \tag{5.16}$$

where

$$\mathcal{L}_i \left(\nabla \mathbf{u}_i + \nabla \mathbf{u}_i^T \right) = \lambda_i \left(\nabla \mathbf{u}_i + \nabla \mathbf{u}_i^T \right) + \nu_i (\nabla \cdot \mathbf{u}_i) \mathbf{I},$$

and λ_i and ν_i are viscosities.

In the case of two-component mixtures, a local approximation of this model may be achieved by taking $\mathbf{u}_0 = -(\varphi_1/\varphi_0)\mathbf{u}_1$ and p to be constant, setting the viscosity to zero, and discarding Eq. (5.15) to get $\mathbf{u}_1 = -M \varphi_i \nabla \mu_i$; see [147]. Viscosity can be easily included, and this yields a nonlocal equation analogous to the mixture model derived in Byrne and Preziosi [105] in one dimension; the inviscid model is similar to that considered earlier by De Angelis and Preziosi. Interestingly, the local approximation can be shown to converge to a classical single-phase sharp interface model of the type considered in previous sections as the component interactions $\epsilon_{ij} = \epsilon$ tend to zero (see Cristini et al. [147]). A discussion of the model without the above approximation may also be found in [147].

For multicomponent mixtures consisting of more than two components, one may take yet another approximation by supposing that the velocities of the solid components are given by $\mathbf{u}_i = M_i \varphi_i \nabla \mu_i$ and that the velocity of the liquid is $\mathbf{u}_0 = (-1/\varphi_0) \sum_{i=1}^N \varphi_i \mathbf{u}_i$. This is analogous to the closure laws described in De Angelis and Preziosi and in Ambrosi and Preziosi [22].

5.5   Liquid–liquid mixture model II

Another approach to modeling multicomponent tumors was recently developed by Wise et al. [697] and Frieboes et al. [229]. The mass balance equations are the same as Eq. (5.13), and the constitutive laws for the mechanical fluxes $\mathbf{J}_i$ and a generalized Darcy's law for the cell velocities $\mathbf{u}_i$ are derived using an energy variation argument utilizing the mixture energy

$$E = \sum_{i=0}^N \int \Psi_i \, dx, \qquad (5.17)$$

where the energy density is given by Eq. (5.12). As derived in [697], thermodynamically consistent fluxes may be taken to be the generalized Fick's law,

$$\mathbf{J}_i = -\bar{M}_i \nabla \left(\frac{\delta E}{\delta \varphi_i} - \frac{\delta E}{\delta \varphi_N} \right), \qquad 1 \leq i \leq N - 1, \qquad (5.18)$$

and $\mathbf{J}_N = -\sum_{i=1}^{N-1} \mathbf{J}_i$; $\bar{M}_i > 0$ is the mobility and the $\delta E / \delta \varphi_i$ are variational derivatives of the total energy E given by

$$\frac{\delta E}{\delta \varphi_i} = \sum_{j=0}^N \left(\frac{\partial \tilde{\Psi}_j}{\partial \varphi_i} - \nabla \cdot \left(\bar{\varepsilon}_{ji}^2 \nabla \varphi_i \right) \right) + \sum_{l=1}^L \chi_{il} c_l, \qquad 1 \leq i \leq N. \qquad (5.19)$$

The velocities of the components may be also determined in a thermodynamically and mechanically consistent way. Assuming that the solid and liquid volume fractions remain constant, i.e., $\sum_{i=1}^N \varphi_i = \tilde{\varphi}_S$ and $\varphi_0 = 1 - \tilde{\varphi}_S$ with φ_S constant in space and time, the resulting generalized Darcy laws for the velocities of the components are given by

$$\mathbf{u}_0 = -\bar{k}_0 \nabla \left(\frac{\delta E}{\delta \varphi_0} + q \right), \qquad (5.20)$$

$$\mathbf{u}_j = -\bar{k} \left(\nabla p - \sum_{i=1}^N \frac{\delta E}{\delta \varphi_i} \nabla \varphi_i \right) - \bar{k}_j \nabla \left(\frac{\delta E}{\delta \varphi_j} - \frac{1}{\tilde{\varphi}_S} \sum_{i=1}^N \varphi_i \frac{\delta E}{\delta \varphi_i} + \frac{p}{\tilde{\varphi}_S} \right), \quad j \geq 1, \qquad (5.21)$$

where q is the water pressure, p is the solid pressure, and $\bar{k}_0$, $\bar{k}$, $\bar{k}_j$ are positive definite motility matrices. The constitutive laws (5.18), (5.20), and (5.21) guarantee that in the absence of mass sources the energy in Eq. (5.17) is nonincreasing in time as the fields evolve.

In related work, Armstrong *et al.* [40], and later Gerisch and Chaplain [271], considered a nonlocal model of adhesion in which $\bar{k}_j = 0$, $p = 0$, and the variational derivatives in the velocity $\mathbf{u}_j$ are replaced (in two dimensions) by

$$\mathcal{A}[\varphi](\mathbf{x}, t) = \frac{1}{R} \int_0^R r \int_0^{2\pi} \mathbf{n}(\theta) \cdot \mathbf{O}(r) g(\varphi(t, \mathbf{x} + r\mathbf{n}(\theta))) \, d\theta dr, \qquad (5.22)$$

where $\mathbf{n}$ is the normal vector to a ball of radius R centered at $\mathbf{x}$, which is termed the sensing region; R is the sensing radius. The functions $\mathbf{O}$ and g determine the strengths of the interactions among the different cell types (and the extracellular matrix) and $\varphi = (\varphi_1, \ldots \varphi_N)$ is the vector of volume fractions. As the sensing radius R tends to zero, the nonlocal model converges to a local reaction–diffusion–taxis system of partial differential equations [40, 271]. This adhesion velocity may actually be derived using energy variation arguments starting with a fully nonlocal version of the energy E such as

$$E = \frac{1}{2} \sum_{i,j} \int \int J_{ij}(\mathbf{x}, \mathbf{y}) \varphi_i(\mathbf{x}) \varphi_j(\mathbf{y}) \, d\mathbf{x} d\mathbf{y}, \qquad (5.23)$$

where J_{ij} is an appropriately defined interaction kernel. See the appendix of [697] for a description of the procedure.

In other related work, Khain and Sander [372] developed a model of cell–cell adhesion for a single cell species similar to that described in the liquid–liquid model I. In particular, Khain and Sander used a generalized Cahn–Hilliard [107] (GCH) equation to study tumor invasion. Below a critical level of adhesion, the tumor invades as a propagating front. Above the critical level, a second peak is found to appear in the tumor fraction (i.e., density) curve behind the leading front. The results of the GCH model compare well with a stochastic discrete-cell model recently developed by Khain *et al.* [373].

Note that only hydrostatic stresses in the tumor and host tissues are simulated in the approach by Wise *et al.* [697] and Frieboes *et al.* [229]. This is highly simplified. An important research direction is the incorporation of more realistic models of soft-tissue mechanics. These include elastic, poroelastic, and viscoelastic models, e.g., [20, 35, 62, 248, 328, 352, 426, 574]. These effects may be included in the model by incorporating the relevant energy in the system energy and following the mixture model development by Araujo and McElwain [37, 38].

5.6   A special case of the generalized Darcy's law liquid–liquid mixture model

The model given in Eqs. (5.18)–(5.21) may be simplified [697] by assuming (i) that tumor cells prefer to adhere to one another rather than to the host (as observed experimentally [40, 41]), and (ii) that no distinction is made between the adhesive properties of different cell species (e.g., between viable and dead cells). Accordingly, in Eq. (5.12) we may take $\tilde{\Psi}_i(\varphi_0, \ldots, \varphi_N) = \varphi_i f(\varphi_T)$, where $\varphi_T = \sum_{i=1}^{N-1} \varphi_i$ is the solid fraction of the tumor tissue and $\varphi_N = \varphi_H$ is the volume fraction of the host tissue ($\varphi_T + \varphi_H = \tilde{\varphi}_S$). Further, taking $\epsilon_{ij} = 0$ for $i, j < N$ and $\epsilon_{NN} = \bar{\epsilon}$, the total adhesion energy (5.17) becomes

$$E = \int_\Omega \left(f(\varphi_T) + \frac{\bar{\epsilon}^2}{2} |\nabla \varphi_T|^2 \right) d\mathbf{x}. \tag{5.24}$$

This form of the energy arises also in the classic theory of phase transitions (e.g., [107]). For example, f may be written as the difference of two convex functions,

$$f(\varphi_T) = f_c(\varphi_T/\tilde{\varphi}_S) - f_e(\varphi_T/\tilde{\varphi}_S), \tag{5.25}$$

where one may take

$$f_c(\varphi) = \tfrac{1}{4} \bar{E} \alpha_1 \left(\left(\varphi - \tfrac{1}{2} \right)^4 + 1 \right) \qquad \text{and} \qquad f_e(\varphi) = \tfrac{1}{4} \bar{E} \alpha_2 \left(\varphi - \tfrac{1}{2} \right)^2 \tag{5.26}$$

and where α_1 and α_2 describe the strength of adhesion (attraction) of tumor cells to the host tissue and to each other, respectively, and $\bar{E}$ is an overall energy scale. Setting $\alpha_1 = \alpha_2$ yields a double-well energy f with minima at $\varphi_T = \tilde{\varphi}_S$ and $\varphi_T = 0$ and gives rise to a well-delineated phase separation of the tumor ($\varphi_T \approx \tilde{\varphi}_S$) and host tissues ($\varphi_T \approx 0$). Since φ_T is continuous, it is necessary that $0 < \varphi_T/\tilde{\varphi}_S < 1$ in the interfacial region dividing the tumor and host domains. However, the states $\varphi_T > \tilde{\varphi}_S$ or $\varphi_T < 0$ are not physical, and the interaction energy tends to prevent their formation by increasing the energy of these states. Note that taking an interaction energy with logarithmic terms would explicitly prevent the formation of these states [198]. By varying the ratio α_1/α_2 the relative tendency of the tumor cells to aggregate can be modified. For example, increasing the ratio results in an increasingly diffuse tumor mass.

The thickness of the diffuse interface between the tumor and host tissue depends on the relative sizes of $\bar{\epsilon}$, $\bar{E}$, and α_1/α_2. Specifically, for fixed $\bar{E}$ the smaller the constant $\bar{\epsilon}$, the less diffuse the interfacial region. If $\bar{E} \sim 1/\epsilon$ then it can be shown that this system converges as $\epsilon \to 0$ to a classical sharp-interface single-phase tumor model [697]. If the tumor contains different species that have different adhesion properties then the energy (5.24) can be modified to account for the different cell–cell interactions, following the more general approach described earlier (see also [381]).

5.7 Fluxes and velocities

From the flux constitutive equation (5.18) and the adhesion energy equation (5.24), the adhesion fluxes may be determined. Recalling that the densities of the components are matched and taking the mobilities $\bar{M}_i = \bar{M}\varphi_i$, where $\bar{M}$ is a positive constant, the fluxes are obtained [697] as

$$\mathbf{J}_i = -\bar{M}\varphi_i \nabla \frac{\delta E}{\delta \varphi_T}, \tag{5.27}$$

where $i = 1, \ldots, N - 1$ and $\mathbf{J}_N = \mathbf{J}_H = -\sum_{i=1}^{N-1} \mathbf{J}_i$, where the energy is taken not to depend explicitly on φ_H. The variational derivative is given by

$$\frac{\delta E}{\delta \varphi_T} = f'(\varphi_T) - \bar{\varepsilon}^2 \nabla^2 \varphi_T. \tag{5.28}$$

Setting $\bar{k}_i = 0$ for $i > 0$, which is consistent with assuming that the cells are tightly packed and that they move together with the mass-averaged velocity, the component velocities become [697]

$$\mathbf{u}_0 = \mathbf{u}_W = -\bar{k}_W \nabla q, \tag{5.29}$$

$$\mathbf{u}_i = -\bar{k} \left(\nabla p - \frac{\delta E}{\delta \varphi_T} \nabla \varphi_T \right), \tag{5.30}$$

assuming that the energy does not depend explicitly on φ_0 or φ_N. In Eq. (5.30), the term dependent on $\delta E / \delta \varphi_T$ represents the excess force due to adhesion and arises from cell–cell interactions. The quantities q, p are the water pressure and the solid (oncotic) pressure, respectively. The coefficients $\bar{k}_W$ and $\bar{k}$ are motilities that reflect the responses of the water and cells, respectively, to pressure gradients. These coefficients may depend on the volume fractions and other variables since the individual components may respond to the pressure and adhesive forces differently, but mixed components tend to move together. The cell motilities contain the combined effects of cell–cell and cell–matrix adhesion. The constitutive choices (5.27), (5.30), and (5.29) guarantee that, in the absence of mass sources ($S_i = 0$), the adhesion energy is nonincreasing as the fields evolve while the total tumor mass is conserved.

5.8   Mass-exchange terms

As a first approximation, viable tumor cells may be assumed to necrose when the local cell substrate (e.g. oxygen, nutrient) concentration n falls below the cell viability limit $\bar{n}_{N,i}$ which may be different for different cell types. Cells may be assumed to consist entirely of water; in terms of volume fraction, this is a reasonable first approximation. Cell mitosis may be assumed to be proportional to the amount of oxygen and nutrients present and, as mitosis occurs, an appropriate amount of water is converted into cell mass. Conversely, the lysing of cells represents a mass sink as cellular membranes are degraded and the mass converts completely into water (neglecting residual solid cellular components). Mitosis is neglected in the host domain, as the proliferation rate for tumor cells is much larger. Accordingly, defining φ_D to be the volume fraction of unlysed dead tumor cells, we may take [697], for viable species i not including dead tumor cells D or nutrient N,

$$S_i = \bar{\lambda}_{M,i} \frac{n}{\bar{n}_\infty} \varphi_i - \bar{\lambda}_{A,i} \varphi_i - \bar{\lambda}_{N,i} \mathcal{H}(\bar{n}_{N,i} - n)\varphi_i, \qquad N > i > 0 \text{ and } i \neq D. \tag{5.31}$$

Furthermore,

$$S_{\mathrm{D}} = \sum_{\substack{i>0 \\ i \neq \mathrm{D}}}^{N-1} \left[\bar{\lambda}_{\mathrm{A},i} + \bar{\lambda}_{\mathrm{N},i} \mathcal{H}(\bar{n}_{\mathrm{N},i} - n) \right] \varphi_i - \bar{\lambda}_{\mathrm{L}} \varphi_{\mathrm{D}}, \tag{5.32}$$

$$S_N = S_{\mathrm{H}} = 0, \tag{5.33}$$

$$S_{\mathrm{W}} = -\sum_{i=1}^{N} S_i = -\sum_{\substack{i=1 \\ i \neq \mathrm{D}}}^{N-1} \bar{\lambda}_{\mathrm{M},i} \frac{n}{n_\infty} \varphi_i + \bar{\lambda}_{\mathrm{L}} \varphi_{\mathrm{D}}, \tag{5.34}$$

where $\bar{\lambda}_{\mathrm{M},i}$, $\bar{\lambda}_{\mathrm{A},i}$, $\bar{\lambda}_{\mathrm{N},i}$, and $\bar{\lambda}_{\mathrm{L}}$ are the rates of volume gain or loss due to cellular mitosis, apoptosis, necrosis, and lysing, respectively, and where $\bar{n}_\infty$ is the far-field level of the cell substrates. The source terms S_i, S_{D}, S_N, S_{H}, S_{W}, refer to the various species: viable, dead, nutrient, host and water. Finally, $\mathcal{H}$ is the Heaviside function. For simplicity, we have omitted mass-exchange terms corresponding to genetic or epigenetic events that transform one cell species into another.

5.9 Diffusion of cell substrates

Following the single-phase approach discussed in Section 3.2, the host tissue is modeled at equilibrium as a first approximation, in which the net cell substrate uptake is regarded as negligible compared with the uptake by tumor cells. Substrates uptaken by the host tissue are assumed to be replaced by supply from the normal vasculature. This may not be the case in the tumor, where not only does the uptake exceed the supply but the uptake may also be much higher than that of the host tissue [556, 207]. Using the quasi-steady approximation described earlier for single-component tumor models, the cell-substrate transport equation may be written as

$$0 = \nabla \cdot (D(\varphi_1, \ldots, \varphi_N) \nabla n) + T_{\mathrm{C}} - \sum_{\substack{i=1 \\ i \neq \mathrm{D}}}^{N-1} v_i^{\mathrm{u}} \varphi_i, \tag{5.35}$$

where the sum excludes dead cells; D is the substrate diffusivity and can vary depending on the medium and cell type, e.g., it can be an interpolated function from the host to the tumor, with value 1 inside the tumor and value D_{H} in the host medium; the quantity T_{C} is a source of cell substrates from a pre-existing uniform vasculature or a vasculature newly formed through angiogenesis; and the v_i^{u} are the uptake rates of the different cell types.

5.10 Mutation of tumor-cell species

The phenotypic transformation of one viable tumor-cell species φ_i into another φ_j, can be modeled by including a transfer function $\mathcal{M}^{i \to j}$ as part of Eq. (5.31); it transfers a

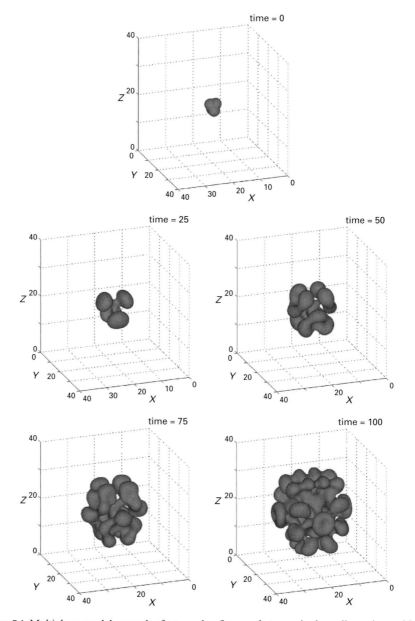

Figure 5.1 Multiphase model: growth of a two-plus-four mode tumor in three dimensions with cell adhesion $\gamma = 0$. The $\varphi_T = 0.5$ isosurface is shown. The parameter values are as in Table 5.1 The model predicts that this tumor morphological instability, which increases the overall surface-area-to-volume ratio, enables the tumor to increase its access to cell substrates from the surrounding host vasculature. Reprinted with permission from Frieboes *et al.* [228].

Table 5.1. Nondimensional parameters (as defined in [697]) used in the simulation shown in Figure 5.1 (from [228])

$v_P^H = \bar{v}_P^H / \bar{v}_U$	0.5	$v_P^T = \bar{v}_P^T / \bar{v}_U$	0.0
$D_H = \bar{D}_H / \bar{D}_T$	1.0	$n_N = \bar{n}_N / \bar{n}_\infty$	0.4
$\lambda_A = \bar{\lambda}_A / \bar{\lambda}_M$	0.0	$\lambda_N = \bar{\lambda}_N / \bar{\lambda}_M$	3.0
$\lambda_L = \bar{\lambda}_L / \bar{\lambda}_M$	1.0	$\gamma \approx \tilde{\gamma}/(6\sqrt{2})$	0.0
M	10.0	ε	0.1

corresponding partial volume fraction from S_i to S_j [228]. Speaking loosely, we use the term "mutation" to denote this phenotypic change, which may be of genetic or epigenetic origin. The term $\mathcal{M}^{i \to j}$ is subtracted from S_i and added to S_j. The choice of mutation event $(\mathbf{x}^*, t^*)$ has a random component and is generated under a temporal frequency f^* at viable cells, where the density $\varphi_V > 0.5$. A successful mutation covers a circular space of radius r^* and lasts for a time δt^*. The mutation transfer function is chosen to be [228]

$$\mathcal{M}^{V \to M}(\mathbf{x}, t, \varphi_i) = g(\mathbf{x}, \mathbf{x}^*) a(t, t^*) \varphi_V(\mathbf{x}, t), \qquad (5.36)$$

where g is a Gaussian distribution function spatially located at $\mathbf{x}^*$ and a is a time activation function starting at time t^* and lasting δt^*. Precisely,

$$g(\mathbf{x}, \mathbf{x}^*) = C \exp\left(-\frac{||x - x^*||^2}{(r^*)^2}\right), \qquad (5.37)$$

$$a(t, t^*) = \frac{4b}{1 - b}, \qquad b = \frac{t - t^*}{(\delta t)^*}, \qquad (5.38)$$

where $C > 0$ is a rate constant.

The effects of phenotypic transformations are coupled nonlinearly to the tumor scale through the numerical calculation of cellular-pressure and cell-substrate gradients in the microenvironment and of their effect on cell proliferation and apoptosis [228].

5.11 Nonlinear results: avascular growth

5.11.1 Morphological instability as an invasive mechanism

Following [697], we consider the growth of a tumor consisting of viable cells, with volume fraction φ_V, and dead cells, with volume fraction φ_D, i.e., $\varphi_T = \varphi_V + \varphi_D$. The dead-cell population includes the tumor cells that have undergone apoptosis or necrosis. Dead cells are assumed not to consume cell substrates. There is also a host-cell species and a liquid component. Using the generalized Darcy's law mixture model described above, the growth of an initially perturbed spherical tumor in three dimensions with cell adhesion $\gamma = 0$ is shown in Figure 5.1. The isosurface $\varphi_T = 0.5$ is plotted. The parameters are given in Table 5.1 (as defined in [697]).

In this simulation, the substrate gradients contribute only to variable tumor-cell proliferation and necrosis; they do not induce the migration of cells. Consistently with linear analyses [149, 416], the tumor is unstable owing to the limited supply of cell substrates and the corresponding tumor mass loss due to necrosis in the tumor interior. The simulation shows that in the nonlinear regime a tumor emerges with repeated budding and recursive growth. By acquiring this complex morphology, the tumor has effectively increased its surface-area-to-volume ratio and so gains better access to the cell substrates from the surrounding host vasculature.

5.11.2   Chemotaxis as an invasive mechanism

Cristini *et al.* [147] considered the effects of chemotaxis (directed cell movement along a chemical concentration gradient). Tumor evolution in substrate-rich and substrate-poor cell tissues was investigated by performing spherically symmetric and fully two-dimensional nonlinear numerical simulations. It was demonstrated that a tumor may suffer from taxis-driven fingering instabilities, which are most dramatic when the cell proliferation is low as predicted by linear stability theory. This is also observed in experiments [230]. This work shows quantitatively that taxis may play a role in tumor invasion and that when cell substrates play the role of chemoattractant, diffusional instability is exacerbated by the substrate gradients. The model is thus capable of describing some of the complex invasive patterns observed in experiments.

Briefly, cell-substrate-driven taxis in Cristini *et al.* [147] is included by adding the term $\chi_n n \varphi_T$, where n is the substrate concentration and χ_n is the corresponding taxis coefficient, to the total energy in Eq. (5.24), following the general fomulation given in Eq. (5.12). Equation (5.28) then becomes

$$\frac{\delta E}{\delta \varphi_T} = f'(\varphi_T) - \varepsilon^2 \nabla^2 \varphi_T + \varepsilon \chi_n n. \tag{5.39}$$

Using the liquid–liquid mixture model I (see Section 5.4), Cristini *et al.* simulated the growth of a solid tumor in a cell-substrate-poor microenvironment. The proliferation rate was taken to be low and cell apoptosis was neglected [147]. In Figure 5.2, left the tumor evolution for low cell proliferation is shown using an adaptive algorithm (the mesh is shown at time $t = 30$). Note that if $\chi_n = 0$, i.e., when there is no taxis, the evolution is stable. The solid curve represents the tumor interface ($\varphi_T = 0.5$) and the broken-and-dotted curve represents linear results from the sharp-interface model (here, the time scaling used is different from the one described in Chapters 3 and 4; see [147] for further details).

Figure 5.2 shows that invasive fingers develop around time $t = 10$ and get stretched at time $t = 20$, forming long and slim protrusions and thereby increasing the tumor surface area and enabling better access to cell substrates. At later times the fingers continue to stretch and start to bend inward. There is good agreement between the linear and nonlinear results at early times, but there is significant deviation at later times owing to the strong nonlinearity. The right-hand column show similar simulations of tumor evolution for a larger proliferation rate. All other parameters are the same as in

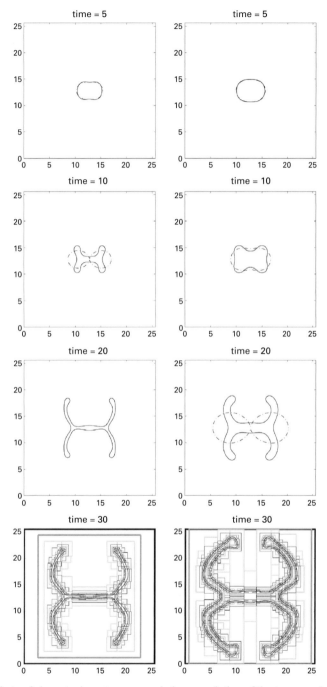

Figure 5.2 Effects of chemotaxis on tumor morphology: evolution of the tumor surface without apoptosis for $D_H = 1$ and $\chi_n = 5$. The initial tumor surface is given by $(x(\alpha) - 12.8, y(\alpha) - 12.8) = (2 + 0.1 \cos 2\alpha)(\cos \alpha, \sin \alpha)$, where $\alpha \in [0, 2\pi]$. Left-hand column, low proliferation; right-hand column, the proliferation is $5\times$ that of the left-hand column. Solid lines, nonlinear simulation; faint broken- and -dotted lines, linear results. The last row shows the details of the mesh development. With kind permission from Springer Science and Business Media: *J. Math. Biol.*, Cristini *et al.*, vol. 58, p. 750, © 2009 Springer.

the left-hand column. The fingers are thicker for the case with larger proliferation. The spread of the fingering into the surrounding tissue at early times is more pronounced for smaller proliferation than for larger proliferation. As before, there is good agreement between the linear and nonlinear results at early times, before nonlinear effects dominate the evolution. Figure 5.3 shows tumor evolution using the same parameter values as in Figure 5.2 but with an elliptical initial tumor-surface shape. A dramatic fingering, indicating invasive behavior, is also observed in this case.

The numerical results reveal that a tumor may possess a cell substrate-taxis-driven invasiveness that is most dramatic when the cell proliferation is low (see [149]). This suggests that substrate taxis may play a role in tumor invasion and that diffusional instability is exacerbated by cell-substrate gradients. This process resembles the invasive patterns observed experimentally in tumor spheroids grown in hypoxic conditions (see [528]).

5.12   Modeling angiogenesis in three dimensions

5.12.1   Background

The modeling of tumor vasculature described in Section 3.3.2 enabled a two-dimensional representation of vessel topology. Here, we consider an extension of the angiogenesis model to three dimensions that is independent of the underlying computational grid (i.e., it is lattice-free). In this approach, continuum conservation laws are introduced to describe the dynamics of the vessel densities and angiogenic factors; this model does not provide morphological or blood-flow information about the vasculature. We then couple this algorithm with the three-dimensional multiphase model described in Section 5.5. The angiogenesis model is a refinement of earlier work [534, 535], which had shown that dendritic structures could be created that are consistent with experimentally observed tumor capillaries [408, 622]. This random-walk model describes the proliferation and migration of vascular endothelial cells in response to soluble angiogenic regulators such as VEGF (which induce chemotaxis) and bound matrix macromolecules such as fibronectin (which induce haptotaxis). Vessels that connect (anastomose) may provide a source of cell substrates in the tissue [45] and may further undergo spontaneous shutdown and regression during tumor growth [320]. The VEGF level is represented by a single continuum variable that reflects the excess of pro-angiogenic over anti-angiogenic regulators. Perinecrotic tumor cells and host tissue cells close to the tumor boundary are assumed to be the source of these regulators, stimulating the endothelial cells to proliferate and begin to form vessels towards the tumor [343].

Endothelial cells near the sprout tips proliferate, and their migration can be described by chemotaxis and haptotaxis. For simplicity we model only the leading endothelial cells since proliferation occurs at the cells close to the leading tip and the trailing cells follow passively. The vasculature architecture, i.e., its interconnectedness and anastomoses, is captured via a set of rules. A leading endothelial cell has a fixed probability of branching at each time step while anastomosis may occur if a leading endothelial cell crosses a

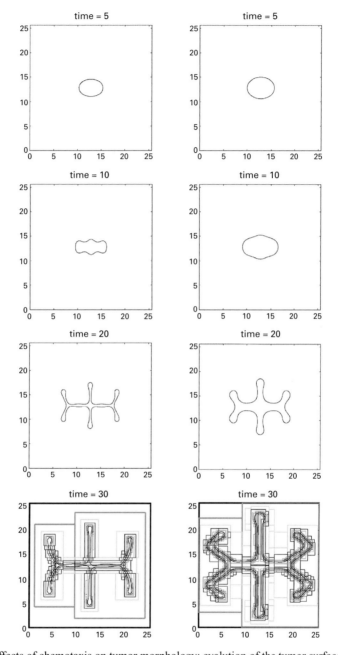

Figure 5.3 Effects of chemotaxis on tumor morphology: evolution of the tumor surface without apoptosis for $D_H = 1$ and $\chi_n = 5$. The initial tumor surface is given by $((x - 12.8)/2.1)^2 + ((y - 12.8)/1.9)^2 = 1$. Left-hand column, low proliferation; right-hand column, the proliferation is $5\times$ that of the left-hand column. The last row shows the details of the mesh development. With kind permission from Springer Science and Business Media: *J. Math. Biol.*, Cristini *et al.*, vol. 58, p. 753, © 2009 Springer.

vessel's trailing path. Tumor vessels are typically more tortuous than normal vessels [87]. This can be quantified by various means, including a "sum of angles metric" (SOAM) that sums the total curvature along a space curve and normalizes by the path length, thus indicating high-frequency low-amplitude sine waves or coils [87].

In the model described here [57, 228, 229], the tumor-induced vasculature does not initially conduct blood, as the vessels need to form loops first. As observed experimentally, the neovasculature model may also account for increasing vessel diameters and the spontaneous shutdown and consecutive regression of initially functioning tumor vessel segments or whole microvascular areas [321]. Here, functioning anastomosed vessels are assumed to provide a source of cell substrates in the tissue in inverse proportion to the local (tumor) pressure.

5.12.2   Biased circular random walk

In the implementation [57, 228, 229] vessels are generated using the biased circular random walk model of Plank and Sleeman [534, 535]. The release of tumor angiogenic factors (e.g. VEGF) is governed by the diffusion–reaction equation (3.36). The trajectories of the endothelial cells (ECs) at the tips of the vessels move with a random walk on the unit circle. In two-dimensional space a tip location (x, y) can be characterized by its speed s and direction angle θ:

$$\frac{dx}{dt} = s \cos \theta, \qquad \frac{dy}{dt} = s \sin \theta. \tag{5.40}$$

The tip cell has a probability at each time step k of turning through a small finite angle δ, clockwise or counterclockwise. This probability is generated through the normalized transition rate,

$$\tau(\theta, t) = \frac{1}{\sigma^2} \exp\left(2 \int \frac{\mu}{\sigma^2} d\theta\right), \tag{5.41}$$

where μ and σ^2 are the expectation (mean) value and the variance of the turning rate, which can be adjusted according to the statistical analysis of the capillary network. Both are functions of orientation θ and time t:

$$\sigma = \sigma(\theta, t), \qquad \mu = \mu(\theta, t). \tag{5.42}$$

Chemotaxis and haptotaxis can be modeled by taking the mean turning rate μ such that the tips tend to reorient themselves to move up the local gradients of the species. The mean turning rate is:

$$\mu = \sum_i -d_i \sin(\theta_i - \theta_i^*), \tag{5.43}$$

where θ_i^* is the preferred orientation along the gradient of each taxis species i and d_i is its corresponding turning coefficient, which indicates the cell's ability to reorient itself due to taxis and is hence proportional to the magnitude of the gradient.

The model can be extended to three dimensions by adding a second angle, φ, as the polar angle and letting θ represent the azimuthal angle. The tip location (x, y, z) is tracked by the equations of motion:

$$\frac{dx}{dt} = s \cos\theta \cos\varphi, \quad \frac{dy}{dt} = s \cos\theta \sin\varphi, \quad \frac{dz}{dt} = s \sin\theta, \quad (5.44)$$

where the transition rate of φ can be independently defined analogously to Eq. (5.41).

At each time step the local gradients of the taxis species are measured near the tip cell. For each θ and φ the transition rate is computed with respect to the gradients, and these transition rates are used to determine whether the cell stays at the current angle or makes a turn. The new location of the cell is updated once the two angles are determined. As in previous work (e.g., [534, 535]), it is assumed that the previously generated cells trail after the leading cells and do not change position after their formation.

5.12.3 Branches and anastomoses

For each tip cell, the capillary has a fixed probability of branching at each time step. When branching occurs a cell splits into two leading cells, and the new cells reorient by a fixed angle of 30 degrees. These two cells then continue to migrate and proliferate into new vessels.

If the leading cell of one vessel crosses the trail of another vessel then anastomosis may occur. This process forms a closed vessel loop and blood begins to circulate, flowing from the root of one vessel to that of the other. The fluid dynamics of the blood are not modeled; thus, changing the direction of the flow has no effect. Simplified models of the fluid dynamics of blood flow through capillary networks have been developed (e.g, Pries *et al.* [550], McDougall *et al.* [458, 459], Stephanou *et al.* [634, 635], Wu *et al.* [702], Zhao *et al.* [721], Sun and Munn [641], and Pries and Secomb [549]), In [458, 549, 550, 635], shear-stress-induced vessel remodeling is also simulated. Here, we do not describe this level of detail; instead, we will assume that anastomosed vessels whose branches arise from different roots are capable of releasing cell substrates.

When anastomosis occurs the looped vessel starts to supply nutrients and oxygen, as described by the source term λ_{neo}^n in Eq. (3.40).

5.13 Nonlinear results: vascularized tumor growth

We now present an example of vascularized tumor growth in the context of human glioma (a form of brain cancer). Gliomas may evolve into glioblastoma multiforme (GBM), which is the most aggressive human brain tumor. Unlike lower-grade gliomas, this cancer is highly vascularized (Preusser *et al.* [542]), and typical post-diagnosis survival is less than 12 months. To study this highly invasive cancer, the mixture model presented above was coupled with the lattice-free model of angiogenesis [534, 535] described in Section 5.12, and the simulation results were compared with clinical brain tumor data (Frieboes *et al.* [229]).

Parameters were estimated from *in vitro* [230] and *in vivo* data [229], while the extent of neovascularization and cell-substrate supply due to blood flow can be

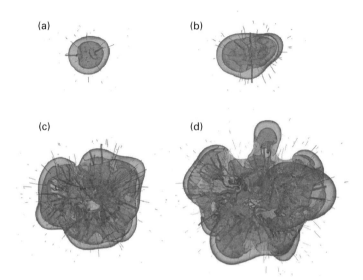

(a)

(b)

(c)

(d)

Figure 5.4 Simulation of a human glioblastoma using the multiscale three-dimensional mixture model [229]. The 3-month time sequence along frames (a) through (d) shows that the morphology is affected by successive cycles of starvation, neovascularization, and vessel cooption [45, 320, 666]. Viable and necrotic tissue regions are shown in light and dark gray, respectively. The vasculature displays new nonconducting vessels (thin lines) and mature blood-conducting vessels (thick lines).

estimated in part from dynamic-contrast-enhanced magnetic resonance imaging (DCE-MRI) observations in patients [501] (see [229]). Because the simulation was based upon the hypothesized relationships between cell adhesion, motility, death, and proliferation described in [229], the model enables a quantitative analysis of these relationships in virtual patients. This undertaking could thus drive the development of morphological and immuno-histopathological criteria for *in vivo* tumor invasion.

The calibrated model correctly predicts the gross tumor morphology, including interior regions of necrosis surrounded by viable tumor tissue as well as a tortuous neovasculature (Figure 5.4), as observed *in vivo* [87]. The vessels migrate towards the tumor–host interface since perinecrotic tumor cells and host-tissue cells close to the tumor boundary release tumor angiogenic factors and other such regulators. The tumor eventually co-opts and engulfs the vessels.

The model simulates the details of the tumor morphology, by quantifying the cells' metabolic response to spatial gradients in substrate levels caused by spatiotemporal heterogeneity in proliferation and by the abnormal neovasculature. Excellent agreement was found when comparing the tumor "virtual histopathology" with clinical pathology observations (Figure 5.5) on sections of tumor tissue 100–200 μm thick encircling neovessels and surrounded by necrotic tumor cells and vessels that have been shut down due to either age or mechanical intratumoral pressure (see [513]). This vessel shutdown enhances the hypoxic gradients seen clinically and predicted by the model. The overall tumor shape depends strongly upon the vascular patterning, a result supported by animal

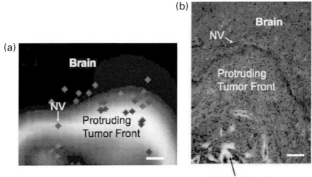

Figure 5.5 Study of human glioma using a multiscale three-dimensional mixture model [229]. (a) Detail of a virtual tumor histology showing the invasive tumor front (white) moving up towards extratumoral conducting neovessels (NV). The aged vessels in the tumor have thicker walls and are assumed to provide fewer cell substrates than the thin-walled vessels at the tumor periphery [513]. Conducting vessels, light diamonds; nonconducting vessels, dark diamonds. The bar is of length 100 μm. (b): H & E-stained patient glioblastoma histopathology sections viewed by fluorescence microscopy [229]. The tumor (bottom) is invading the normal tissue (top). Note the demarcated margin between the tumor and brain parenchyma (middle top), the fluorescent outlines of large aged, vessels deep in the tumor, and the new functional vessels NV at the tumor boundary. The bar is of length 100 μm. Reprinted with permission from Frieboes *et al.*, *NeuroImage*, vol. 37, p. S66, ⓒ 2007 Elsevier.

[58, 392, 398, 528, 576] and clinical imaging (e.g., [703]) as well as histological data. The virtual histopathology also suggests that a tumor may rely upon vessels in the nearby host tissue [50, 542] and proliferate towards them; this is suggested in the brain histopathology in Figure 5.5.

The model enables a quantitative analysis; e.g., the viable region thickness, of about 100–200 μm, and the extent of necrosis seen in Figure 5.5 are shown to be strongly dependent on diffusion gradients of cell substrates in the microenvironment, in agreement with previous experiments [230, 311]. Further, the model predicts that the tumor boundary moves at a rate of about 50–100 μm per week, presenting a mass of diameter of about 5 cm in one year (these data are not shown). The results are supported by well-known clinical observations of glioblastoma (e.g., [489]). As the tumor grows and engulfs the vessels in its vicinity, it may compress them [513] and disrupt the flow of cell substrates, leading to further necrosis and even temporary mass and vascular regression [321, 714]. In addition, chaotic angiogenesis leads to heterogeneous perfusion in the tumor and this may also cause the regression of parts of the vascular network and the necrosis of tumor cells [112, 524]. This enhances variable tumor-cell proliferation.

These results support the hypothesis that diffusion gradients of cell substrates drive collective tumor-cell infiltration into the healthy tissue in addition to playing a major role in determining the tumor's morphological structure. The movement of tumor fronts towards the sources of cell substrates may thus strongly influence glioblastoma invasiveness.

6 Discrete cell modeling[1]

With P. Macklin and M. E. Edgerton

In this chapter, we introduce discrete cancer-cell modeling, assess the strengths and weaknesses of the available discrete cell modeling approaches, sample the major discrete cell modeling approaches employed in current computational cancer modeling, and introduce a discrete agent-based cell modeling framework. This framework currently being developed by the present authors and collaborators will be used to implement the next-generation multiscale cancer-modeling framework detailed in Chapter 7.

6.1   A brief review of discrete modeling in cancer biology

Thus far we have discussed continuum modeling, in which cancer is modeled at the *tissue* scale and the effects of individual cells are averaged out. We now turn our attention to *discrete* models, in which the behavior of one or more individual cells as they interact with one another and the microenvironment is addressed.

Discrete modeling has enjoyed a long history in applied mathematics and biology, dating as far back as the 1940s when John von Neumann applied lattice crystal models to study the necessary rule sets for self-replicating robots [674]. Perhaps the most famous early example of discrete biological modeling is John Conway's 1970 "game of life," a two-dimensional rectangular lattice of "cells" that changed color according to rules based upon the colors of the neighboring cells [260]. Even simple rules can lead to complex emergent behavior, and Conway's model was later shown to be Turing complete [595]. Today, discrete cell modeling has advanced to study a broad swath of cancer biology, spanning carcinogenesis, tumor growth, invasion, and angiogenesis.

Discrete, or individual-based, models are generally divided into two categories: lattice-based (including cellular automata) and lattice-free (agent-based). Both approaches track and update individual cells according to a set of biophysical rules. Typically these models involve a composite[2] discrete–continuum approach in the sense that the microenvironmental variables (glucose, oxygen, extracellular matrix, growth factors, etc.) are described using continuum fields while the cells are discrete. See the reviews by

---

[1] The agent model presented in this chapter is an extension of Macklin *et al.* (2009) [433], and includes work in preparation by Macklin *et al.* in [434].

[2] Many workers refer to this as a hybrid model, but we reserve the term "hybrid" for models that simultaneously combine discrete and continuum representations for cells.

Aber *et al.* [11], Moreira and Deutsch [475], Drasdo [182], Araujo and McElwain [34], Quaranta *et al.* [553, 552], Hatzikirou *et al.* [307], Nagy [490], Abbott *et al.* [2], Byrne *et al.* [96], Fasano *et al.* [209], Galle *et al.* [252], Drasdo and Höhme [186], Thorne *et al.* [656], Anderson *et al.* [28], Deisboeck *et al.* [169], Anderson and Quaranta [29], and Zhang *et al.* [720]. We now review these two main approaches and give some key examples from the literature. A further review can be found in [425].

6.1.1 Lattice-based models

In lattice-based modeling the cells are confined to a regular two-dimensional or three-dimensional lattice. Each computational mesh point is updated in time according to deterministic or stochastic rules derived from physical conservation laws and biological constraints. Some models use a high-resolution mesh to discretize the cells and the surrounding microenvironment at subcellular resolution, allowing a description of the cells' finite sizes, morphologies, and biomechanical interactions. Cellular automata (CA) models, which describe each cell by with a single computational mesh point, can be viewed as a special case of the lattice-based approach.

The simple spatial arrangement of lattice-based models is relatively easy to implement, with less need for advanced computational expertise, rendering them broadly accessible beyond the traditional scientific computing communities. The regular structure imposed by the computational mesh also eliminates the need for elaborate interaction-testing between the discrete cells. This is particularly true for CA models on rectangular meshes, where cell–cell interaction involves only the immediate neighboring mesh points. It is also straightforward to couple lattice-based methods directly to the microenvironment, by assigning continuum variables to every mesh point (in the case of CA methods), or to a coarsened sub-mesh (in the more general case).

The uniform spacing of lattice-based methods can be a weakness. While computationally efficient, the low-resolution lattices used in CA methods can impose artificial constraints on the arrangement, orientation, and interaction of the cells, which can sometimes be observed as square or diamond-like artifacts. Such low-resolution lattice-based models cannot capture the non-lattice cell patternings often found in both normal tissue (e.g., in hepatic lobules) and cancer (e.g., in cribriform ductal carcinoma *in situ* (or DCIS)). Because the cells are limited to just a few orientations, these models can only crudely treat cell polarization, and cell mechanics can only be modeled with great difficulty. Hence, low-resolution models are poorly suited to a rigorous exploration of the balance of cell–cell adhesion, cell–BM adhesion, and cell mechanics (e.g., cell deformability). High-resolution, subcellular, meshes can better approximate cell mechanics but are much more computationally expensive, impeding their ability to describe large systems of cells and their microenvironment.

6.1.2 Lattice-free models

Lattice-free models, frequently referred to as agent-based models, do not restrict the cells' positions and orientations in space. This allows more complex and accurate coupling between the cells and their microenvironment and imposes fewer artificial constraints on the behavior of multicellular systems. The cells are treated as distinct objects or agents and are allowed to move, divide, and die individually according to biophysically based rules. For example, many models apply free-body force diagrams to the individual cells, allowing a mechanistic description of the cell–cell and cell–ECM interactions. A model's level of detail in the cell size, volume, and morphology can vary from simple (e.g., the cells are infinitesimal points in three-dimensional space, such as in [1, 57, 228, 425]) to complex (e.g., the cells are evolving deformable spheres that develop cusps during mitosis, as in [187]). The agent interpretation of the cells makes modern object-oriented programming languages (e.g., C++ and Java) very suitable for implementing these models.

Agent-based models are ideal for situations of freely wandering and nonuniformly arranged cells, such as in angiogenesis, carcinogenesis, immune system attacks on tumor cells, and metastasis. The level of detail of the agents can be tailored to the simulation. Each cell agent can be assigned individual protein and surface-receptor signaling networks, a model of cell cycle progression, and genotypic and phenotypic characteristics; this makes agent-based modeling a powerful tool in multiscale frameworks. (Some agent-based models are restricted to motion on a regular lattice to save on computational cost; see [718, 719] for some examples.) Their flexibility in detail, at times even down to the biochemical level, can make agent-based models easier to calibrate to biological data.

This flexibility comes at a price, however. Because each individual cell can be made almost arbitrarily complex, the computational cost can be very high for non-lattice, agent-based, models, limiting this approach to small systems of cells. The lack of a regular cell arrangement also makes cell–cell-interaction testing computationally expensive. In the worst case, where there are n cell agents that may each interact with the other $n - 1$ cells and there are no constraints on cell placement, one must test $\mathcal{O}(n^2)$ possible cell–cell interactions for each computational time step. In such a case, the computational cost increases rapidly as more cells are added to the multicell system, rendering large-scale simulations infeasible.

6.1.3   Comparison with continuum methods

Continuum modeling can be too coarse-scaled to capture the spatial intricacy of tissue microarchitecture when the cells are polarized (i.e., they have visible apex and base distributions, and anisotropic surface-receptor distributions) or during individual cell motility. Furthermore, continuum models tend to lump multiple physical properties into one or two phenomenological parameters. For example, the models in [229, 437] lump cell–cell, cell–BM, and cell–ECM adhesion, motility, and ECM rigidity into a single "mobility" parameter, as well as into some of the conditions modeling forces on the tumor boundary. While this eases mathematical analysis of the physical systems, it can be difficult to directly match such lumped parameters to physical measurements. Some key patient-specific measurements occur at the molecular (e.g., immunohistochemical (IMC)) and cellular scales, i.e., at finer scales than those of continuum models.

In contrast, discrete cell models can in some cases be directly matched to such measurements. For example, Zhang, Athale, and Deisboeck have been very successful in tying intracellular signaling models to individual cell phenotypes and motility in brain cancer [718]. Their work also made key advances in using molecular and cellular data to inform and calibrate the cell-scale model. Several groups (e.g., [716]) have made considerable advances in the linking of molecular- and cell-scale models, often with calibration to data at the appropriate scales. This is promising in the context of patient-specific cancer simulation (see Chapter 10) and multiscale modeling, since it allows molecular-, cellular-, and tissue-scale data to be matched at their "native" scales; a bidirectional data flow subsequently propagates this information to all the scales in the model. See Sections 7.1 and 10.4.3 and the Conclusion following Chapter 10 for more on this topic.

Discrete models have some drawbacks when compared with continuum approaches. Because they rely upon the behavior of individual cells to determine emergent system properties, they can be difficult to analyze. In addition, the computational cost of the methods increases rapidly with the number of cells modeled, the lattice resolution (for lattice-based methods), and the complexity of each cell object (for agent models). This can make such models difficult or impossible to apply to large systems, even with parallel programming.

Other difficulties relate to model calibration. Nonlocal effects such as biomechanics are often better described by continuum variables, making calibration more feasible at the continuum scale since it can be based upon macroscopic measurements. In hybrid modeling (Chapter 7) one can address this issue by applying and calibrating discrete and continuum models alongside one another, and then incorporating a rigorous information flow between the scales.

Some model parameters, even if clearly related to cell-scale phenomena, may be difficult to measure in controlled experiments whereas macroscopic measurements based upon the averaged behavior of many cells are simple to measure accurately and to match to lumped parameters in continuum models. However, we note that the analysis of volume-averaged agent models can sometimes provide further insight on how to motivate and interpret such matching. See Section 6.5 for such an example.

6.1.4 Some discrete modeling examples

We now discuss some discrete modeling examples that sample the range of cell-scale tumor modeling. Our survey does not exhaust the immense amount of discrete modeling in cancer cell biology; the reader is encouraged to examine the reviews listed in Section 6.1 as well as [425].

Composite cellular automata modeling

In an illustrative example of cellular automata modeling, Anderson *et al.* developed a composite discrete–continuum model of solid tumor growth with microenvironmental interactions [25, 27]. The microenvironmental variables, i.e., the ECM density and matrix metalloproteinases (MMPs) are represented as continuous concentrations and the tumor cells are described as cellular automata. Cells move via a biased random walk on a Cartesian lattice; the movement probabilities are generated by discretizing an analogous continuum model of the tumor cell density. The transition probabilities are similar in spirit to those in the chemotaxis model developed earlier by Othmer and Stevens [510]. The cells respond haptotactically to the ECM density and produce MMPs that degrade the matrix. Cell–cell adhesion is not considered. The model predicts more extensive local tumor invasion in a heterogeneous ECM than is predicted by the analogous continuum model.

Building on this work, Anderson *et al.* extended their model to include cell–cell adhesion by weighting the probability of motion by the number of desired neighbors [25]. Different cell phenotypes are modeled by varying the number of desired neighbors, the proliferation rate, and the nutrient uptake rate. The microenvironment plays an important role in the model: the nutrient supply is assumed to be proportional to the ECM in order to model the pre-existing vasculature, and so matrix degradation disrupts the nutrient supply as well. This model enabled an evaluation of how individual cell–cell and cell–ECM interactions may affect the tumor shape. Anderson and his co-workers extended their model further to provide a theoretical and experimental framework to characterize tumor invasion quantitatively as a function of microenvironmental selective factors [31]. In the extended model they considered random mutations among 100 different phenotypes. In agreement with the findings of Cristini and co-workers [145, 149, 230, 722], hypoxia and a heterogenous ECM were found to induce invasive fingering in the tumor margins, with selection of the most aggressive phenotypes.

Gerlee and Anderson simplified this model to investigate complex branched cell-colony growth patterns arising under nutrient-limited conditions [273]. In agreement with earlier stability analyses (e.g., [149]), the stability of the growth was found to depend on how far the nutrient penetrates into the colony. For low nutrient consumption rates the penetration distance was large, and this stabilized the growth; for high consumption rates the penetration distance was small, leading to unstable branched growth. After incorporating a feed-forward neural network to model the decision-making mechanisms governing the evolution of the cell phenotype, Gerlee and Anderson demonstrated how the oxygen concentration may significantly affect the selection pressure, cell-population diversity, and tumor morphology [272, 274]. They further extended this model to study the emergence of a glycolytic phenotype. Their results suggest that this phenotype is most likely to arise in hypoxic tissue with a dense ECM. Recently, the group explored these themes further, with Quaranta and Rejniak [552]. Similar discrete modeling work on hypoxia has been pioneered in various collaborations between Gatenby, Smallbone, Maini, Gillies, Gavaghan, and others (e.g., [263, 264, 267, 624–627]).

Lattice-gas cellular automata models

Dormann and Deutsch developed a lattice-gas cellular automaton method for simulating the growth and size saturation of avascular multicell tumor spheroids [181, 174]. Unlike traditional cellular automaton methods, where at most one cell can be at a single grid point (this is known as volume exclusion), lattice-gas models accommodate variable cell densities by allowing multiple cells per mesh point, with separate channels of movement between mesh points. The channels specify direction and velocity magnitude and may include zero-velocity resting states. Lattice-gas models typically require channel exclusion, i.e., only one cell may occupy a channel at any time. Hatzikirou *et al.* [306] later used this approach to study traveling fronts characterizing glioma cell invasion. Very recently they performed a mean field analysis of a lattice-gas model of moving and proliferating cells to demonstrate that certain macroscopic information, e.g. scaling laws, can be accurately obtained from a microscopic model [306]. Accurate predictions of other macroscopic quantities that depend sensitively on higher-order correlations are more difficult to obtain.

Immersed boundary model

In [560, 561], Rejniak developed a highly detailed lattice-based approach to the modeling of solid tumor growth. Each individual cell is modeled using the immersed boundary method [529, 530] on a regular computational grid. The cell is represented as the interior of an elastic membrane and the nucleus is represented as an interior point. Cell–cell adhesion and cell contractile forces are modeled using linear springs to mimic a discrete set of membrane receptors, adhesion molecules, and the effect of the cytoskeleton on the cell mechanics. The cytoplasm and interstitial fluid are modeled as viscous incompressible fluids. The elastic, adhesive, and contractive forces impart singular stresses to the fluids. Cell growth is modeled with an interior volume source; once the cell grows to a threshold volume, contractile forces on opposite sides of the cell create a neck that pinches off to produce two approximately equal-sized daughter cells. The nutrient supply is modeled using continuum reaction–diffusion equations, with uptake localized in the cells. The method can describe individual cell morphology but is computationally expensive, thus restricting simulations to about 100 cells. Rejniak and Dillon recently extended the model to better represent the lipid bilayer cell membrane structure as two closed curves connected by springs [564]; sources and sinks, placed in the simulated bilipid membrane, model water channels.

The immersed boundary-cell model has been applied to study pre-invasive intraductal tumors, as well as the formation and stability of epithelial acini (single-layered spherical shells of polarized cells attached to a BM) [562–564]; genetic mutations that disrupt cell polarity could lead to abnormal acini and ductal carcinoma. The model has also been used to study the interaction between nutrient availability, metabolism, phenotype, and the growth and morphological stability of avascular tumors [30, 180]. Morphological instabilities are in qualitative agreement with the continuum [149, 416, 437, 722] and cellular automata [25, 31, 272, 273] modeling results discussed earlier.

An extended Q-Potts model

A less detailed lattice-based method has been developed using an extended Q-Potts model, which originated in statistical physics to study surface-diffusion grain growth in materials science, as in [33]. Graner and Glazier adapted the Q-Potts model to simulate cell sorting through differential cell–cell adhesion [284, 290]. In this approach, now referred to as the GGH (Graner–Glazier–Hogeweg) model, each cell is treated individually and occupies a finite set of grid points within a Cartesian lattice; space is divided into distinct cellular and extracellular regions. Each cell is deformable with a finite volume. Cell–cell adhesion is modeled with an energy functional. A Monte Carlo algorithm is used to update each lattice point and hence change the effective shape and position of a cell. Although the description of the cell shape is less detailed than in the immersed boundary approach described above, finite-size-cell effects are incorporated.

Later work incorporated nutrient-dependent proliferation and necrosis in the GGH model to simulate the growth of benign multicellular avascular tumors to a steady state [639]; nutrient transport was modeled with a continuum reaction–diffusion equation. The steady-state configuration consisted of a central necrotic core, a surrounding band of quiescent cells, and an outermost shell of proliferative cells; the parameters were determined by matching the thickness of these regions to experimental measurements. Reference [664] extended the GGH model, by incorporating ECM–MMP dynamics, haptotaxis, and adhesion-controlled proliferation, to study tumor invasion. The GGH model has also been extended to account for chemotaxis [591], cell differentiation [319], and cell polarity [715]. Others modified the GGH model to include a subcellular protein signaling model to tie the cell cycle to growth promoters and inhibitors through continuum reaction–diffusion equations [351]. In that work, parameter values were selected such that model produced avascular tumors that quantitatively replicated experimental measurements of *in vitro* spheroids.

The GGH model has been used to simulate vasculogenesis and tumor angiogenesis [463–465] in heterogeneous tumor microenvironments [53] as well as the role of the ECM in glioma invasion [575]. Very recently, Poplawski and co-workers used the GGH method to study the morphological evolution of two-dimensional avascular tumors [538]; the work developed a phase diagram characterizing the tumor morphology and the stability of the tumor–host interface with respect to critical parameters characterizing nutrient limitations and cell–cell adhesion. In particular, they found that morphological stability depends primarily on the diffusional limitation parameter, whereas the morphological details depend on cell–cell adhesion. The results are consistent with previous continuum [145, 149, 416, 437] and discrete [30, 272, 273, 560] modeling results.

Some semi-deformable agent-based models

Drasdo and co-workers developed an agent-based model that incorporates finite cell size to study epithelial cell–fibroblast–fibrocyte aggregates in connective tissue [187]. In their approach, simplified cells are modeled as a roughly spherical space containing a central region. The cells are slightly compressible and are capable of migration, growth, and division. A non-dividing cell is taken to be spherical. As a cell mitoses it deforms into a dumbbell shape until its volume roughly doubles, and then it divides into two daughter cells. The adhesion and repulsion (arising from limitations on cell deformation and compressibility) among cells are modeled using an interaction energy that describes nearest-neighbor interactions. Mitosis and migration may induce pressure on neighboring cells. The cells respond by changing their mass or orientation to minimize the total interaction energy via a stochastic Metropolis algorithm [467]. Interaction potentials have also been used in agent models by Ramis-Conde and co-workers [557, 558]; in the model, cells move down the gradient of the potential, analogously to minimizing the interaction energy. Others have modeled cells as deformable viscoelastic ellipsoids [161, 519].

Drasdo and Höhme [184] adapted their approach to early-stage avascular tumor spheroids, where growth is not primarily limited by oxygen or nutrient supply but rather by volume exclusion arising from limited cell compressibility. The biomechanical and kinetic parameters were estimated by comparison with tumor spheroid experiments from [225]. Drasdo and Höhme extended the model to account for glucose-limited growth, necrosis, and cell lysis, in order to simulate the spatiotemporal growth dynamics of two-dimensional tumor monolayers and three-dimensional tumor spheroids, with biophysical and kinetic parameters drawn from the experimental literature [185]. The results suggested that biomechanical growth inhibition is responsible for the transition from exponential to subexponential growth that is observed experimentally for sufficiently large tumors; glucose deprivation was found to be a primary factor determining the size of the necrotic core but it was found not to affect the size of the tumor. Galle *et al.* extended the model to incorporate the effect of BM contact on cell cycle progress and apoptosis (see Section 2.1.1) and studied epithelial cell growth in monolayers [254]. They found that the inactivation of BM-dependent cell cycle progression and apoptosis or the removal of contact-mediated growth inhibition (e.g., see Section 2.1.5) could lead to epithelial tumor growth. In similar monolayer simulations, Drasdo showed how the agent model can be used to determine the rules for a simpler cellular automaton (CA) model, which was then used to derive a continuum model with contact inhibition by averaging the CA behavior on a coarser Cartesian grid (coarse-graining). This provided a link between different scales and biophysical processes [183]. Byrne and Drasdo performed further analysis of the continuum model [102].

Recent examples of subcellular modeling

Individual cell agents can readily be endowed with subcellular models, making them ideal multiscale modeling platforms. Recently, Ramis-Conde and colleagues incorporated E-cadherin/β-catenin dynamics in an agent-based approach to obtain a more realistic model of cell–cell adhesion mechanics [558]; β-catenin binds the membrane-bound E-cadherin to the cytoskeleton. Their detailed model could describe the detachment of cells from a primary tumor and the corresponding epithelial–mesenchymal transition. Galle *et al.* incorporated cell–ECM interactions and ECM contact-dependent cell regulation as well as cell differentiation [253]. They studied the effect of cancer stem-cell organization on tumor growth, finding that tumors invade the host tissue much more rapidly when stem cells are on the tumor periphery rather than confined to the interior.

Wang *et al.* developed a multiscale model of non-small-cell lung cancer within a two-dimensional microenvironment, implementing a specific intracellular signal transduction pathway between the epidermal growth factor receptor (EGFR) and extracellular receptor kinase (ERK) at the molecular level [678]. Phenotypic changes at the cellular level were triggered through dynamical alterations of these molecules. The results indicated that, for this type of cancer, downstream EGFR–ERK signaling may be processed more efficiently in the presence of a strong extrinsic chemotactic stimulus, leading to a migration-dominant cell phenotype and an accelerated rate of tumor expansion. Zhang *et al.* presented a three-dimensional multiscale agent-based model to simulate the cellular decision process to either proliferate or migrate, in the context of brain tumors [718]. Each cell was equipped with an EGFR gene–protein interaction network module that also connected to a simplified cell-cycle description. The results show that the proliferative and migratory cell populations directly impact the spatiotemporal expansion patterns of the cancer. Zhang and co-workers later refined their model to incorporate mutations representing a simplified tumor-progression pathway [719].

6.2   An agent-based cell modeling framework

To illustrate these concepts we now introduce a discrete, cell-scale, modeling framework that combines and extends the major features of the models introduced in Section 6.1. Our main objective is to develop a model that is sufficiently mechanistic that cellular and multicellular behavior manifest themselves as *emergent phenomena* of the modeling framework, rather than through computational rules that are imposed *a priori*. An additional design goal is that the model is modular (both in software and mathematics), allowing "submodels" (describing, e.g., molecular signaling or cell morphology) to be expanded, simplified, or replaced outright, as necessary.

We use a lattice-free, agent-based, approach to allow more accurate cell mechanics and treat the cells (the agents) as physical objects subject to biophysically justified forces. Cell–cell and cell–BM mechanical interactions are modeled using interaction-potential functions, in a similar way to [102, 183–185, 187]. The balance of these forces explicitly determines the cell's velocity. The cells have a nonzero, finite size [183–185, 187]. We model explicitly the mechanical interactions between the cells and the basement membrane, in a similar way to the work in [567]. As with many of the discrete models discussed in Section 6.1.4, each cell has a phenotypic state, the transitions between those phenotypic states being governed by stochastic processes. We note that the same model is used for both cancerous and noncancerous cells. The cells differ primarily in the values of their proliferation, apoptosis, and other coefficients; this is analogous to the modeling of altered oncogenes and tumor suppressor genes [300].

We incorporate the essential molecular biology through carefully chosen constitutive relations. In particular, the mechanics, time duration, and biology of each phenotypic state is modeled as accurately as our data will allow; this should facilitate the model's calibration to molecular and cellular data. As in the preceding models (e.g., [27]), the microenvironment is incorporated using a composite discrete–continuum approach. Thus, it is modeled as a set of field variables (e.g., oxygen concentration, ECM density) governed by continuum equations that can be altered by the discrete cells. Cell agents interact with this microenvironment, both mechanically and biochemically, through surface receptors that are part of a molecular-scale signaling model.

This agent model introduces several new features to discrete modeling as part of our design philosophy. The cell states are chosen specifically to facilitate calibration to immunohistochemistry in patient-specific simulations. We explicitly link the phenotypic state transitions to the microenvironment and signaling models through functional relationships in the stochastic parameters. Our model differentiates between apoptosis and necrosis and introduces a model for necrotic cell calcification. To facilitate a mechanistic understanding of the model and to match it to experimental biology, we separate the forces into different potential functions rather than lump them together into a single function.

We use interaction potential functions with compact support (i.e., they are zero outside some finite maximum interaction distance) to model the finite interaction distances between cells and their neighbors more realistically. The adhesion model can differentiate between homophilic and heterophilic adhesion, and it separates the effects of cell–cell, cell–BM, and cell–ECM adhesion.

In this discussion cells are not polarized and, in particular, we assume an isotropic distribution of cell surface receptors; model extensions to address this were discussed in [434]. Also, we do not currently focus on stem cell dynamics, although this can readily be added by identifying cells as stem cells, progenitor cells, or differentiated cells and assigning each cell class different proliferation and other phenotypic characteristics. We treat the cells as mostly rigid spheres (cells are allowed to partly overlap), with growth in both two and three dimensions; basement membranes are currently modeled as sharp boundaries using level set functions [229, 435, 438, 439]. We will present applications of the model to breast cancer in Section 6.6 and Chapter 10.

6.2.1 A brief review of exponential random variables and Poisson processes

Because we are going to model transitions between cell states as stochastic processes, we will begin with a brief review of the necessary preliminaries. This discussion necessarily only introduces the key concepts and does not explore the full richness of measure-theory-based probability and stochastic processes. The interested reader can find out more in references such as [504, 609].

A random variable T is *exponentially distributed* with parameter α if, for any $t > 0$, the probability $\Pr(T < t)$ is given by

$$\Pr(T < t) = 1 - e^{-\alpha t}. \tag{6.1}$$

Also, T has expected value $\mathrm{Ex}[T] = 1/\alpha$ (i.e., the mean $\langle T \rangle$ is $1/\alpha$) and variance $\mathrm{Var}[T] = 1/\alpha^2$. This simple relationship between the mean $\langle T \rangle$ and α is useful when calibrating with limited data. Exponential random variables are *memoryless*: for any t, $\Delta t \geq$, the probability that $T > t + \Delta t$ given that $T > t$ is

$$\Pr(T > t + \Delta t | T > t) = \Pr(T > \Delta t), \tag{6.2}$$

i.e., if the event T has not occurred by time t then the probability that the event occurs within an additional Δt units of time is unaffected by the earlier times, and so we can "reset the clock" at time t. This is useful for modeling cell decision processes that depend upon the current subcellular and microenvironmental state and not on states at previous times. Even if the current cell decisions do depend upon past states, that information can be built into the time evolution of the internal cell state.

A *stochastic process* N_t is a series of random variables indexed by the "time" t. In particular, N_t is a *counting process* if:

1. $N_0 \geq 0$. (The initial number of events N_0 is at least zero.)
2. $N_t \in \mathbb{Z}$ for all $t \geq 0$. (The current number of events N_t is an integer.)
3. If $s < t$ then $N_t - N_s \geq 0$. (N_t is the cumulative number of events, and the number of events occuring within the interval $(s, t]$ is $N_t - N_s$.)

A *Poisson process* X_t is a particular type of counting process satisfying:

1. $X_0 = 0$. (The initial count is 0.)
2. If $[s, s + \Delta s]$ and $[t, t + \Delta t]$ are nonoverlapping $X_{s+\Delta s} - X_s$ and $X_{t+\Delta t} - X_t$ are independent random variables. (What happens in the interval $[t, t + \Delta t]$ is independent of what happened in $[s, s + \Delta s]$.)
3. For any $0 \leq s < t$, the distribution of $X_t - X_s$ depends only upon the length of the interval $[s, t]$ (the process has stationary increments) and, in particular, for $n \in \mathbb{N}$,

$$\Pr(X_t - X_s = n) = \frac{\mathrm{e}^{-\alpha(t-s)}\alpha(t-s)}{n!}. \tag{6.3}$$

Poisson processes have a useful property upon which we rely in the model. If $A_n = \inf\{t : P_t = n\}$ is the *arrival time* of the nth event (i.e., the first time at which $X_t = n$), and $T_{n+1} = A_{n+1} - A_n$ is the *interarrival time* between the nth and $(n + 1)$th events for $n \in \mathbb{N}$ then T_n is an exponentially distributed random variable with parameter α, and

$$\Pr(X_{t+\Delta t} - X_t \geq 1) = \Pr(A_n \in (t, t + \Delta t] | A_n > t)$$
$$= \Pr(T_n < \Delta t) = 1 - \mathrm{e}^{-\alpha \Delta t}. \tag{6.4}$$

Thus, for a sequence of events governed by a Poisson process, the times between consecutive events are simple exponentially distributed random variables. In the context of stochastic cell models, if X_t is a Poisson process that gives the cumulative number of phenotypic state changes experienced by the cell by time t, the time until the next phenotypic state change is exponentially distributed.

Lastly, we note that if $\alpha = \alpha(t)$ varies in time then X_t is a *nonhomogeneous* Poisson process, with interarrival times given by

$$\Pr(X_{t+\Delta t} - X_t = n) = \frac{\left\{\exp\left[-\int_t^{t+\Delta t} \alpha(s)\,ds\right]\right\} \left(\int_t^{t+\Delta t} \alpha(s)\,ds\right)^n}{n!},$$
$$\Pr(T_n < \Delta t) = \Pr(X_{t+\Delta t} - X_t \geq 1)$$
$$= 1 - \exp\left[-\int_t^{t+\Delta t} \alpha(s)\,ds\right]$$
$$\approx 1 - \exp[-\alpha(t)\Delta t], \qquad \Delta t \downarrow 0.$$

In our work the Poisson processes are nonhomogeneous owing to their dependencies upon microenvironmental and intracellular variables that vary in time. However, on small time intervals $[t, t + \Delta t]$ these can be approximated by homogeneous processes as above [433, 434].

6.2.2 A family of potential functions

As in [102, 183–185, 187], we shall use potential functions to model biomechanical interactions between cells and the microenvironment. We now introduce the family of interaction potential functions $\varphi(\mathbf{x}; R, n)$ used in the agent model. Parameterized by R and n, they satisfy

$$\varphi(\mathbf{x}; R, n) = \begin{cases} -\dfrac{R}{n+2} \left(1 - \dfrac{|\mathbf{x}|}{R}\right)^{n+2} & \text{if } |\mathbf{x}| < R, \\ 0 & \text{otherwise,} \end{cases} \tag{6.5}$$

$$\varphi'(x; R, n) = \begin{cases} \left(1 - \dfrac{x}{R}\right)^{n+1} & \text{if } x < R \\ 0 & \text{otherwise,} \end{cases} \tag{6.6}$$

$$\nabla\varphi(\mathbf{x}; R, n) = \begin{cases} \left(1 - \dfrac{|\mathbf{x}|}{R}\right)^{n+1} \dfrac{\mathbf{x}}{|\mathbf{x}|} & \text{if } |\mathbf{x}| < R \\ \mathbf{0} & \text{otherwise.} \end{cases} \tag{6.7}$$

where R is the *maximum interaction distance* of φ and n is the *exponent* or *power* of the potential. We use this form of potential function because:

1. The potential and its derivatives have *compact support*: they are zero outside a closed bounded set (in this case, the closed ball $\overline{B}(\mathbf{0}, R)$). This models finite cell–cell and cell–BM interaction distances.
2. For any R and n, and for any $0 < |\mathbf{x}| < R$, we have

$$0 = \varphi'(R; R, n) < \varphi'(|\mathbf{x}|; R, n) < \varphi'(0; R, n) = 1. \tag{6.8}$$

The baseline case, $n = 0$, is a linear ramping and for higher n the function tapers off to zero gradient smoothly.

A good discussion of the use of potential functions to mediate cell–cell adhesion and interaction for individual-based models can be found in [102].

6.2.3 Cell states

We model the cells' biological function by endowing each agent with a state $\mathcal{S}(t)$ in the state space $\{\mathcal{Q}, \mathcal{P}, \mathcal{A}, \mathcal{H}, \mathcal{N}, \mathcal{C}, \mathcal{M}\}$. Quiescent cells ($\mathcal{Q}$) are in a "resting state" (G0, in terms of the cell cycle); this is the "default" cell state in the agent framework. We model the transitions between cell states as stochastic events governed by exponentially distributed random variables. (The transition events are interarrival times modeling the time elapsed between proliferation and apoptosis events. See [434] for a discussion of the mathematical theory of this modeling construct.) Quiescent cells can become proliferative ($\mathcal{P}$), apoptotic ($\mathcal{A}$), or motile ($\mathcal{M}$). Cells in any state can become hypoxic ($\mathcal{H}$); hypoxic cells can recover to their previous state or become necrotic ($\mathcal{N}$), and necrotic cells are degraded and replaced by calcified debris ($\mathcal{C}$). See Figure 6.1. The subcellular scale is built into this framework by making the random exponential variables depend upon the microenvironment and the cell's internal properties. Cell cycle models which

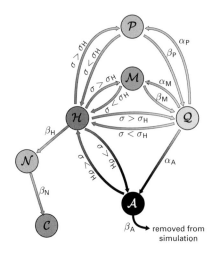

Figure 6.1 Flow between the cell states in the agent-based model. Reprinted with permission from [434]. The arrows are labeled by oxygen levels σ and the parameters α, β discussed in Section 6.2.1–6.2.3.

can regulate the $\mathcal{P} \to \mathcal{Q}$ transition [2, 718], and protein signaling networks to regulate the $\mathcal{Q} \to \mathcal{P}$, $\mathcal{Q} \to \mathcal{A}$, and $\mathcal{Q} \to \mathcal{M}$ transitions, have been developed. These can be integrated directly into the agent framework presented here by modifying the stochastic parameters; see Section 6.3 below. Further discussions of agent-based modeling with subcellular signaling components are to be found in Section 6.1.4 and in the excellent publications [44, 131, 132, 169, 175, 345, 376, 402, 678, 688, 689, 694, 716, 718, 720].

Proliferation ($\mathcal{P}$)

Quiescent cells $\mathcal{Q}$ enter the proliferative state $\mathcal{P}$ with a probability that depends upon the microenvironment. We model the probability that a quiescent cell enters the proliferative state in the time interval $(t, t + \Delta t]$ as an exponential interarrival time with parameter $\alpha_P(\mathcal{S}, \bullet, \circ)$; $\bullet$ represents the cell's internal (genetic and proteomic) state and $\circ$ denotes the local microenvironmental conditions. Hence,[3]

$$\Pr\big(\mathcal{S}(t + \Delta t) = \mathcal{P}\big|\mathcal{S}(t) = \mathcal{Q}\big) = 1 - e^{-\alpha_P \Delta t} \approx \alpha_P \Delta t. \tag{6.9}$$

Assuming a correlation between the microenvironmental oxygen level σ (nondimensionalized by σ_∞, the far-field oxygen level in non-diseased, "well-oxygenated," tissue) and proliferation (See Section 10.4.2), we expect α_P to increase with σ.[4] We model

---

[3] The interarrival time normally gives the probability that there is at least one (rather than precisely one) proliferation event during $(t, t + \Delta t]$. Our α_P in Eq. (6.10) precludes this because $\alpha_P \downarrow 0$ until proliferation is complete.

[4] In Chapters 6 and 10, σ and g denote the oxygen and glucose levels, which are generalized by the substrate variable n in the remainder of the book. In Chapters 6 and 10, n denotes an integer.

this by

$$
\alpha_P = \alpha_P(\mathcal{S}(t), \sigma, \bullet, \circ) =
\begin{cases}
\overline{\alpha}_P(\bullet, \circ)\dfrac{\sigma - \sigma_H}{1 - \sigma_H} & \text{if } \mathcal{S}(t) = \mathcal{Q}, \\
0 & \text{otherwise},
\end{cases}
\tag{6.10}
$$

where σ_H is the threshold oxygen value at which cells become hypoxic and $\overline{\alpha}_P(\bullet, \circ)$ is the cell's $\mathcal{Q} \to \mathcal{P}$ transition rate when $\sigma = 1$ (i.e., in "well-oxygenated" tissue); this transition rate depends upon the cell's genetic profile and protein signaling state ($\bullet$) and on the local microenvironment ($\circ$). For simplicity, we will model $\overline{\alpha}_P$ as constant for, and specific to, each cell type. In Section 6.1.4, we will discuss how to incorporate $\bullet$ (i.e., a cell's internal protein expression) and $\circ$ (as sampled by a cell's surface receptors) into α_P through a molecular signaling model. We note that models have been developed that reduce the proliferation rate in response to mechanical stresses (e.g., see the excellent description in [610]); in the context of the model above a cell samples these stresses from continuum-scale field variables or tensors (i.e., $\circ$) in order to reduce α_P.

Once a cell has entered the proliferative state $\mathcal{P}$ it remains in that state for a time β_P^{-1}, which generally depends upon the microenvironment and the cell's internal state, but which we currently model as fixed at the same value for both tumor and epithelial cells. This models our assumption that both tumor and noncancerous cells use the same basic cellular machinery to proliferate, but with differing frequency owing to altered oncogene expression in the cancerous cells [300]. Once a cell exits the proliferative state we replace it with two identical daughter cells. Both inherit the parent cell's phenotypic properties, are randomly positioned adjacently to one another while conserving the parent cell's center of mass, and placed in the "default" quiescent state.

Apoptosis ($\mathcal{A}$)

Apoptotic cells are undergoing "programmed" cell death in response to internal protein signaling. As with proliferation, we model entry into $\mathcal{A}$ as an exponential interarrival time with parameter $\alpha_A(\mathcal{S}, \bullet, \circ)$. We assume no correlation between apoptosis and oxygen level [194]. Hence $\alpha_A(\mathcal{S}, \bullet)$ is fixed for each cell population:

$$
\Pr\big(\mathcal{S}(t + \Delta t) = \mathcal{A} \big| \mathcal{S}(t) = \mathcal{Q}\big) = 1 - e^{-\alpha_A \Delta t},
\tag{6.11}
$$

where

$$
\alpha_A = \alpha_A\big(\mathcal{S}(t), \bullet, \circ\big) =
\begin{cases}
\overline{\alpha}_A(\bullet, \circ) & \text{if } \mathcal{S}(t) = \mathcal{Q}, \\
0 & \text{otherwise}.
\end{cases}
\tag{6.12}
$$

Here $\circ$ does not include the oxygen level σ but may include other microenvironmental stimuli such as chemotherapy or continuum-scale mechanical stresses that increase α_A, as in [610]. Cells remain in the apoptotic state for a fixed amount of time, β_A^{-1}. Cells leaving the apoptotic state are deleted from the simulation, to model the phagocytosis of apoptotic bodies by the surrounding epithelial cells. The previously occupied volume is made available to the surrounding cells to model the release of the cells' water content after lysis.

Hypoxia ($\mathcal{H}$)

Cells enter the hypoxic state if $\sigma < \sigma_H$. Hypoxic cells have an exposure-dependent probability of becoming necrotic:

$$\Pr\big(\mathcal{S}(t + \Delta t) = \mathcal{N}\,|\,\mathcal{S}(t) = \mathcal{H}\big) = 1 - e^{-\beta_H \Delta t}, \tag{6.13}$$

where at present we model β_H as constant. If $\sigma > \sigma_H$ (i.e., normoxia is restored) at time $t + \Delta t$ and the cell has not become necrotic, it returns to its former state ($\mathcal{Q}$, $\mathcal{P}$, $\mathcal{A}$, or $\mathcal{M}$) and resumes its activity. For example, if the cell transitioned from $\mathcal{P}$ to $\mathcal{H}$ after spending a time τ in the cycle and normoxic conditions are restored before the cell transitions to $\mathcal{N}$ then it returns to $\mathcal{P}$ after a time τ has elapsed in its cell-cycle progression. Because $\Pr(\mathcal{S}(t + \Delta t) = \mathcal{N}\,|\,\mathcal{S}(t) = \mathcal{H}) \approx \beta_H \Delta t$, the probability that a cell succumbs to hypoxia is approximately proportional to $\ell\,(t : \mathcal{S}(t) = \mathcal{H})$, its cumulative exposure time to hypoxia. This construct could model the cell's response to other stressors (e.g., chemotherapy), in the same way as "area under the curve" models (e.g., [196, 197]).

Necrosis ($\mathcal{N}$)

In our model, a hypoxic cell has a probability of irreversibly entering the necrotic state; this simulates the depletion of its ATP store. We can simplify the model and neglect the hypoxic state by letting $\beta_H \to \infty$.

We assume that cells remain in the necrotic state for a fixed amount of time β_N^{-1}, during which time their surface receptors and subcellular structures degrade, they lose their liquid volume, and their solid component is calcified (replaced by calcium deposits). We define β_{NL}^{-1} to be the length of time for the cell to swell, lyse, and lose its water content, β_{NS}^{-1} the time for all surface adhesion receptors to degrade and become functionally inactive, and β_C^{-1} the time for the cell to fully calcify. Generally, $\beta_{NL}^{-1} \leq \beta_{NS}^{-1} < \beta_C^{-1} = \beta_N^{-1}$. In [433] we found that a simplified model with $\beta_N^{-1} = \beta_{NS}^{-1} = \beta_{NL}^{-1} = \beta_C^{-1}$ could not reproduce certain aspects of the breast cancer microarchitecture.

If τ is the time spent in the necrotic state, we model the degradation of the surface receptor species S (scaled by the non-necrotic expression level) by exponential decay with rate constant $\beta_{NS} \log 100$; the constant is chosen so that $S(\beta_{NS}^{-1}) = 0.01\,S(0) = 0.01$, i.e., virtually all the surface receptor is degraded by time $\tau = \beta_{NS}^{-1}$.

For a preliminary model of the necrotic cell's volume change, we neglect its early swelling and instead model its volume change after lysis:

$$V(\tau) = \begin{cases} V & \text{if } 0 \leq \tau < \beta_{NL}^{-1}, \\ V_S & \text{if } \beta_{NL}^{-1} < \tau, \end{cases} \tag{6.14}$$

where V_S is the cell's solid volume.

Lastly, we assume a constant rate of cell calcification, such that the necrotic cell is 100% calcified by time β_C^{-1}. If C is the nondimensional degree of calcification then $C(t) = \beta_C \tau$.

Calcified debris ($\mathcal{C}$)

Cells leaving the necrotic state $\mathcal{N}$ irreversibly enter the calcified debris state $\mathcal{C}$. Lacking functional adhesion receptors, these cells adhere only to other calcified debris. This is a

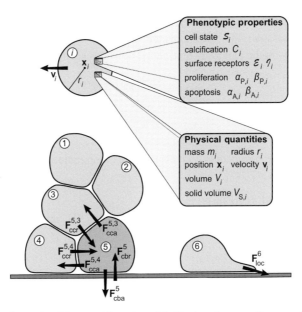

Figure 6.2 Basic schematic of the agent-based model. The key forces acting on cell 5 are labeled. Reprinted with permission from [434].

simplified model of the crystalline bonding between calcium phosphate and/or calcium oxalate molecules in the microcalcification.

Motility ($\mathcal{M}$)

The transition of cells to the motile state (via α_M) is a complex "decision" that depends upon the cell's microenvironment, as communicated to the cell by its surface receptors and its internal signaling network. Once the cell enters $\mathcal{M}$, its speed (and hence β_M) and direction of motion depend upon its interaction with the microenvironment. Depending upon the sophistication of the model, the duration of motility can be fixed, by taking β_M to be constant, or determined by the motility model; for example, we can set $\beta_M = 0$ until the cell reaches its destination, at which time we "force" an immediate $\mathcal{M} \to \mathcal{Q}$ transition. We discuss this further in Section 6.2.4.

6.2.4   Forces acting on the cells

Each cell is subject to competing forces that determine its motion within the microenvironment. Cells adhere to other cells (via the cell–cell adhesion force $\mathbf{F}_{cca}$), to the extracellular matrix (via the cell–ECM adhesion force $\mathbf{F}_{cma}$), and to the basement membrane (via the cell–BM adhesion force $\mathbf{F}_{cba}$); calcified debris adheres to other calcified debris (via the debris–debris adhesion force $\mathbf{F}_{dda}$); cells and calcified debris resist compression by other cells and debris (via the cell–cell repulsion force $\mathbf{F}_{ccr}$); and the basement membrane resists its penetration and deformation by cells and debris (via the cell–BM repulsion force $\mathbf{F}_{cbr}$). Motile cells experience a net locomotive force $\mathbf{F}_{loc}$ along the direction of intended travel. See Figure 6.2, where we show the forces acting on cell 5. In addition, moving cells and debris experience a drag force $\mathbf{F}_{\text{drag}}$ from the luminal and interstitial fluids, which we model by $\mathbf{F}_{\text{drag}} = -\nu\mathbf{v}_i$. We relate the balance of forces on the cell to its motion by Newton's second law:

$$m_i\dot{\mathbf{v}}_i = \sum_j \left(\mathbf{F}_{\text{cca}}^{ij} + \mathbf{F}_{\text{ccr}}^{ij} + \mathbf{F}_{\text{dda}}^{ij}\right)$$

$$+ \mathbf{F}_{\text{cma}}^i + \mathbf{F}_{\text{cba}}^i + \mathbf{F}_{\text{cbr}}^i + \mathbf{F}_{\text{loc}}^i + \mathbf{F}_{\text{drag}}^i. \tag{6.15}$$

Here, the sum is over all cells j in the computational domain.

Cell–cell adhesion ($\mathbf{F}_{\text{cca}}$)

Cell–cell adhesion can be both homophilic [521] and heterophilic [630, 654, 427].

Homophilic adhesion

Adhesion molecules with receptor expression $\mathcal{E}$ on the cell surface bond with the same type of molecule on neighboring cells. Hence the strength of the adhesive force between the cells is proportional to the product of their respective expressions $\mathcal{E}$. Furthermore, the strength of adhesion increases as the cells are drawn more closely together, bringing more surface area and hence more adhesion molecules into direct contact. We model this adhesive force between cells i and j by

$$\mathbf{F}_{\text{cca}}^{ij} = \alpha_{\text{cca}}\mathcal{E}_i\mathcal{E}_j\nabla\varphi(\mathbf{x}_j - \mathbf{x}_i; R_{\text{cca}}^i + R_{\text{cca}}^j, n_{\text{cca}}), \tag{6.16}$$

where $\mathcal{E}_i$ and R_{cca}^i are cell i's (nondimensionalized) adhesion receptor expression and maximum cell–cell adhesion interaction distance, respectively, r_i is the cell's radius, and n_{cca} is the cell–cell adhesion power for our potential function family; see Section 6.2.2. We typically set $R_{\text{cca}}^i > r_i$ to approximate the cell's ability to deform before all adhesive bonds are broken, with the bond strength decreasing as the cells are separated.

Heterophilic adhesion

Adhesion molecules A with receptor expression $\mathcal{I}_A$ on the cell surface bond with different molecules B having receptor expression $\mathcal{I}_B$ on neighboring cells. Hence the strength of the adhesive force between the cells is proportional to the product $\mathcal{I}_A\mathcal{I}_B$ of their receptor expressions. Furthermore, as before the strength of adhesion increases as the cells are drawn more closely together, bringing more surface area and hence more receptors into direct contact. We model this adhesive force between cells i and j by

$$\mathbf{F}_{\text{cca}}^{ij} = \alpha_{\text{cca}}(\mathcal{I}_{A,i}\mathcal{I}_{B,j} + \mathcal{I}_{B,i}\mathcal{I}_{A,j})\nabla\varphi(\mathbf{x}_j - \mathbf{x}_i; R_{\text{cca}}^i + R_{\text{cca}}^j, n_{\text{cca}}), \tag{6.17}$$

where $\mathcal{I}_{A,i}$ and $\mathcal{I}_{B,i}$ are cell i's (nondimensionalized) $\mathcal{I}_A$ and $\mathcal{I}_B$ receptor expressions, R_{cca}^i is the cell's maximum cell–cell adhesion interaction distance, and n_{cca} is the cell–cell adhesion power as before. As with homophilic cell–cell adhesion, we typically set

$R^i_{\text{cca}} > r_i$ to approximate the ability of cells to deform before all adhesive bonds are broken, with the bond strength decreasing as the cells are separated.

Cell–ECM adhesion ($\mathbf{F}_{\text{cma}}$)

Integrins with receptor expression $\mathcal{I}_E$ on the cell surface form heterophilic bonds with suitable ligands $\mathcal{L}_E$ in the ECM. We assume that the ligands $\mathcal{L}_E$ are distributed proportionally to the (nondimensional) ECM density E. If we also assume that $\mathcal{I}_E$ is distributed uniformly across the cell surface and that E varies slowly relative to the spatial size of a single cell, then cells at rest encounter a uniform pull from cell–ECM adhesive forces $\mathbf{F}_{\text{cma}}$ in all directions, resulting in a zero net cell–ECM adhesive force. For cells in motion, the forces $\mathbf{F}_{\text{cma}}$ resist that motion; they act as a drag force due to the energy required to overcome integrin–ligand bonds:

$$\mathbf{F}_{\text{cma}} = -\alpha_{\text{cma}} \mathcal{I}_{E,i} E \mathbf{v}_i, \tag{6.18}$$

where $\mathcal{I}_{E,i}$ is the integrin receptor expression of cell i. If E or $\mathcal{L}_E$ varies with a higher spatial frequency, or if $\mathcal{I}_E$ is not uniformly distributed, then the finite half-life of integrin–ligand bonds will lead to net haptotactic-type migration up gradients of E [583]. We currently describe this effect only by including it in the cell's (active) locomotive force $\mathbf{F}_{\text{loc}}$.

Cell–BM adhesion ($\mathbf{F}_{\text{cba}}$)

Integrin molecules on the cell surface form heterophilic bonds with specific ligands $\mathcal{L}_B$ (generally laminin and fibronectin [90]) on the basement membrane (with density $0 < B < 1$). We assume that the ligands $\mathcal{L}_B$ are distributed proportionally to the (nondimensional) BM density B. Hence, the strength of the cell–BM adhesive force is proportional to the cell's integrin surface receptor expression and B. Furthermore, the strength of the adhesion increases as the cell approaches the BM, bringing more cell adhesion receptors in contact with corresponding ligands on the BM. We model this adhesive force on cell i by

$$\mathbf{F}^i_{\text{cba}} = \alpha_{\text{cba}} \mathcal{I}_{B,i} B \nabla \varphi \left(d(\mathbf{x}_i); R^i_{\text{cba}}, n_{\text{cba}} \right), \tag{6.19}$$

where d is the distance from cell i to the basement membrane, $\mathcal{I}_{B,i}$ and R^i_{cba} are cell i's (nondimensionalized) integrin receptor expression and maximum cell–BM adhesion interaction distance, respectively, and n_{cba} is the cell–BM adhesion power. (see Section 6.2.2). As with cell–cell adhesion, we typically set $R^i_{\text{cba}} > r_i$ to approximate the cell's limited capacity to deform before breaking all its adhesive bonds.

Calcified-debris–calcified-debris adhesion ($\mathbf{F}_{\text{dda}}$)

We model adhesion between calcified debris particles similarly to homophilic cell–cell adhesion; hence calcite crystals in the interacting calcified debris particles remain strongly bonded as part of the microcalcification. We model this adhesive force between the calcified debris particles i and j by

$$\mathbf{F}^{ij}_{\text{dda}} = \alpha_{\text{dda}} C_i C_j \nabla \varphi \left(\mathbf{x}_j - \mathbf{x}_i; R^i_{\text{dda}} + R^j_{\text{dda}}, n_{\text{dda}} \right), \tag{6.20}$$

where C_i and R_{dda}^i are cell i's (nondimensionalized) degree of calcification and maximum debris–debris adhesion interaction distance and n_{dda} is the debris–debris adhesion power. The coefficient α_{dda} can be interpreted as the adhesive force between two fully calcified debris particles.

Cell–cell repulsion (including calcified debris) ($\mathbf{F_{ccr}}$)

Cells resist compression by other cells, owing to the structure of their cytoskeletons, the incompressibility of their cytoplasm (fluid), and the surface tension of their membranes. We introduce a cell–cell repulsive force that is zero when cells are just touching and then increases rapidly as the cells are pressed together. We approximate any pressure-induced cell deformation by allowing some overlap between cells. We model $\mathbf{F}_{\text{ccr}}$ by

$$\mathbf{F}_{\text{ccr}}^{ij} = -\alpha_{\text{ccr}} \nabla \varphi(\mathbf{x}_j - \mathbf{x}_i; r_i + r_j, n_{\text{ccr}}), \qquad (6.21)$$

where n_{ccr} is the cell–cell repulsion power (Section 6.2.2) and α_{ccr} is the maximum repulsive force, when the cells are completely overlapping.

Cell–BM repulsion (including calcified debris) ($\mathbf{F_{cbr}}$)

We model the basement membrane as rigid and nondeformable by virtue of its relative stiffness and strength. Hence, it resists deformation and penetration by cells and debris particles. We model the corresponding force by

$$\mathbf{F}_{\text{cbr}}^{i} = -\alpha_{\text{cbr}} B \nabla \varphi \left(d(\mathbf{x}_i); r_i, n_{\text{cbr}} \right), \qquad (6.22)$$

where n_{cbr} is the cell–BM repulsion power, d is the distance from the cell to the BM, and α_{cbr} is the maximum repulsive force when the cell's center is embedded in the basement membrane. Cells can secrete matrix metalloproteinases (MMPs) that degrade the BM (Section 2.2.3) and hence reduce $\mathbf{F}_{\text{cbr}}$; we model this effect by making the cell–BM repulsive force proportional to the remaining BM density B. This is discussed further in Section 6.4.

Motile locomotive force ($\mathbf{F_{loc}}$)

If an agent cell is not motile ($\mathcal{S} \neq \mathcal{M}$) then $\mathbf{F}_{\text{loc}} = \mathbf{0}$. Otherwise, we can model the locomotive force with various levels of detail, as follows.

Imposed chemotaxis and haptotaxis

For a simple motility model, we choose a deterministic direction of motion based upon biological hypotheses such as chemotaxis in response to growth factors f and haptotaxis along gradients in the ECM density E:

$$\mathbf{F}_{\text{loc}} = \alpha_1 \nabla f + \alpha_2 \nabla E. \qquad (6.23)$$

The coefficients α_1 and α_2 can be either fixed or made variable to model energetic factors such as the oxygen availability, the level of receptor activation, and the expression of

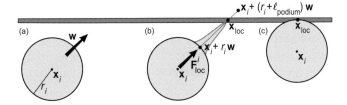

Figure 6.3 To simulate motion along a basement membrane, we (a) choose a random unit vector **w**, (b) test for intersection with the BM along that direction and, if there is an intersection point, apply $\mathbf{F}_{\mathrm{loc}}$ towards it until (c) the cell membrane reaches the BM. Reprinted with permission from [434].

adhesive ligands necessary for traction. For instance, one could set

$$\alpha_1 = \overline{\alpha}_1(\sigma, g) f f_{\mathrm{r}} \mathcal{I}_{E,i} E, \tag{6.24}$$

where g is the cell's internal glucose level, $\overline{\alpha}_1(\sigma, g)$ models the rate of locomotion as a function of oxygen and glucose, f_{r} is the cell's surface expression of receptors to the growth factor f. Note that $f f_{\mathrm{r}}$ denotes the overall activation of f_{r}. Similarly, $\mathcal{I}_{E,i} E$, where $\mathcal{I}_{E,i}$ is the cell receptor expression, gives the overall level of binding of the cell adhesion receptors to their ligands in the ECM. We apply $\mathbf{F}_{\mathrm{loc}}$ for a fixed amount of time β_{M}^{-1}; afterwards, we set $\mathcal{S} = \mathcal{Q}$ and $\mathbf{F}_{\mathrm{loc}} = \mathbf{0}$. One could use similar functional forms for α_2.

Biased random motion
Increasing the complexity somewhat, we choose a random direction of motion when the cell enters the motile state and fix it for the duration of motility, with the distribution of directions dependent upon the gradients of microenvironmental factors such as oxygen level σ, growth factors f, and ECM density E. One such method is to write, for the vector of travel **w**,

$$\mathbf{w} = r_1 \nabla \sigma + r_2 \nabla f + r_3 \nabla E, \qquad -1 \leq r_1, r_2, r_3 \leq 1, \tag{6.25}$$

$$\mathbf{F}_{\mathrm{loc}} = \frac{\alpha_{\mathrm{loc}}}{|\mathbf{w}|} \mathbf{w}, \tag{6.26}$$

where the distributions of the random variables r_1, r_2, and r_3 are chosen according to the desired weighting of oxygen taxis, chemotaxis, and haptotaxis. Random weightings can be used to model more complex signaling dynamics, where the cell must "choose" among competing signals but the internal decision process is unknown. We apply the vector of travel for a pseudotime $0 \leq \tau \leq \beta_{\mathrm{M}}^{-1}$, after which $\mathcal{S} = \mathcal{Q}$ and $\mathbf{F}_{\mathrm{loc}} = \mathbf{0}$.

Motion along the BM
We can model motility along the basement membrane due to the extension and contraction of lamellipodia as follows. At any time t when the cell state changes to $\mathcal{M}$ we choose a random direction **w** (with $|\mathbf{w}| = 1$; see Figure 6.3(a)), and test the line segment between $\mathbf{x} + r\mathbf{w}$ (on the cell membrane) and $\mathbf{x} + (r + \ell_{\mathrm{podium}})\mathbf{w}$ (the maximum

extension of a lamellipodium) for intersection with the BM. The intersection point is labeled $\mathbf{x}_{loc}$ in Figure 6.3(b)). If there is no intersection we set $\mathcal{S} = \mathcal{Q}$. We assume that the cell adheres to the BM by means of a lamellipodium at $\mathbf{x}_{loc}$ and pulls towards this location until either (i) the cell's boundary reaches $\mathbf{x}_{loc}$ (Figure 6.3(c)) or (ii) the cell "gives up its effort" at a maximum time $\beta_{M,max}^{-1}$. Until then, we model the lecomotive force as

$$\mathbf{F}_{loc} = \alpha_{loc} \frac{\mathbf{x}_{loc} - \mathbf{x}}{|\mathbf{x}_{loc} - \mathbf{x}|}, \tag{6.27}$$

where α_{loc} gives the cell's speed of lamellipodium contraction. To avoid "double-counting" cell–BM adhesion, we set $\mathbf{F}_{cba} = \mathbf{0}$ during motility. Once motility is complete we set $\mathbf{F}_{loc} = \mathbf{0}$ and $\mathcal{S} = \mathcal{Q}$.

Mechanistic motility

In a more rigorous motility model, we recognize locomotion as the combined effect of directed actin polymerization (which drives the protrusion of the cell membrane) and differential cell–ECM adhesion [403]. Suppose that $a(\theta, \varphi, g, \sigma, f, f_r)$, where $0 \le a \le 1$, gives the distribution of actin polymerization activity across the cell's surface, and f and f_r also vary with position (θ, φ) on the cell surface. Let $\mathcal{I}(\theta, \varphi)$ and $E(\theta, \varphi)$ denote the distributions of surface-adhesion receptor and nearby ligands, respectively, where these also vary between 0 and 1. Then the distribution of successful adhesions of cell membrane protrusions to the ECM can be modeled by

$$M(\theta, \varphi) = a(\theta, \varphi, g, \sigma, f, f_r)\mathcal{I}(\theta, \varphi)E(\theta, \varphi). \tag{6.28}$$

We set

$$\mathbf{F}_{loc} = \alpha_{loc} \nabla M / |\nabla M|. \tag{6.29}$$

If $\nabla M = \mathbf{0}$ then we choose $\mathbf{F}_{loc}$ randomly. We then apply this force until the maximum motility duration β_M^{-1} is reached.

This approach could replace the explicit treatment of cell–cell, cell–BM, and cell–ECM adhesion. We are investigating it in a modified agent model that explicitly discretizes cell membrane extension and retraction [583].

This general form should capture oxygen and glucose taxis, chemotaxis, and haptotaxis as emergent properties, as well as the effects of ECM anisotropy: if the rate of actin polymerization is assumed to correlate with local ATP availability (which is proportional to the glucose and oxygen levels) then ∇a should correlate with ∇g and $\nabla \sigma$, thereby recovering nutrient taxis. If actin polymerization is assumed to occur as the result of an internal signaling response to a chemoattractant f and if the receptor f_r is roughly uniformly distributed then ∇a should correlate with ∇f, giving chemotaxis as an emergent behavior. If a is uniform but the ECM has a local gradient, then $\mathbf{F}_{loc} \parallel \nabla E$, thus recovering haptotaxis. By incorporating a signaling model into a that connects with actin polymerization activity (e.g., as in [215, 367, 454, 473, 590, 704]), we can make the model more mechanistic and less phenomenological, thereby improving its predictivity. This is a key strength of multiscale modeling.

"Inertialess" assumption

We assume that the forces equilibrate quickly, and so $|m_i \mathbf{v}_i| \approx 0$. Hence, we make the approximation $\sum \mathbf{F} = \mathbf{0}$ and solve for the cell velocity from Eq. (6.15); given that $\mathbf{F}_{\text{drag}} = -\nu \mathbf{v}_i$:

$$\mathbf{v}_i = \frac{1}{\nu + \alpha_{\text{cma}} \mathcal{I}_{E,i} E} \left(\sum_j \left(\mathbf{F}_{\text{cca}}^{ij} + \mathbf{F}_{\text{ccr}}^{ij} + \mathbf{F}_{\text{dda}}^{ij} \right) + \mathbf{F}_{\text{cba}}^i + \mathbf{F}_{\text{cbr}}^i + \mathbf{F}_{\text{loc}}^i \right), \quad (6.30)$$

where ν is the drag coefficient. This has a convenient interpretation: each term

$$\frac{1}{\nu + \alpha_{\text{cma}} \mathcal{I}_{E,i} E} \mathbf{F}_{\square}$$

is the "terminal" (equilibrium) velocity of the cell when fluid drag, cell–ECM adhesion, and $\mathbf{F}_{\square}$ are the only forces acting upon it. (Here, "$\square$" represents any individual force in (6.30) after the summations have been carried out, e.g., cba, cca, etc., and the summation is over all cells j in the computational domain.) This should be useful when calibrating cell motility in future work, as motility is generally measured as a cell velocity (e.g., [302]).

6.3 Subcellular modeling

We can incorporate a molecular-scale signaling model to improve the cell agent's "decision process." For example, in the context of the EGFR pathway, Deisboeck *et al.* modeled the PLCγ and ERK proteins to drive cell decisions on quiescence, proliferation, and motility using thresholding on the time derivatives of these proteins [43, 44, 131, 168, 678, 718, 720].

Taking an analogous approach, we may derive a generalized signaling model by introducing a set of proteins $\mathbf{P} = \left[P_1, P_2, \ldots \right]^T$ governed by a nonlinear system of ODEs. Because a cell's protein state depends upon its sampling of the microenvironment, we define a "stimulus" vector $\mathbf{S} = \left[S_1, S_2, \ldots \right]$, where the S_i may be oxygen, receptor ligands, and so forth. Furthermore, the signaling network topology depends upon the cell's genetic makeup, which we generically refer to as $\mathbf{G}$. Hence:

$$\dot{\mathbf{P}} = \mathbf{f}\left(\mathbf{G}, \mathbf{P}, \mathbf{S} \right). \quad (6.31)$$

The phenotypic transition probabilities then depend upon $\mathbf{P}$ and $\dot{\mathbf{P}}$.

We will illustrate this idea with a rudimentary E-cadherin/β-catenin signaling model. Let P_1 be the cell's unligated E-cadherin, P_2 the ligated E-cadherin (bound to E-cadherin, S_1, on neighboring cells, considered as an external stimulus), P_3 be the free cytoplasmic β-catenin, and P_4 be the ligated E-cadherin/β-catenin complexes. Then a basic system of nonlinear ODEs that includes protein synthesis, binding, dissociation, and proteolysis is given by [434]

$$\dot{P}_1 = \overbrace{c_1}^{\text{synthesis}} - \overbrace{c_2 S_1 P_1}^{\text{homophilic binding}} + \overbrace{c_3 P_2}^{\text{dissociation}} - \overbrace{c_4 P_1}^{\text{proteolysis}}, \tag{6.32}$$

$$\dot{P}_2 = c_2 S_1 P_1 - c_3 P_2 - \overbrace{c_5 P_2}^{\text{proteolysis}} - \overbrace{d_2 P_2 P_3}^{\beta\text{-catenin binding}} + \overbrace{d_3 P_4}^{\beta\text{-catenin dissociation}}, \tag{6.33}$$

$$\dot{P}_3 = \overbrace{d_1}^{\text{synthesis}} - d_2 P_2 P_3 + d_3 P_4 - \overbrace{d_4 P_3}^{\text{proteolysis}}, \tag{6.34}$$

$$\dot{P}_4 = d_2 P_2 P_3 - d_3 P_4 - \overbrace{d_5 P_4}^{\text{proteolysis}}. \tag{6.35}$$

The first two equations describe E-cadherin *receptor trafficking*, and the second two give the interaction of cytoplasmic β-catenin with ligated E-cadherin. We assume that the transcription of downstream targets of β-catenin is proportional to P_3 (see Section 2.1.5) and incorporate this molecular signaling into the phenotypic transformations by writing the $\mathcal{Q} \rightarrow \mathcal{P}$ transition parameter α_P (see Eq. (6.10)) in terms of a modified parameter $\bar{\alpha}_P$:

$$\alpha_P = \bar{\alpha}_P f_P(P_3) \frac{\sigma - \sigma_H}{1 - \sigma_H}. \tag{6.36}$$

Here, f_P satisfies $f_P' > 0$ (transcription increases with P_3), $f_P(1) = 1$ (β-catenin is not a limiting factor if there is no E-cadherin binding), and $f_P(0) = 0$ (there is no downstream transcription if all the β-catenin is bound).

6.4   Dynamic coupling with the microenvironment

As discussed in Section 6.2.3, a cell agent's behavior is inexorably linked with the microenvironment. We now integrate the agent model with the microenvironment as part of a discrete–continuum composite model. We do this by introducing field variables for the key microenvironmental components (e.g., oxygen, signaling molecules, extracellular matrix, etc.), which are updated according to continuum equations. The distribution of these variables affects the cell agents' evolution as already described; simultaneously, the agents impact the evolution of the continuum variables. We give several key examples to illustrate this concept.

Oxygen
All cell agents uptake oxygen needed for metabolism. At the macroscopic scale, this is modeled by

$$\frac{\partial \sigma}{\partial t} = \nabla \cdot (D \nabla \sigma) - \lambda \sigma, \tag{6.37}$$

where σ is the oxygen level, D is the oxygen diffusion constant, and λ is the (spatiotemporally variable) uptake rate by the cells.[5] Suppose that tumor cells uptake oxygen at a rate λ_t and host cells at a rate λ_h and that elsewhere the oxygen level "decays" (by interacting chemically with the molecular landscape) at a low background rate λ_b. Suppose that, in a small neighborhood B (a ball $B_\epsilon(\mathbf{x})$ of radius ϵ centered at $\mathbf{x}$), the tumor cells, host cells, and stroma (non-cells) respectively occupy fractions f_t, f_h, and f_b of B, where $f_t + f_h + f_b = 1$. Then the uptake rate $\lambda(\mathbf{x})$ is given by

$$\lambda(\mathbf{x}) \approx f_t\lambda_t + f_h\lambda_h + f_b\lambda_b, \tag{6.38}$$

i.e., by averaging the component uptake rates with weighting according to the combination of cell types and stroma present near $\mathbf{x}$. Note that we could further decompose f_t and f_h according to cell phenotype, if the component uptake rates are expected to vary. In a numerical implementation, λ is calculated first on a high-resolution mesh (e.g., 1 mesh point is 1 cubic micron in three dimensions) and subsequently averaged to obtain λ at the continuum scale (e.g., 1000-cubic-micron mesh points) [434]. In this formulation the cell uptake rate varies with the tumor microstructure, which, in turn, evolves according to the nutrient and oxygen availability.

Extracellular matrix

Cell agents adhere to the extracellular matrix and require the ECM for certain types of cell locomotion (Section 6.2.4). Cells can also affect the ECM by secreting matrix metalloproteinases, which degrade the ECM, and possibly by depositing new ECM (Section 2.2.3). If E is the concentration of the ECM and M is the concentration of MMPs secreted by the cells then we model this at the continuum scale as in our prior work in [439]:

$$\frac{\partial E}{\partial t} = \lambda_{\text{production}}^E (\mathbf{x}) - \lambda_{\text{degradation}}^E E M, \tag{6.39}$$

$$\frac{\partial M}{\partial t} = \nabla \cdot (D_M \nabla M) + \lambda_{\text{production}}^M (\mathbf{x}) - \lambda_{\text{decay}}^M M, \tag{6.40}$$

where $\lambda_{\text{production}}^\square$ is a production rate ($\square = E, M$), $\lambda_{\text{degradation}}^E$ is the rate of ECM degradation by MMPs, λ_{decay}^M is the MMP decay rate, and D_M is the MMP diffusion constant (which is generally low).

The production rates $\lambda_{\text{production}}^\square$ are upscaled from the cell scale in the same way as the oxygen uptake rate and are generally functions of E and M. The authors in [439] used production rates of the form $\bar{\lambda}_{\text{production}}^\square (1 - \square)$, $\square = E, M$, for constant values of $\bar{\lambda}$ within the tumor viable rim. This was an approximation of subcellular signaling under the assumption that ECM and MDE production decrease as E and M approach an equilibrium value, here scaled to 1. In a multiscale composite modeling context,

---

[5] In Chapters 6 and 10, σ and g denote the oxygen and glucose levels, which are generalized by the substrate level n in the remainder of the book; in these chapters, n denotes an integer.

however, these assumptions are not required as the phenomena can emerge directly from the subcellular signaling models.

Basement membrane

In Section 6.2.4, we discussed how the basement membrane impacts the cell agents through a balance of adhesive and repulsive forces. The cells can also impact the BM by deforming and degrading it (Section 2.2.3). We model the BM as a sharp, heterogeneous, deformable interface. The cell agents require information on (i) their distance from the basement membrane, and (ii) the properties of the BM at that location. We satisfy these requirements using an augmented level set approach. For simplicity, we describe this method for simulations in two dimensions; the three-dimensional approach is analogous.

Let $\{\mathbf{x}^{\mathrm{BM}}(s) : 0 \leq s \leq 1\}$ parameterize the basement membrane, let $\mathbf{n}(s)$ be the normal to the BM at $\mathbf{x}^{\mathrm{BM}}(s)$ for all s ($\mathbf{n}(s)$ is taken to face towards the epithelial side of the BM), and let $\mathbf{b}(s) = \left(b^1(s), b^2(s), \ldots\right)$ be a set of properties along the basement membrane at position $\mathbf{x}(s)$. In our work, b^1 is the BM density and b^2 is the integrin concentration. For any point $\mathbf{x}$ in the computational domain, define

$$s(\mathbf{x}) = s \text{ that minimizes } \left\{ \left| \mathbf{x} - \mathbf{x}^{\mathrm{BM}}(s) \right| : 0 \leq s \leq 1 \right\}, \tag{6.41}$$

that is, $\mathbf{x}^{\mathrm{BM}}\left(s\left(\mathbf{x}\right)\right)$ is the closest point on the BM to $\mathbf{x}$. Note that the minimum value in Eq. (6.41) is equal to $|d\left(\mathbf{x}\right)|$. For notational simplicity, let $s_i = s\left(\mathbf{x}_i\right)$, $\mathbf{x}_i^{\mathrm{BM}} = \mathbf{x}^{\mathrm{BM}}(s_i)$, and $\mathbf{n}_i = \mathbf{n}(s_i)$ for any cell i.

Here, $d(\mathbf{x})$ is the signed distance function (first mentioned in Section 6.2.4), satisfying:

$$\begin{aligned} d &= 0 \quad \text{on the basement membrane;} \\ d &> 0 \quad \text{on the epithelial side of the BM;} \\ d &< 0 \quad \text{on the stromal side of the BM;} \\ |\nabla d| &\equiv 1 \quad \text{(since } d \text{ is a distance function).} \end{aligned}$$

We now modify the cell–BM adhesive and repulsive forces to account for the heterogeneity introduced by $\mathbf{b}(s)$. The modified cell–BM adhesive force acting on cell i is

$$\mathbf{F}_{\mathrm{cba}}^i = \alpha_{\mathrm{cba}} \mathcal{I}_{B,i} b^2(s_i) \nabla \varphi \left(d\left(\mathbf{x}_i\right); R_{\mathrm{cba}}^i, n_{\mathrm{cba}}\right) \tag{6.42}$$

and the modified cell–BM repulsive force on cell i is

$$\mathbf{F}_{\mathrm{cbr}}^i = -\alpha_{\mathrm{cbr}} b^1(s_i) \nabla \varphi \left(d\left(\mathbf{x}_i\right); r_i, n_{\mathrm{cbr}}\right); \tag{6.43}$$

note that we have assumed that the BM stiffness is proportional to its density.

The MMPs secreted by the cells (with concentration $M(\mathbf{x})$; see the previous section) degrade the BM. We model this by writing

$$\frac{db^1(s)}{dt} = -\lambda_{\mathrm{degradation}}^E b^1(s) M\left(\mathbf{x}^{\mathrm{BM}}(s)\right) \qquad \text{for each } 0 \leq s \leq 1. \tag{6.44}$$

Cells in contact with the BM impart forces that deform it. In the simplest deformation model, the membrane acts like a semi-plastic material whose deformations remain even if the cell-imparted stresses are removed. Any cell contacting the BM with velocity directed towards it contributes locally to the membrane's normal velocity $\mathbf{v}^{\mathrm{BM}}(s)$. This membrane velocity varies according to the heterogeneous membrane stiffness. We model this by first defining the BM normal velocity wherever the cells are touching it:

$$\mathbf{v}^{\mathrm{BM}}(s_i) = \frac{-\max\left(-\mathbf{v}_i \cdot \mathbf{n}_i, 0\right)}{b^1(s)} \qquad \text{for any } i \text{ such that } d(\mathbf{x}_i) < r_i. \tag{6.45}$$

Then we smoothly interpolate this function to obtain $\mathbf{v}^{\mathrm{BM}}$ for other values of s. The boundary position is updated using this extended normal velocity (e.g., by extending $\mathbf{v}^{\mathrm{BM}}$ away from the basement membrane and using level set techniques, as in [229, 435–439]).

More sophisticated models can also be applied to BM deformation. We could discretize $\mathbf{x}^{\mathrm{BM}}$ and $\mathbf{b}$, connect the discrete membrane points with springs, and balance the forces at each membrane mesh point to model an elastic boundary; then, gradually reducing the strain in each virtual spring could model relaxation in a viscoelastic material.

In a still more sophisticated model, we could upscale the cells' velocities $\mathbf{v}_i$ to obtain the mean mechanical pressure in a coarsened spatial grid using Darcy's law $\langle \mathbf{v} \rangle = \mu \nabla p$, as in Chapter 3. In combination with proper material properties and boundary conditions, we could then solve for the BM velocity. Ribba and co-workers used similar approaches for continuum descriptions of membrane deformations under viscoelastic stresses [567]. Such an approach would be very well suited to hybrid models, where both discrete and continuum representations of the cell velocities are already present. In fact membrane deformations are better suited to continuum descriptions, which have been highly developed in solid material mechanics, while membrane heterogeneity is best characterized by localized alterations by the discrete cells; this is a good example of where a hybrid model would be stronger than either a discrete or continuum approach alone. Chapter 7 will explore hybrid models in greater detail.

6.5 A brief analysis of the volume-averaged model behavior

Let us fix a volume Ω contained within the viable rim (i.e., including any cell i in Ω that satisfies $\mathcal{S}_i \notin \{\mathcal{H}, \mathcal{N}, \mathcal{C}\}$). We analyze the population dynamics in a simplified $\mathcal{Q}$–$\mathcal{A}$–$\mathcal{P}$ cell-state network (assuming no motility); this analysis is the basis of the model calibration in Section 10.3. Let $P(t)$, $A(t)$, and $Q(t)$ denote respectively the number of proliferating, apoptosing, and quiescent cells in Ω at time t. Let $N(t) = P + A + Q$ be the total number of cells in Ω. If $\langle \alpha_{\mathrm{P}} \rangle(t) = |\Omega|^{-1} \int_{\Omega} \alpha_{\mathrm{P}} \, dV$ is the mean value of α_{P} at time t throughout Ω, then the net number of cells entering state $\mathcal{P}$ in the time interval $[t, t + \Delta t)$ is approximately

$$\begin{aligned} P(t + \Delta t) &= P(t) + \Pr\big(\mathcal{S}(t + \Delta t) = \mathcal{P} \,\big|\, \mathcal{S}(t) = \mathcal{Q}\big) Q(t) - \beta_{\mathrm{P}} P(t) \Delta t \\ &\approx P(t) + \left(1 - \mathrm{e}^{-\langle \alpha_{\mathrm{P}} \rangle \Delta t}\right) Q(t) - \beta_{\mathrm{P}} P(t) \Delta t, \end{aligned} \tag{6.46}$$

whose limit as $\Delta t \downarrow 0$ becomes, after some rearrangement,

$$\dot{P} = \langle \alpha_P \rangle Q - \beta_P P. \tag{6.47}$$

Similarly

$$\dot{A} = \alpha_A Q - \beta_A A, \tag{6.48}$$

$$\dot{Q} = 2\beta_P P - (\langle \alpha_P \rangle + \alpha_A) Q. \tag{6.49}$$

Summing these, we obtain

$$\dot{N} = \beta_P P - \beta_A A. \tag{6.50}$$

Next, define PI $= P/N$ and AI $= A/N$ to be the *proliferative* and *apoptotic indices*, respectively. We can express the equations above in terms of AI and PI by dividing by N and using Eq. (6.50) to treat $(d/dt)(P/N)$ and $(d/dt)(A/N)$ correctly. After simplifying, we obtain a nonlinear system of two ODEs for PI and AI:

$$\dot{PI} = \langle \alpha_P \rangle (1 - AI - PI) - \beta_P(PI + PI^2) + \beta_A AI \cdot PI \tag{6.51}$$

$$\dot{AI} = \alpha_A (1 - AI - PI) - \beta_A(AI - AI^2) - \beta_P AI \cdot PI. \tag{6.52}$$

It is much easier to compare these equations, rather than (6.48), (6.49), with immuno-histochemical measurements, which are generally given in terms of AI and PI.

Lastly, let us nondimensionalize the equations by letting $t = \hat{t}\,\bar{t}$, where $\hat{t}$ is dimensionless. Then if $f' = (d/d\hat{t})f$ we have

$$\frac{1}{\bar{t}}PI' = \langle \alpha_P \rangle (1 - AI - PI) - \beta_P(PI + PI^2) + \beta_A AI \cdot PI \tag{6.53}$$

$$\frac{1}{\bar{t}}AI' = \alpha_A(1 - AI - PI) - \beta_A(AI - AI^2) - \beta_P AI \cdot PI. \tag{6.54}$$

The cell cycle length β_P^{-1} is on the order of 1 day (e.g., as in [512]), and in Section 10.4.1 we will determine that β_A is of similar magnitude. Thus, if we choose $\bar{t} \sim \mathcal{O}(10 \text{ days})$ or greater then we can assume that $(1/\bar{t})PI' = (1/\bar{t})AI' \approx 0$ and conclude that the local cell-state dynamics reach a steady state after after 10–100 days. This is significant, because it allows us to calibrate the population dynamic parameters (α_A, α_P, β_A, and β_P) without the inherent difficulty of estimating temporal derivatives from often noisy *in vitro* and immunohistochemistry data.

6.6   Numerical examples from breast cancer

We now present numerical examples of the agent model as applied to ductal carcinoma in situ, a type of breast cancer where the cells are confined to growth and motion in a breast duct. We focus on the simulation results and defer a biological discussion of breast cancer and model calibration to Chapter 10.

For simulation in two dimensions, consider cells growing in a fluid-filled domain Ω (representing a rigid-walled duct) of length ℓ (1 mm in our simulations) and width

$2R$ (340 µm in our simulations). We "cap" the left edge of the simulated duct with a semicircle of radius R. Cells are removed from the simulation if they cross the right-hand edge of the computational boundary. We represent the duct wall with a signed distance function d satisfying $d > 0$ inside the duct.

Cell states (see Figure 6.1) include the proliferative ($\mathcal{P}$), quiescent ($\mathcal{Q}$), apoptotic ($\mathcal{A}$), hypoxic ($\mathcal{H}$), necrotic ($\mathcal{N}$), calcified debris ($\mathcal{C}$), and motile ($\mathcal{M}$) states. We use the simplified model of motility along a basement membrane in randomly selected directions; see Figure 6.3. For simplicity, we neglect membrane degradation, membrane deformation, and molecular-scale signaling, which allows us to instead focus upon the effects of the various cell states and forces. We assume there is no ECM in the duct (i.e., $\mathbf{F}_{cma} = \mathbf{0}$) and that cells adhere to one another with E-cadherin (homophilic adhesion) and to the BM with integrins. We assume that the surface adhesion receptors and BM adhesion ligands are distributed uniformly on the cell surfaces and the BM, respectively.

We model oxygen transport within the duct by

$$\begin{cases} \dfrac{\partial \sigma}{\partial t} = D\nabla^2 \sigma - \lambda \sigma & \text{if } \mathbf{x} \in \Omega, \\ \sigma = \sigma_B & \text{if } \mathbf{x} \in \partial\Omega, \end{cases} \tag{6.55}$$

where λ is the locally averaged oxygen uptake rate discussed in Section 6.4. We use the Neumann condition $\partial\sigma/\partial n = 0$ on the right-hand side of the duct. The parameter values for the simulations are given in Table 6.1 The duct size, population parameters (α_A, $\bar{\alpha}_P$, β_A, and β_P), the balance of cell–cell adhesion and repulsion (α_{cca} versus α_{ccr}), and oxygen boundary value σ_B are calibrated to the patient data presented in Chapter 10; more detail is given there. Further detail on the numerical implementation can be found in Section 10.2.4 and [434].

6.6.1 Baseline calibrated run

We first simulate ductal carcinoma in situ (DCIS) with simplified hypoxia ($\beta_H^{-1} = 0$) and no motility ($\alpha_M = 0$). We apply a simplified model of necrosis, in which $\beta_{NS}^{-1} = \beta_{NL}^{-1} = \beta_{NC}^{-1}$ and set $\lambda_b = 0.01\lambda_t$. The dynamic simulation is presented in Figure 6.4.

In the simulation, a small initial population (two cells) begins proliferating into the duct (at 0 days). As the tumor grows along the duct, oxygen uptake by the tumor cells leads to the formation of an oxygen gradient (not shown). At time 5.04 days, the oxygen level drops below σ_H in the center of the duct near the leading edge of the tumor, causing the first instance of necrosis (light gray cells). By 7 to 14 days, a viable rim of nearly uniform thickness (equal to approximately 80 µm) can be observed, demonstrating an overall oxygen gradient decrease from σ_B at the duct boundary to σ_H at the edge of the necrotic core. (See also Section 10.3.3.) This is consistent with the prediction in Section 6.5 that the cell-state dynamics reach a local steady state by 10 to 100 days.

Consistently with the assumed functional form of the $\mathcal{Q} \to \mathcal{P}$ transition, proliferating cells (diagonally striped) are most abundant near the duct wall, where the oxygen level is highest, with virtually no proliferation at the perinecrotic boundary. Because oxygen

Table 6.1. Main parameters used for the agent-based model DCIS simulations

Parameter	Physical meaning	Value
$2R_{duct}$	duct diameter	340 μm
L_{duct}	duct length	1 mm
R	cell radius	9.953 μm
σ_H	hypoxic oxygen threshold	0.2
σ_B	duct boundary O_2 value	0.263 717 if $\lambda_b = 0.001$
		0.277 491 if $\lambda_b = 0.01$
		0.386 095 if $\lambda_b = 0.1$
$\langle\lambda\rangle$	mean oxygen uptake rate	0.1 min^{-1}
λ_t	tumor cell O_2 uptake rate	0.1 min^{-1}
λ_b	background O_2 decay rate	0.001, 0.01, or 0.1 min^{-1}
$\langle\lambda\rangle L$	oxygen diffusion length scale	100 μm
α_A^{-1}	mean time to apoptosis	786.61 hours
$\overline{\alpha}_P^{-1}$	mean time to proliferation (when $\sigma = 1$)	115.27 min if $\lambda_b = 0.001$
		151.21 min if $\lambda_b = 0.01$
		434.53 min if $\lambda_b = 0.1$
β_A^{-1}	time to complete apoptosis	8.6 hours
β_P^{-1}	time to complete cell cycle	18 hours
β_{NC}^{-1}	time to complete calcification	15 days
β_H^{-1}	mean survival time for hypoxic cells	0 or 5 hours
β_{NL}^{-1}	time for necrotic cells to lyse	2, 24, 120, or 360 hours
β_{NS}^{-1}	necrotic-cell surface receptor degradation time	5 or 15 days
V_S/V	solid cell fraction	10%
$\alpha_{cca}, \alpha_{cba}, \alpha_{dda}$	cell–cell, cell–BM, and debris adhesive forces	0.391 531v (μm min^{-1})
α_{ccr}	cell–cell repulsive force	8v (μm min^{-1})
α_{cbr}	cell–BM repulsive force	5v (μm min^{-1})
α_{loc}	locomotive (motile) force	5v (μm min^{-1})
α_M^{-1}	mean time between migration events	120, 300, or ∞ min
$\beta_{M,max}^{-1}$	maximum motility time	15 min
ℓ_{podium}	maximum lamellipodium length	15 or 20 μm
$n_{cca}, n_{cba}, n_{dda}$	adhesion-potential parameters	1
n_{ccr}, n_{nbr}	repulsion-potential parameters	1
$R_{cca}, R_{ccr}, R_{cba}, R_{cbr}$	maximum interaction distances	12.083 μm

can diffuse into the tumor from the duct lumen (with a low decay rate λ_b), viable tumor cells are also observed along the tumor's leading edge near the center of the duct. See Figure 6.4 at times 7, 14, and 21 days.

Cells begin to lyse at 20.04 days and, because $\beta_{NC} = \beta_{NL}$, they immediately calcify (dark gray cells). By 28 days, another characteristic length emerges: the trailing edge of the microcalcification maintains a distance of approximately 500 μm from the end of the duct. Several features combine to cause this. The model does not include contact

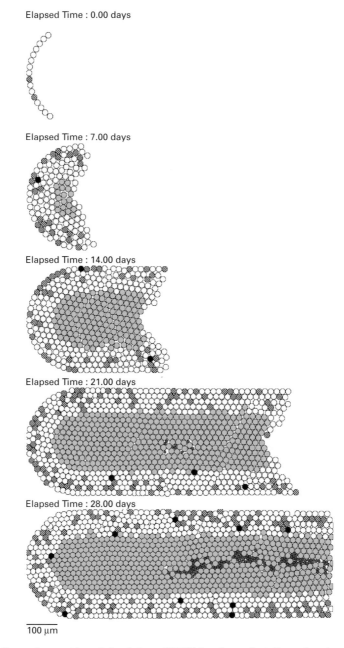

Elapsed Time : 0.00 days

Elapsed Time : 7.00 days

Elapsed Time : 14.00 days

Elapsed Time : 21.00 days

Elapsed Time : 28.00 days

100 μm

Figure 6.4 Dynamic agent-based simulation of DCIS in a 1 mm duct. Legend: quiescent cells are white, apoptotic cells are black, proliferating cells are diagonally striped, necrotic cells are light gray, and calcified cell debris (in the middle right of the lowest panel) is dark gray. Reprinted with permission from [434].

inhibition, and so tumor cells at the end of the duct continue to proliferate and push other cells towards the tumor's leading edge. Because the end of the duct has reached a local dynamic equilibrium by this time (see the discussion above and [437]), a steady flux of tumor cells into the necrotic region has emerged. Because the lysis and calcification times (β_{NL}^{-1} and β_{NC}^{-1}) are fixed, the cells are pushed a fixed distance along the necrotic core before lysing and calcifying, leading to the observed "standing wave" pattern.

The viable rim of the tumor reaches the end of the simulated 1 mm stretch of the duct by time 22.5 days, and the necrotic core reaches this computational boundary by approximately 24 days. Cells that exit the computational domain are removed from the simulation, leading to an artificial drop in the mechanical resistance to growth, particularly as cells from the trailing edge of the tumor continue to proliferate and push into the lumen. Thus, the simulation is more valuable for examining the radial tumor dynamics than the rate of progression through the duct after this time.

6.6.2   Impact of hypoxic survival time

We next examined the impact of changing β_{H}^{-1}, the mean time for which hypoxic cells survive before necrosing. We found that increasing β_{H}^{-1} from 0 to 5 hours has a minimal impact on the simulation; the principal effect is to delay calcification, because cells wait a few more hours before necrosing. (The results are not shown.)

This minimal impact of β_{H} on the current model behavior is not surprising: hypoxic cells are not motile in this model implementation, and the overall flux of cells is from the proliferating rim towards the center of the duct. Hence, there is no opportunity for hypoxic cells to take advantage of their increased survival time to return to the viable rim. Thus, increasing β_{H}^{-1} merely delays necrosis (in this model formulation). Furthermore, the hypoxic cells are not glycemic in this model, and so they do not lower the pH (acidosis) in the nearby viable rim; hence, the increased hypoxic survival time does not improve the survivability and behavior of cells in the viable rim. We expect that β_{H} would have a greater impact if:

1. hypoxic cells were allowed to switch to anaerobic glycolysis, thus making the viable rim microenvironment acidic and giving the hypoxic cells a competitive advantage over nearby nonresistant tumor cell strains;
2. hypoxic cells were allowed to migrate out of the hypoxic region, thereby propagating their phenotypic adaptations into the tumor viable rim;
3. hypoxic cells that had migrated out of the hypoxic region were allowed to reach the basement membrane and initiate invasive carcinoma; and
4. hypoxic cells were allowed to experience increased genetic instability due to the harshness of the environment. This would tend to accelerate the previous points.

Hypoxia would be expected to increased tumor invasiveness if these aspects, taken together, were incorporated in the model.

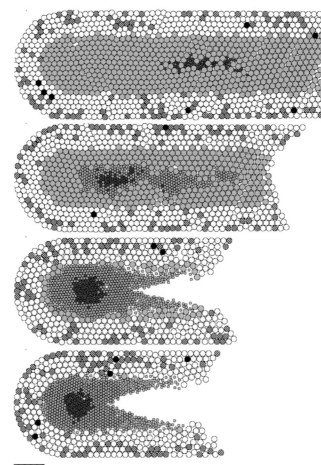

100 µm

Figure 6.5 Impact of the necrotic-cell lysis timescale: the DCIS progression is shown at time 25 days for $\beta_{NL}^{-1} = 15$ days (top), 5 days (second from top), 1 day (second from bottom), and 2 hours (bottom). Reprinted with permission from [434].

6.6.3 Impact of the cell lysis time

Next, we investigate the effect of variations in the necrotic cell lysis time β_{NL}^{-1} on the overall tumor evolution. We used the values 15 days, 5 days, 1 day, and 2 hours. Because surface receptor degradation is likely to be a faster process than cell calcification we set β_{NS}^{-1} to 5 days. However, we found that varying β_{NS} had little impact on the mechanical behavior, owing to the dense cell packing in the simulated tumors.

In Figure 6.5, we plot the DCIS simulations at time 25 days for each of these lysis parameter values. The parameter has a great effect on the rate of tumor advance through the duct: as the lysis time is decreased, the rate of tumor advance through the duct slows. The reason is that the necrotic core acts as a volume sink when the cells lyse, thereby relieving mechanical pressure. In turn, more of the cell flux is directed towards the center of the duct rather than forward towards the tumor's leading edge. Once $\beta_{\text{NL}}^{-1} < 1$ day, further reductions have little additional impact on the tumor progression, because the time spent by the cells in an unlysed state is much smaller than the growth time scale.

In addition, the size of the microcalcification grows as β_{NL}^{-1} decreases, because more cells are pushed into the necrotic core rather than forward for small values of β_{NL}^{-1}; consequently a greater number of cells have been in the necrotic region longer than the calcification time β_{NC}^{-1}. Furthermore, the microcalcification appears to have a rounder morphology for the shorter lysis times. This phenomenon is still under investigation but it appears to be due to the more uniform packing of the lysed necrotic cells, owing to their occupying a greater percentage of the necrotic core for small values of β_{NL}^{-1}. Furthermore, the microcalcification appears now to occupy a larger percentage of the duct's cross section owing to the slower rate of tumor advance through the duct.

Similarly to the "standing wave" calcification pattern in the baseline simulation (and for similar reasons), a characteristic length emerges, that between the end of the duct and the start of the lysed-cell region. This length is shorter than the distance from the end of the duct to the trailing edge of the calcification. These lengths both decrease as β_{NL}^{-1} decreases, owing to the resulting faster progression from viable cells to necrotic cells to lysed cells to calcified cells.

In a detailed comparison of these simulation results with breast histopathology data, we found that 1 day $\leq \beta_{\text{NL}}^{-1} \leq 5$ day yields the best match between the simulated and actual necrotic-core morphological features, including the rough distributions of lysed and unlysed cells and the general appearance of the necrotic cross sections [434].

6.6.4   Impact of background oxygen decay rate

We varied the background oxygen decay rate λ_{b}, using the values 0.001 min^{-1}, 0.01 min^{-1}, and 0.1 min^{-1}, to investigate its impact on the simulation results. Aside from eliminating the viable cells from the center of the tumor's leading edge (see the baseline run, Section 6.6.1), there was very little impact on the simulations. The reason is that the tumor necrotic centers were densely packed for all values of λ_{b} and, since necrotic cells uptake oxygen at a rate $\lambda_{\text{t}} = \langle \lambda \rangle$, the oxygen uptake rate is equal throughout the bulk of the computational domain. Thus, the oxygen profile was similar for all three cases, leading to comparable amounts of cell proliferation in all cases; i.e., the extra viable cells in the leading edge did not contribute substantially to the tumor's advance through the duct. However, λ_{b} may have a greater impact in less dense tumors such as cribriform DCIS (see Section 10.1.2).

6.6.5   Impact of cell motility

We next allowed a random cell motility along the basement membrane, as described in Section 6.2.4. We set $\beta_{\text{NL}}^{-1} = 2$ hours, $\beta_{\text{H}}^{-1} = 5$ hours, and investigated the role of the mean time to migration, using $\alpha_{\text{M}}^{-1} = 300$ min or 120 min, and that of the maximum

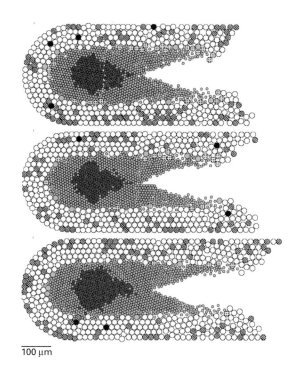

100 μm

Figure 6.6 Impact of cell motility: as the mean time to migration is decreased from infinity (top) to 300 min (middle) to 120 min (bottom), the rate of the tumor's advance through the duct is slightly increased, with the leading edge more spread out. The hypoxic cells are light gray with a pattern of black dots. Reprinted with permission from [434].

lamellipodium length, using 15 or 20 μm. We set the maximum migration speed α_{loc}/ν to 5 μm min^{-1} and the maximum duration of migration to 15 min. In all simulations we set $\lambda_{\text{b}} = 0.001$ min^{-1}.

In Figure 6.6 we plot the simulations at time 30 days for 15 μm maximum lamellipodium length, for $\alpha_{\text{M}} = 0$ (top), $\alpha_{\text{M}}^{-1} = 300$ min (middle), and $\alpha_{\text{M}}^{-1} = 120$ min (bottom). The legend for Figure 6.4 is enlarged to include hypoxic cells, represented as light-gray circles containing a pattern of black dots. As α_{M}^{-1} is decreased from infinity (top) to 120 min (bottom), the tumor leading-edge morphology becomes less blunt and the tumor advances slightly further along the duct. Thus, the introduction of even random motility can increase the rate of a tumor's advance through a breast duct. We expect that this effect will be more pronounced if the cells instead move in a directed manner (e.g., chemotactically, as in Paget's disease [165]).

We also increased the maximum lamellipodium length ℓ_{podium} from 15 μm to 20 μm for $\alpha_{\text{M}}^{-1} = 120$ min. Similarly to decreasing the α_{M}^{-1} value, increasing ℓ_{podium} accentuates the effect of motility because now the motile cells can travel further per migration event (this simulation is not shown). Note that these effects are subtle, and so further analysis and simulations, with different random seeds, are required to quantify them fully and confirm their statistical significance.

6.7   Conclusions

In this chapter, we have reviewed the major discrete modeling approaches currently employed in mathematical cancer cell biology, with particular focus on a generalized agent-based cell modeling framework. After a brief analysis of the model's population dynamics, we presented numerical examples of the model as applied to breast cancer; this application is explored in greater depth in Chapter 10. The specific modeling of the necrotic volume loss can have a major impact on the rate of tumor advance through the duct. Furthermore, even random cell motility along the basement membrane can speed a tumor's spread through the duct system, and the effect should be much more pronounced in cases of directed motility such as Paget's disease (in which there is chemotactic motion along the ducts towards chemoattractants released by keratinocytes in the breast nipple [165]). However, if hypoxia is modeled merely as a delayed progression to necrosis then it has little impact on DCIS progression; instead glycolysis, acidosis, and motility must be considered. This finding is consistent with previous modeling results by Gatenby and co-workers (e.g., [263, 264, 267, 624–627]).

7   Hybrid continuum–discrete tumor models

With F. Jin and Y.-L. Chuang

Recently, hybrid modeling techniques have been proposed that combine the advantages of the continuum and the discrete approaches in order to simulate multiscale multibody problems; these provide more realistic descriptions of microscopic mechanisms while efficiently evolving the entire system to obtain macroscopic observations. Solid-tumor growth is one example of such multiscale problems, where the cellular and subcellular scale pathways have been intensively studied and are fairly well understood while the tissue-scale tumor morphology is of interest in clinical applications. We presented a sharp-interface model in Chapter 3 and a Cahn–Hilliard mixture model in Chapter 5, using the continuum approach to simulate the tissue-scale macroscopic morphology of tumors. The dynamics of how tumor cells interact with the microenvironment were studied in Chapter 6 by formulating agent-based models at the cellular and subcellular scales using the discrete approach. By combining the continuum and the discrete approaches, we now present a hybrid modeling framework that utilizes knowledge collected at the cellular and subcellular scales in the study of tissue-scale tumor evolution. Figure 7.1 illustrates the basic components constituting a hybrid model; each component will be detailed in this chapter.

We first review, in Section 7.1, some earlier related work that led to the development of hybrid models. Next, in Section 7.2, we discuss how the fundamental conservation laws are used to establish the general concept of hybrid tumor models. Then the hybrid coupling mechanisms through mass and momentum exchange will be rigorously formulated in Section 7.3. We close the chapter by presenting two examples from the published literature to demonstrate the general framework. A hybrid tumor model that describes the growth of *multicellular tumor spheroids* (MCTSs) developed by Kim *et al.* [382] will be considered in Section 7.4; this is followed in Section 7.5 by a discussion of the hybrid vascularized tumor growth model presented by Bearer *et al.* [57] and Frieboes *et al.* [228], which incorporates continuum and discrete tumor growth models as well as a dynamic vascular system to simulate tumor-induced angiogenesis.

7.1   Background

The concept of hybrid modeling is founded on the study of the connection between continuum and discrete modeling expressions. The development of the general theory

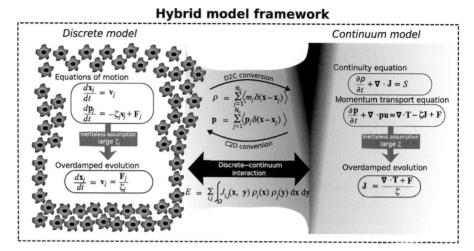

Figure 7.1 Illustration of the framework of a continuum–discrete hybrid model. The discrete and the continuum model components are respectively represented on the left and on the right. The general forms of the constitutive equations for each model are listed, with the dynamic description of the model in the upper part and the kinematic description, with an inertialess assumption, in the lower part. The coupling mechanism between the two models is shown, including a D2C (discrete-to-continuum) conversion, a C2D (continuum-to-discrete) conversion, and a mechanical continuum–discrete interaction.

can be traced back to a 1950 paper of Irving and Kirkwood [331], where the continuum quantities in fluid mechanics, such as density, flux, strain, and stress, were derived as ensemble-averaged quantities of discrete particles using classical statistical mechanics. For biological problems, earlier work of Othmer *et al.* [509] had investigated the behaviors of two major types of classical random walk models, which are often adopted to describe the dispersal of discrete cells in a biological system. By taking the models to the limit of zero jump-step size and infinite jump frequency, the two types of random walks, position-jump processes and velocity-jump processes, respectively converged to continuum descriptions in terms of diffusion equations and damped wave equations. Othmer and Stevens [510] later studied some general types of reinforced random walk models, i.e., random walks with nonhomogeneous transition probabilities, which are commonly chosen to describe taxis-driven cell dispersal. The continuum limit of the reinforced random walk models resulted in various classes of advection–diffusion equations. Asymptotic analysis of the continuum limit suggested that a variety of dynamics, such as aggregation, blowup, or collapse, could occur in a cell population. The investigation was further supported by a follow-up study of Horstmann *et al.* [324], where rigorous connections were established between solutions of the reinforced random walk models and the corresponding continuum equations. An application to a more sophisticated problem can be found in Erban and Othmer [203, 204], where the micro- and macro-scale descriptions of bacterial chemotaxis were connected by taking a continuum limit of a microscopic individual-cell-behavior model. In the microscopic model, the velocity of each bacterium jumped according to a nonhomogeneous transition rate which depended on an internal state assigned to the bacterium. The internal signal transduction network that governed the variation of the bacterium's internal state was modeled by a system of ordinary differential equations forced by a time- and/or space-dependent external signal. The continuum limit of the microscopic model yielded a system of partial differential equations of moments with respect to the slow component of the internal state, which could be reduced further to the classical chemotaxis equation; the chemotactic sensitivity is determined directly from the parameters of the internal dynamics. Solutions of the macroscopic equations agreed well with Monte Carlo simulations of individual cell movement.

The development of equation-free methods has also provided a route for connecting the microscopic descriptions to macroscopic evolutions that is alternative to deriving the continuum limit of microscopic models explicitly. Equation-free methods were proposed and developed by Kevrekidis and co-workers [270, 370, 445, 446] (see also the review by Li *et al.* [415]) to solve the dynamics of a huge microscopic system that does not converge to a closed-form expression in the continuum limit. The idea is to evolve the slow variables of the problem by projecting it to a slow manifold, where the fast variables rapidly relax to a quasi-static state and are "enslaved" by the slow "master" variables. The slow variables are usually a few lower moments of some statistical distributions of the microscopic variables; hence, integrating only the slow variables greatly reduces the dimension of the problem, improving the efficacy of the simulation and simplifying the analysis. Such equation-free methods require the identification of a set of intrinsic slow variables of a problem and consist of the following elementary steps:

1. running the full microscopic model for a short period of time to obtain essential information for a coarse time stepper, to be used later;
2. using a restriction operator to restrict the microscopic states of the system to the slow variables at the macroscopic scale;
3. projecting the slow variables over a longer period of time, using the coarse time stepper constructed from the information obtained at step 1;
4. using a lifting operator to generate microscopic realizations that are consistent with the current state of the slow variables; and
5. repeating steps 1–4 until the simulation is complete.

In order to achieve statistically meaningful results, multiple microscopic ensembles should be generated at steps 1 and 4. Hence, the relative efficiency of these equation-free methods in comparison with the full microscopic simulations depends on whether the computations saved by the coarse time stepper overcome the additional computations needed to generate multiple microscopic samples during the short-burst runs.

Setayeshgar *et al.* [601] utilized the equation-free method to address multiscale characteristics in biological problems, where the temporal evolution usually occurs on widely disparate time scales. For such problems the intrinsic slow variables can usually be expressed in terms of a small number of moments of an underlying probability distribution. As a demonstration, Setayeshgar *et al.* [601] applied these methods to simulate the evolution of noninteracting bacteria in the presence of a chemoattractant. A microscopic model described the flagellar motor-driven cell motility as a random walk process coupled with a system of deterministic ODEs to formulate the internal signal transduction network. Monte Carlo methods were used to simulate short bursts of microscopic runs. Using error analysis the authors demonstrated that the projective coarse time integration of the moments could be carried out on a time scale much longer than that of the microscopic dynamics, thus overcoming the additional computations required to generate multiple microscopic ensembles.

As mentioned earlier, Erban *et al.* [202] applied these equation-free methods to a simplified version of the chemotactic bacteria model in [203, 204], in which the closed-form continuum limit of the model was derived. For the purposes of analysis, the continuum model was simulated using the explicit projective integration method introduced by Gear and Kevrekidis [269], which is similar in spirit to the equation-free approach. Then the authors showed that the kinetic Monte Carlo simulations of the microscopic model could be accelerated by a factor as much as 1000 using the equation-free approach. It was also suggested that, even when a closed-form continuum limit exists, equation-free discrete simulations could assist the continuum simulations of the chemotaxis model in regard to high-chemical-gradient regions, which generate localized strong chemotactic signals and may otherwise require an adaptive-mesh scheme (as in Chapter 8) to solve the problem. While lower-order moments of density and momentum flux were used in [601, 202] for applications of the equation-free methods, there exist alternative choices for the intrinsic slow variables, which do not directly correspond to physical observables. Bold *et al.* [74] expanded a microscopic model of coupled heterogeneous biological oscillators using generalized polynomial chaos (gPC) expansions. A few leading-order gPC coefficients were selected as the intrinsic slow variables and advanced by the coarse time stepper of the equation-free methods. Then microscopic realizations were generated from the evolved gPC expansions to study the synchronization of such oscillators on the population level.

Hybrid modeling is a further extension of the previous studies. The coupled continuum and discrete models in the hybrid modeling framework may or may not describe the same cell species or cells of the same phenotype. Hence, the continuum model is not necessarily the continuum limit of the discrete system. Nevertheless, a closed-form continuum model usually exists for such a modeling framework and one does not need to rely on the equation-free methods to accelerate the macroscopic simulations. Instead, the continuum–discrete connection established by studying the continuum limit of discrete models sets up the foundation of hybrid modeling, in which the continuum and discrete components exchange mass and momentum during the course of a simulation; this requires a rigorous translation between these two descriptions. Furthermore, the mass and momentum exchange in a hybrid model is operated dynamically and thus a dynamic continuum–discrete coupling mechanism is necessary, which is very similar to the restriction operator and the lifting operator of the equation-free methods. In the next section, we present the general concept of hybrid modeling as well as the basic conservation laws involved.

7.2   General concept and conservation laws

In a hybrid continuum–discrete tumor growth model, discrete and continuum systems are coupled via mass and momentum exchange. As shown in Figure 7.1, the coupling mechanism can further be divided into a continuum-to-discrete (C2D) conversion, a discrete-to-continuum (D2C) conversion, and mechanical interactions between the discrete and the continuum elements. Through the C2D and the D2C conversions, the discrete and the continuum models trade mass as well as the momentum carried by the traded mass, while the mechanical interactions enable the models to modify each other's momentum. Like all physical systems, the hybrid tumor growth model, as well as its components, should obey the fundamental conservation laws, notably the conservation of mass and momentum. Mass conservation is deduced from energy conservation as a basic conservation law in the limit of classical mechanics. Taking everything into account, the total mass must not change. For example, during the cell-proliferation process, cells uptake amounts of nutrient and water equivalent to their masses in order to divide into two. Momentum conservation includes the conservation of linear momentum and the conservation of angular momentum, respectively corresponding to the linear transportation of and the rotation of cells. In tumor growth problems, the cell rotation is usually ignored either because of the assumption of a point cell mass or because of the assumption of quick relaxation to a quasi-static cell orientation owing to a strongly dissipative environment. Hence, in this chapter, we will discuss only the conservation of linear momentum, not angular momentum, and simply use the term "momentum" to refer to linear momentum.

We will review the conservation laws of a typical continuum model and a typical discrete model in Sections 7.2.1 and 7.2.2. In Section 7.2.3 we introduce the inertialess assumption, often made for tumor growth models designed to approximate a dissipative system. Then, consistently with the conservation of mass and momentum, theoretical connections between continuum and discrete variables are derived in Section 7.2.4 that pave the way toward formulating a hybrid coupling mechanism to link the continuum and the discrete models in Section 7.3.

7.2.1   Conservation laws in continuum models

In general, a typical continuum model describes the macroscopic evolution of a population density and is essentially constructed on the basis of the fundamental conservation laws. Here, we briefly review the construction of continuum models via the conservation of mass and momentum.

In a continuum model, mass is conserved through the *continuity equation*

$$\frac{\partial \rho}{\partial t} + \nabla \cdot \mathbf{J} = S, \tag{7.1}$$

where ρ is the density field, $\mathbf{J}$ is the density flux, and S represents sources and sinks of mass. Integrating Eq. (7.1) over a domain Ω gives us

$$\frac{d}{dt} \int_\Omega \rho \, d\mathbf{x} = - \oint_{\partial\Omega} \mathbf{J} \cdot \mathbf{n} \, d\mathbf{s} + \int_\Omega S \, d\mathbf{x}, \tag{7.2}$$

where $\partial\Omega$ denotes the boundary of Ω and $\mathbf{n}$ is a unit normal vector on $\partial\Omega$. Equation (7.2) states that the total sources within a domain and the flux through the boundary of the domain add up to the net change of total mass inside the domain.

The momentum of a continuum model is defined as $\mathbf{p} \equiv \rho\mathbf{u}$, in which $\mathbf{u} = \mathbf{J}/\rho$ is the velocity field of ρ. Momentum is conserved through the *momentum transport equation*

$$\frac{\partial\mathbf{p}}{\partial t} + \nabla\cdot(\mathbf{pu}) = \nabla\cdot\mathbf{T} + \mathbf{F}, \tag{7.3}$$

where $\mathbf{T}$ and $\mathbf{F}$ are respectively the stress tensor and the force. We may similarly integrate Eq. (7.3) over a domain Ω and obtain

$$\frac{d}{dt}\int_\Omega \mathbf{p}\,d\mathbf{x} = -\oint_{\partial\Omega}\mathbf{p}(\mathbf{u}\cdot\mathbf{n})\,ds + \oint_{\partial\Omega}\mathbf{T}\cdot\mathbf{n}\,ds + \int_\Omega\mathbf{F}\,d\mathbf{x}. \tag{7.4}$$

Equation (7.4) states that the change in total momentum within Ω results from the sum of the integrated surface force $\mathbf{T}\cdot\mathbf{n}$ over $\partial\Omega$, the total force $\mathbf{F}$ in the interior of Ω, and the momentum flux $\mathbf{pu}$ flowing through $\partial\Omega$ into Ω.

7.2.2   Conservation laws in discrete models

In Chapter 6 we presented a discrete agent-based model of tumor growth, which directly tracks the movement of each individual agent, allowing cellular level biological mechanisms to be programmed explicitly into the model. Like continuum models, discrete models should obey the conservation of mass and density.

In a pure discrete model, cells are treated as rigid bodies and mass is implicitly conserved, as all the existing mass is explicitly tracked as individual agents. Furthermore, the addition and the subtraction of individual cells are governed by biologically mass-conserving cell-proliferation and cell-death processes that also satisfy the conservation of mass globally. The conservation of momentum is satisfied through Newton's laws of motion,

$$\frac{d\mathbf{x}_j}{dt} = \mathbf{v}_j, \qquad \frac{d\mathbf{v}_j}{dt} = \frac{\mathbf{F}_j}{m_j}; \tag{7.5}$$

the position of cell j, denoted by $\mathbf{x}_j$, evolves according to its velocity $\mathbf{v}_j$ while the rate of change of the velocity is called the acceleration and is calculated as the external force $\mathbf{F}_j$ on the cell divided by its constant mass m_j. Velocity itself is not a conserved quantity; therefore, the second equation is often written as

$$\frac{d\mathbf{p}_j}{dt} = \mathbf{F}_j, \tag{7.6}$$

where $\mathbf{p}_j \equiv m_j\mathbf{v}_j$, the momentum of cell j, is a conserved quantity when $\mathbf{F}_j = 0$. Equation (7.6) also shows that force is the rate of momentum variation, which is consistent with Eq. (7.3) for continuum models. Note that Eq. (7.6) represents the general form of possible rotational motion for finite-size cells if $\mathbf{p}_j$ denotes the angular momentum and $\mathbf{F}_j$ the torque.

7.2.3   "Inertialess" assumption

For a strongly dissipative system such as the tumor growth environment, the dissipative force, also known as the drag, is especially important. Objects moving through a fluid with low speeds and stirring up no turbulence experience an approximately linear drag force $\mathbf{F}_{\text{drag},j} = -\zeta_j \mathbf{v}_j$, where ζ_j is the constant damping coefficient of a discrete cell j. Let us rewrite Eq. (7.6) of the discrete model by separating out the drag from the rest of the force:

$$\frac{d\mathbf{p}_j}{dt} = -\zeta_j \mathbf{v}_j + \mathbf{F}_j. \tag{7.7}$$

Now $\mathbf{F}_j$ represents a force that no longer includes the drag. If the dissipation is strong (i.e., ζ_j is large), the force quickly equilibrates and $d\mathbf{p}_j/dt \approx 0$, which is the "inertialess" assumption discussed in Section 6.2.4 for the agent-based model. As a result, the velocity quickly relaxes to a terminal equilibrium velocity

$$\mathbf{v}_j = \frac{1}{\zeta_j} \mathbf{F}_j. \tag{7.8}$$

For details of how the various biologically founded discrete forces $\mathbf{F}_j$ are formulated, readers may refer to the constitution of the agent-based model in Chapter 6. Equation (7.8) is a *kinematic description* of a discrete system, in which the velocities are passively determined by other variables. In contrast, Eq. (7.5) is a *dynamic description* of a discrete system, where the velocities are actively evolved by some accelerations.

Similarly, Eqs. (7.1) and (7.3) in Section 7.2.1 are the dynamic description of a continuum model, from which a kinematic description can be derived for a strongly dissipative system using the inertialess assumption. The continuum form of the drag is $\mathbf{F}_{\text{drag}} = -\zeta \mathbf{J}$, with damping coefficient ζ, and will be justified in Section 7.2.4 after we have established the connection between discrete and continuum variables. By again separating out the drag from the other forces and assuming an incompressible fluid, we can rewrite Eq. (7.3) as

$$\frac{D\mathbf{p}}{Dt} = \nabla \cdot \mathbf{T} - \zeta \mathbf{J} + \mathbf{F}, \tag{7.9}$$

where $D/Dt \equiv \partial/\partial t + \mathbf{u} \cdot \nabla$, called the *advective derivative* as in Eq. (5.4) or the *material derivative* is the derivative along the stream velocity field $\mathbf{u}$. Again for symbolic simplicity, we will continue to use $\mathbf{F}$ to represent a force that no longer includes the drag. Since individual cells move along the stream lines, the material derivative is the continuum form of the total derivative d/dt in Eq. (7.7) with the frame of reference fixed on an individual cell. Then, using the inertialess assumption as in Eq. (7.8), we assume that the force acting on an individual cell quickly equilibrates; hence, $D\mathbf{p}/Dt \approx 0$, and the flux $\mathbf{J}$ can directly be evaluated by

$$\mathbf{J} = \frac{1}{\zeta} (\nabla \cdot \mathbf{T} + \mathbf{F}). \tag{7.10}$$

The continuity equation (7.1) with $\mathbf{J}$ calculated by Eq. (7.10) is the kinematic description of a continuum model. Further biological details of how to construct the flux as well as the source-terms in Eq. (7.1) can be found in Chapter 5 in the derivation of the Cahn–Hilliard phase-field model. Sharp-interface models, such as that presented in Chapter 3, can be obtained as a further simplification of the general form in which the dynamics of continuum fields is reduced to the evolution of their interfaces; the kinematic form gives the velocity that advances the interfaces while the dynamic form describes the force balance along the interfaces.

7.2.4   Linking discrete and continuum variables

We reviewed continuum models and discrete models respectively in Sections 7.2.1 and 7.2.2. To couple satisfactorily the two types of models to form a hybrid continuum–discrete model, in this section we use the conservation of mass and momentum to establish a physically founded connection between discrete and the continuum variables.

Connecting discrete mass and continuum density

Following the classical theory of statistical mechanics, the density field ρ of a continuum model can be defined as an ensemble-averaged local tumor mass of a discrete model [331]:

$$\rho(\mathbf{x}, t) = \left\langle \sum_{j=1}^{N} m_j \delta(\mathbf{x} - \mathbf{x}_j(t)) \right\rangle, \tag{7.11}$$

where m_j, $\mathbf{x}_j$ respectively represent the mass and the position of cell j, $\delta(.)$ is the Dirac delta function, and N is the total number of cells. Equation (7.11) specifies a relation of equivalent exchange for a continuum–discrete conversion at position $\mathbf{x}$ and time t. Many continuum models, such as those discussed in Chapters 3–5, use a dimensionless volume fraction φ instead of ρ as the continuum variable. Such models assume a constant mass density, as in Section 5.2; then ρ and φ are linearly proportional, with $\rho = \rho_0 \varphi$ where ρ_0 represents a constant mass density per cell volume. Using this linear relation, Eq. (7.11) is readily applicable for such models. To preserve the original meaning of mass conservation, we will continue using the density ρ for the general discussion of hybrid models. The derivation of equivalent expressions for φ is straightforward.

Connecting discrete and continuum momentum

In parallel with Eq. (7.11), the classical statistical mechanics definition of the continuum momentum field is

$$\mathbf{p}(\mathbf{x}, t) = \left\langle \sum_{j=1}^{N} \mathbf{p}_j \delta(\mathbf{x} - \mathbf{x}_j(t)) \right\rangle, \tag{7.12}$$

recalling that $\mathbf{p}(\mathbf{x}, t) = \rho(\mathbf{x}, t)\mathbf{u}(\mathbf{x}, t)$ and $\mathbf{p}_j = m_j \mathbf{v}_j$ are the continuum and the discrete momentum, respectively. Equation (7.12) defines the momentum exchange accompanying the mass exchange during a continuum–discrete conversion event at position $\mathbf{x}$ and time t.

In Section 7.2.3 we postponed the justification of the continuum drag expression $\mathbf{F}_{\text{drag}}^{\text{cont}} = -\zeta\mathbf{J}$. Now that we have established the connection between discrete and continuum variables, we are able to derive the continuum version of the linear discrete drag $\mathbf{F}_{\text{drag},j}^{\text{disc}} = -\zeta_j\mathbf{v}_j$:

$$\mathbf{F}_{\text{drag}}^{\text{cont}} = \sum_j \left\langle \mathbf{F}_{\text{drag},j}^{\text{disc}} \delta(\mathbf{x} - \mathbf{x}_j) \right\rangle = -\sum_j \left\langle \zeta_j\mathbf{v}_j\delta(\mathbf{x} - \mathbf{x}_j) \right\rangle$$

$$= -\sum_j \left\langle \frac{\zeta_j}{m_j}\mathbf{p}_j\delta(\mathbf{x} - \mathbf{x}_j) \right\rangle. \qquad (7.13)$$

Assuming that the individual cells are nearly identical, we may approximate ζ_j and m_j by an average damping coefficient $\bar{\zeta}$ and an average mass $\bar{m}$, which allows us to obtain a closed-form expression for $\mathbf{F}_{\text{drag}}^{\text{cont}}$:

$$\mathbf{F}_{\text{drag}}^{\text{cont}} = -\frac{\bar{\zeta}}{\bar{m}} \sum_j \left\langle \mathbf{p}_j\delta(\mathbf{x} - \mathbf{x}_j) \right\rangle = -\zeta\mathbf{p} = -\zeta\mathbf{J}, \qquad (7.14)$$

where we define the continuum damping coefficient $\zeta \equiv \bar{\zeta}/\bar{m}$, and recall that $\mathbf{p} = \rho\mathbf{u} = \mathbf{J}$.

7.3 Mass and momentum exchange

In Section 7.2 we discussed the discrete and continuum model components of a hybrid modeling framework and physically linked the variables of each model through conservation laws. On this basis we can rigorously derive the exchange mechanism of mass and momentum to construct a hybrid model. The continuum–discrete coupling mechanism of a hybrid model primarily includes a continuum-to-discrete (C2D) conversion, a discrete-to-continuum (D2C) conversion, and mechanical interactions between the discrete and the continuum elements. The C2D and the D2C conversion processes will be derived in Sections 7.3.1–7.3.2 along with appropriate formulations of the mass and momentum exchange terms. In addition to directly converting to each other, the continuum and discrete variables can also interact mechanically through interaction forces, resulting in another source of momentum exchange which will be described in Section 7.3.3.

7.3.1 The continuum-to-discrete (C2D) conversion

When tumor cells break away from the collective motion of a bulk tumor mass and exhibit more individual characteristics, a C2D conversion takes place in a hybrid model. A C2D conversion turns a certain amount of continuum cell density into an equivalent number of discrete cells. In Section 7.2.4 we derived Eqs. (7.11) and (7.12) for the equivalent exchange of mass and momentum between the continuum and discrete models. Assuming that a C2D conversion takes place at $\mathbf{x} = \mathbf{x}_k \in \Omega$ and $t = t_k \in [0, \infty)$, where Ω is the entire domain over which the model applies, we may rewrite Eqs. (7.11) and (7.12) as

$$
\mathcal{C}_k \rho(\mathbf{x}_k, t_k) = \left\langle \sum_{j=1}^{N_c} m_j \delta(\mathbf{x}_k - \mathbf{x}_j(t_k)) \right\rangle, \tag{7.15}
$$

$$
\mathcal{C}_k \mathbf{p}(\mathbf{x}_k, t_k) = \left\langle \sum_{j=1}^{N_c} \mathbf{p}_j \delta(\mathbf{x}_k - \mathbf{x}_j(t_k)) \right\rangle. \tag{7.16}
$$

Here we introduce a coefficient $\mathcal{C}_k$ to specify the proportion of the continuum density being converted and use N_c to denote the number of converted discrete cells. In Eq. (7.15) the mass on the left-hand side is subtracted from the continuum model mass and transformed into the mass given by the right-hand side, to generate new individual cells in the discrete model. Correspondingly, the amount of momentum carried by the converted mass is also transferred from the continuum model to the discrete model, according to Eq. (7.16). In this section we will construct the C2D conversion by breaking it down into three essential steps: (i) subtraction of the mass from the continuum density; (ii) addition of new cells to the discrete model; and (iii) transfer of the corresponding momentum.

For the C2D conversion we subtract mass from the continuum model by introducing a sink term to the continuity equation, i.e.,

$$
S_{\text{C2D}}(\mathbf{x}, t) = \sum_k S_{\text{C2D}}^{(k)}(\mathbf{x}, t), \tag{7.17}
$$

where $S_{\text{C2D}}^{(k)}(\mathbf{x}, t)$ represents the mass subtracted from one C2D conversion event at position $\mathbf{x}_k$ and time t_k and is expressed as

$$
S_{\text{C2D}}^{(k)}(\mathbf{x}, t) = -\mathcal{R}_{\text{C2D}}^{(k)}(\mathbf{x}, t)\rho(\mathbf{x}, t). \tag{7.18}
$$

Here we have introduced a localized C2D conversion-rate function $\mathcal{R}_{\text{C2D}}^{(k)}(\mathbf{x}, t)$ to specify the extraction rate. Theoretically, we may make the choice

$$
\mathcal{R}_{\text{C2D}}^{(k)}(\mathbf{x}, t) = \mathcal{C}_k \delta(\mathbf{x} - \mathbf{x}_k)\delta(t - t_k); \tag{7.19}
$$

then the total subtracted mass, obtained by integrating $-S_{\text{C2D}}^{(k)}$ over $\Omega \times [0, \infty)$, equals Eq. (7.15). However, for numerical implementation we need to approximate the Dirac delta functions in Eq. (7.19) by finite distribution functions and formulate $\mathcal{R}_{\text{C2D}}^{(k)}$ as follows:

$$
\mathcal{R}_{\text{C2D}}^{(k)}(\mathbf{x}, t) = \mathcal{C}_k(\mathbf{x}, t)G_k(\mathbf{x} - \mathbf{x}_k)\Pi_k(t - t_k). \tag{7.20}
$$

The spatial function $G_k(\mathbf{x} - \mathbf{x}_k)$ defines a localized distribution around $\mathbf{x}_k$, which may be chosen to cover a discretization mesh cell or an arbitrary domain of conversion. The temporal function

$$
\Pi_k(t - t_k) = \mathcal{H}(t - t_k) - \mathcal{H}(t - (t + \Delta t_k)) \tag{7.21}
$$

is a rectangular function specifying a short period of time, in which Δt_k is generally a numerical time-step size but can be a longer, arbitrary, duration of conversion if necessary. Note that for more generality we also add spatiotemporal dependence to the coefficient C_k, allowing for a variable conversion weight.

The mass subtracted from the continuum density adds new cells to the discrete model. By assuming nearly identical cells and approximating the individual mass m_i by an average cell mass $\bar{m}$, the number of new discrete cells created in correspondence to Eq. (7.18) for a C2D conversion event can be calculated as

$$N_c^{(k)} = \frac{1}{\bar{m}} \int_0^\infty \int_\Omega S_{\text{C2D}}^{(k)}(\mathbf{x}, t) \, d\mathbf{x} dt. \tag{7.22}$$

These $N_c^{(k)}$ newly created cells may emerge deterministically at position $\mathbf{x}_k$ or stochastically near $\mathbf{x}_k$, using $G_k(\mathbf{x} - \mathbf{x}_k)$ as a probability distribution function. The initial velocities of such cells are set by considering the conservation of momentum, which we will discuss next.

During a C2D conversion, as mentioned earlier the momentum carried by the converted mass is also transferred from the continuum model to the discrete model. The momentum of the converted mass is removed from the continuum model by introducing the following negative force to the momentum transport equation:

$$\mathbf{F}_{\text{C2D}}(\mathbf{x}, t) = \sum_k \mathbf{F}_{\text{C2D}}^{(k)}(\mathbf{x}, t), \tag{7.23}$$

where $\mathbf{F}_{\text{C2D}}^{(k)}(\mathbf{x}, t)$ represents the momentum extraction during the kth C2D conversion:

$$\mathbf{F}_{\text{C2D}}^{(k)}(\mathbf{x}, t) = -\mathcal{R}_{\text{C2D}}^{(k)}(\mathbf{x}, t) \mathbf{p}(\mathbf{x}, t); \tag{7.24}$$

note that the mass and the momentum are converted at the same rate, $\mathcal{R}_{\text{C2D}}^{(k)}(\mathbf{x}, t)$. Momentum conservation demands that the initial momentum of the newly generated $N_c^{(k)}$ cells should add up to the momentum of the mass subtracted from the continuum model:

$$\sum_{j=1}^{N_c^{(k)}} \langle \mathbf{p}_j \rangle = \int_0^\infty \int_\Omega \mathcal{R}_{\text{C2D}}^{(k)}(\mathbf{x}, t) \mathbf{p}(\mathbf{x}, t) \, d\mathbf{x} dt, \tag{7.25}$$

setting a constraint for the initial velocities of the new cells.

Kinematic models

In a kinematic model, which describes a strongly dissipative environment, the momentum transport equation is replaced by an overdamped velocity as in Eq. (7.10), rendering Eq. (7.24) inapplicable. Furthermore, the discrete velocities are directly calculated by Eq. (7.8). Indeed, in kinematic models it is assumed that any fluctuations in momentum quickly dissipate, so that the cell velocity immediately relaxes to an equilibrium state. The momentum transfer in Eq. (7.24), characterized by the pulsating rate functions $\mathcal{R}_{\text{C2D}}^{(k)}(\mathbf{x}, t)$, causes only short bursts of momentum fluctuation and thus is quickly damped.

Lagged cell conversion

Note that the value of $N_c^{(k)}$ obtained from Eq. (7.22) may not be exactly an integer. If the scale of the discrete model is much smaller than that of the continuum model, we typically get $N_c^{(k)} \gg 1$, and so $N_c^{(k)} = [N_c^{(k)}](1 + \mathcal{O}(N_c^{(k)^{-1}})) \approx [N_c^{(k)}]$, where $[N_c^{(k)}]$ is the nearest integer to $N_c^{(k)}$. Hence we may use $[N_c^{(k)}]$ as the number of converted discrete cells, conserving the mass to $\mathcal{O}(\bar{m})$, which is negligible in the continuum scale for such cases. However, if the scale of the discrete model is not much smaller than that of the continuum model, we might have $N_c \sim 1$ or even $N_c < 1$. For such cases, we can use a "lagged cell conversion" method, where we extend the duration Δt_k of the temporal function $\Pi_k(t - t_k)$ in such a way that enough mass can be subtracted from the continuum density to create an integer number of cells. Assuming that a C2D conversion event is detected at time $t_{\text{detect},k}$, we can calculate the total subtracted mass for this event at every time step t as

$$\mathcal{M}_c^{(k)}(t) = \int_{t_{\text{detect},k}}^{t} \int_{\Omega} S_{\text{C2D}}^{(k)}(\mathbf{x}, s) \, d\mathbf{x} ds; \tag{7.26}$$

until eventually we have

$$\mathcal{M}_c^{(k)}(t_{\text{complete},k}) = N_c^{(k)}\bar{m} + \mathcal{T}(s, \bar{m}), \tag{7.27}$$

where $N_c^{(k)}$ is an integer and $\mathcal{T}$ is an error tolerance that may depend on the numerical time-step size s of the continuum solver and the cell mass $\bar{m}$. Then $N_c^{(k)}$ cells have been created at time $t_{\text{complete},k}$ and position $\mathbf{x}_k$ or within the localized region defined by G_k. The duration of conversion Δt_k is $t_{\text{complete},k} - t_{\text{detect},k}$, and hence we may write $\Pi_k(t - t_{\text{detect},k}) = \mathcal{H}(t - t_{\text{detect},k}) - \mathcal{H}(t - t_{\text{complete},k})$ although $t_{\text{complete},k}$ can only be determined *a posteriori* by Eqs. (7.26), (7.27). Such a lagged conversion scheme is adopted in the hybrid vascularized tumor growth model presented in Section 7.5, with further discussions in Section 7.5.6.

7.3.2   The discrete-to-continuum (D2C) conversion

While discrete cells may emerge from a continuum density, they may also aggregate and collectively form a tumor bulk which is better described by a continuum model. Therefore, in addition to the C2D conversion, we need to construct a discrete-to-continuum (D2C) conversion mechanism for a hybrid model to transform discrete cells to a continuum cell density. The D2C conversion also consists of three primary steps: (i) removing the converted cells from the discrete model; (ii) injecting the mass of the removed cells into the continuum model; (iii) transferring the momentum of the removed cells from the discrete model to the continuum model.

While cells are removed from the discrete model by D2C conversion, the masses of these cells are added to the continuum model by introduction of the following source term to the continuity equation:

$$S_{\text{D2C}}(\mathbf{x}, t) = \sum_k S_{\text{D2C}}^{(k)}(\mathbf{x}, t), \tag{7.28}$$

where $S_{\mathrm{D2C}}^{(k)}(\mathbf{x}, t)$ represents the mass added to the continuum model during one D2C conversion event at position $\mathbf{x}_k$ and time t_k. The conservation of mass demands that

$$S_{\mathrm{D2C}}^{(k)}(\mathbf{x}, t) = m_{\mathrm{tot}}^{(k)} \delta(\mathbf{x} - \mathbf{x}_k) \delta(t - t_k), \tag{7.29}$$

where $m_{\mathrm{tot}}^{(k)} = \sum_{j=1}^{N_c^{(k)}} m_j$ is the total mass of the $N_c^{(k)}$ cells removed from the discrete model in this particular D2C conversion event. Correspondingly, the momentum of the converted cells is transferred to the continuum model by adding a force to the momentum transport equation:

$$\mathbf{F}_{\mathrm{D2C}}(\mathbf{x}, t) = \sum_k \mathbf{F}_{\mathrm{D2C}}^{(k)}(\mathbf{x}, t), \tag{7.30}$$

$$\mathbf{F}_{\mathrm{D2C}}^{(k)}(\mathbf{x}, t) = \mathbf{p}_{\mathrm{tot}}^{(k)} \delta(\mathbf{x} - \mathbf{x}_k) \delta(t - t_k), \tag{7.31}$$

where $\mathbf{p}_{\mathrm{tot}}^{(k)} = \sum_{j=1}^{N_c^{(k)}} m_j \mathbf{v}_j$ is the total momentum of the converted cells in one D2C conversion event. As in C2D conversion, Eqs. (7.30), (7.31) are not applicable for a kinematic model, in which excessive momentum is assumed to quickly damp away.

In the numerical implementation we approximate the Dirac delta functions in Eqs. (7.29) and (7.31) by $G_k(\mathbf{x} - \mathbf{x}_k)$ and $\Pi_k(t - t_k)$ as in Eq. (7.20) of the C2D conversion:

$$S_{\mathrm{D2C}}^{(k)}(\mathbf{x}, t) = m_{\mathrm{tot}}^{(k)} \frac{G_k(\mathbf{x} - \mathbf{x}_k)}{\int_\Omega G_k(\mathbf{x} - \mathbf{x}_k)\, d\mathbf{x}} \frac{\Pi_k(t - t_k)}{\int_0^\infty \Pi_k(t - t_k)\, dt}, \tag{7.32}$$

$$\mathbf{F}_{\mathrm{D2C}}^{(k)}(\mathbf{x}, t) = \mathbf{p}_{\mathrm{tot}}^{(k)} \frac{G_k(\mathbf{x} - \mathbf{x}_k)}{\int_\Omega G_k(\mathbf{x} - \mathbf{x}_k)\, d\mathbf{x}} \frac{\Pi_k(t - t_k)}{\int_0^\infty \Pi_k(t - t_k)\, dt}. \tag{7.33}$$

Again the duration Δt_k of the temporal function $\Pi_k(t - t_k)$ is generally chosen as the numerical time-step size of the continuum solver. However, in some circumstances the amount of the injected mass and momentum can be too large for the continuum model to absorb within a time step, destabilizing the continuum solver. For such cases, a lagged cell-conversion technique similar to that described in Section 7.3.1 can be adopted, as follows.

Lagged cell conversion

Assuming that a D2C conversion event is detected at time $t_{\mathrm{detect}}^{(k)}$, the discrete cells labeled to be converted are immediately removed and, simultaneously, the converted mass and momentum are transferred to the continuum component by the following source terms:

$$S_{\mathrm{D2C}}^{(k)}(\mathbf{x}, t) = \mathcal{R}_{\mathrm{D2C}}^{(k)}(\mathbf{x}, t)\, m_{\mathrm{tot}}^{(k)} \frac{G_k(\mathbf{x} - \mathbf{x}_k)}{\int_\Omega G_k(\mathbf{x} - \mathbf{x}_k)\, d\mathbf{x}}, \tag{7.34}$$

$$\mathbf{F}_{\mathrm{D2C}}^{(k)}(\mathbf{x}, t) = \mathcal{R}_{\mathrm{D2C}}^{(k)}(\mathbf{x}, t)\, \mathbf{p}_{\mathrm{tot}}^{(k)} \frac{G_k(\mathbf{x} - \mathbf{x}_k)}{\int_\Omega G_k(\mathbf{x} - \mathbf{x}_k)\, d\mathbf{x}}. \tag{7.35}$$

Here we introduce a D2C conversion-rate function $\mathcal{R}_{\mathrm{D2C}}^{(k)}(\mathbf{x}, t)$ to replace the rectangular temporal function $\Pi_k(t - t_k)$; note that we allow the D2C conversion rate function to have spatial dependence also, for more generality. Then we can retain the stability of the numerical solver by choosing a relatively lower conversion rate but in consequence the mass and momentum injection has to last for multiple time steps. To track the total injected mass, we similarly keep evaluating.

$$\mathcal{M}_{c}^{(k)}(t) = \int_{t_{\text{detect},k}}^{t} \int_{\Omega} S_{\text{D2C}}^{(k)}(\mathbf{x},s)\, d\mathbf{x}ds \tag{7.36}$$

until we reach

$$\mathcal{M}_{c}^{(k)}\left(t_{\text{complete},k}\right) = m_{\text{tot}} + \mathcal{O}(s) \tag{7.37}$$

(s is the continuum time-step size), and the source term in Eq. (7.34) is then switched off. Since the momentum is injected at the same rate, it is straightforward to show that at $t = t_{\text{complete},k}$, the total converted momentum $\mathbf{p}_{\text{tot}}$ is also completely absorbed and thus the forcing term in Eq. (7.35) is simultaneously turned off.

7.3.3　Momentum exchange through continuum–discrete interactions

Interactive forces also play an important role in the collective behaviors of individual cells. Tumor cells may interact with each other and with the surrounding tissues and through these interactions momentum is exchanged between each model component, modifying the cell mobility, cell arrangement, and consequently the macroscopic features of the tumor. We will explicitly formulate pairwise interactions for the discrete models given by Eq. (6.5) which collectively affect the overall evolution of the system. The interactions are incorporated in continuum mixture models by constructing an adhesion energy E as in Eq. (5.23), from which variational derivatives can be taken in order to calculate the corresponding flux in Eq. (5.18). Here, we will formulate the interactions between the components of a hybrid model and show that for the purely discrete and purely continuum cases the formula converges to our previous derivations.

To derive the interaction terms, we first consider a total (nonlocal) energy in consistent with Eq. (5.23) of Chapter 5:

$$E = \tfrac{1}{2} \sum_{i,j} \int J_{i,j}(\mathbf{x},\mathbf{y})\rho_i(\mathbf{x})\rho_j(\mathbf{y})\, d\mathbf{x}d\mathbf{y}, \tag{7.38}$$

where $J_{i,j}$ is an interaction potential and ρ_i, ρ_j are the densities of component i and component j, which may be either continuum or discrete. The density of a discrete cell can be written as

$$\rho_i(\mathbf{x}) = m_i \delta(\mathbf{x} - \mathbf{x}_i), \tag{7.39}$$

where $\delta(.)$ is the Dirac delta function and $\mathbf{x}_i$ is the position of cell i with mass m_i. Let us divide the densities into continuum and the discrete components, e.g., using $\rho_0, \dots,$ ρ_N to represent the continuum fields and $\rho_{N+1}, \dots, \rho_M$ to stand for the discrete cells. Then, we may further rewrite the total energy as:

$$E = E_{\text{cc}} + E_{\text{cd}} + E_{\text{dd}}, \tag{7.40}$$

where the E_{cc} is the continuum–continuum interaction energy, E_{dd} the discrete–discrete interaction energy, and E_{cd} the continuum–discrete interaction energy:

$$E_{cc} = \frac{1}{2} \sum_{i,j \leq N} \int J_{i,j}(\mathbf{x}, \mathbf{y}) \rho_i(\mathbf{x}) \rho_j(\mathbf{y}) \, d\mathbf{x} d\mathbf{y}, \tag{7.41}$$

$$E_{dd} = \frac{1}{2} \sum_{i,j > N} m_i m_j J_{i,j}(\mathbf{x}_i, \mathbf{x}_j), \tag{7.42}$$

$$E_{cd} = \frac{1}{2} \sum_{i \leq N, j > N} m_j \int \left(J_{i,j}(\mathbf{x}, \mathbf{x}_j) + J_{j,i}(\mathbf{x}_j, \mathbf{x}) \right) \rho_i(\mathbf{x}) \, d\mathbf{x}. \tag{7.43}$$

The continuum–continuum interaction energy is the same as the result derived for the multiphase model in Section 5.5 using the nondimensionalized volume fraction φ in place of the density ρ. The discrete–discrete interaction energy reduces to the form of Eq. (7.42) on integrating the Dirac delta functions over space. Note that, for homophilic cell–cell adhesion, $J_{i,j}$ corresponds to the pairwise interaction potential of the discrete model, discussed in Section 6.2.2, if the adhesion receptor expressions $\mathcal{E}_i$ and $\mathcal{E}_j$ in Eq. (6.16) are absorbed into $J_{i,j}$. The other types of discrete–discrete interaction in Section 6.2.4 can also be written in the form of Eq. (7.42), with appropriate assignment of terms. Also, after integrating the Dirac delta function over space the continuum–discrete interaction energy is derived from Eq. (7.43). In many cases the pairwise interaction potential $J_{i,j}$ is symmetric and depends only on the distance between the two interacting subjects. For such cases the potential $J_{i,j}(\mathbf{x}, \mathbf{y}) = J_{i,j}(|\mathbf{x} - \mathbf{y}|)$ is an even function, independent of the permutation of i and j, i.e., $J_{i,j} = J_{j,i}$; then Eq. (7.43) reduces further to

$$E_{cd} = \sum_{i \leq N, j > N} m_j \int J_{i,j}(|\mathbf{x} - \mathbf{x}_j|) \rho_i(\mathbf{x}) \, d\mathbf{x}. \tag{7.44}$$

For simplicity, from now on we will assume a symmetric interaction potential $J_{i,j}$; for asymmetric interactions, we simply replace $J_{i,j}$ with $\frac{1}{2}(J_{i,j} + J_{j,i})$ in Eqs. (7.43) and (7.44).

Interaction force of the discrete model
Given a pairwise interaction potential $J_{i,j}(|\mathbf{x}_i - \mathbf{x}_j|)$, the interaction force exerted by one discrete cell m_j on another discrete cell m_i is

$$\mathbf{F}_{(j \text{ on } i)} = -m_i m_j \nabla_{\mathbf{x}_i} J_{i,j}(|\mathbf{x}_i - \mathbf{x}_j|);$$

$\nabla_{\mathbf{x}_i}$ means the gradient with respect to $\mathbf{x}_i$. Hence, the total force exerted on the discrete mass m_i ($i > N$) by all the other discrete cells is

$$\mathbf{F}_i^{dd} = -m_i \nabla_{\mathbf{x}_i} \left(\sum_{j=N+1}^{M} m_j J_{i,j}(|\mathbf{x}_i - \mathbf{x}_j|) \right) = -\frac{\delta E_{dd}}{\delta \mathbf{x}_i}, \tag{7.45}$$

where $\delta E_{dd}/\delta \mathbf{x}_i$ is the variational derivative of E_{dd}, Eq. (7.42), with respect to $\mathbf{x}_i$. Similarly, the total force acting on a discrete mass m_i ($i > N$) due to all the continuum densities ρ_j is

$$\mathbf{F}_i^{\text{cd}} = -m_i \nabla_{\mathbf{x}_i} \left(\sum_{j=1}^{N} \int J_{i,j}(|\mathbf{x}_i - \mathbf{y}|) \rho_j(\mathbf{y}) \, d\mathbf{y} \right) = -\frac{\delta E_{\text{cd}}}{\delta \mathbf{x}_i}, \tag{7.46}$$

where E_{cd} is defined in Eq. (7.44). Combining Eqs. (7.45) and (7.46) and the fact that $\delta E_{\text{cc}}/\delta \mathbf{x}_i = 0$, we may formulate the interaction force acting on a discrete mass m_i ($i > N$) as follows:

$$\mathbf{F}_i = -m_i \nabla_{\mathbf{x}_i} \left(\sum_{j=1}^{M} \int J_{i,j}(|\mathbf{x}_i - \mathbf{y}|) \rho_j(\mathbf{y}) \, d\mathbf{y} \right) = -\frac{\delta E}{\delta \mathbf{x}_i}. \tag{7.47}$$

Here, we have used Eq. (7.39) to express the density of discrete cells ($N < j \le M$), and the total energy E is as defined in Eq. (7.38). If the discrete model describes a strongly dissipative system using a kinematic description, the overdamped cell velocity is then given by

$$\mathbf{v}_i = -\frac{1}{\zeta_i} \frac{\delta E}{\delta \mathbf{x}_i}. \tag{7.48}$$

Note that the derivation of the discrete interaction force is similar in spirit to the directional cell–cell interaction mechanism presented in Ramis-Conde et al. [557], in which the total interaction potential of cell i is calculated as the summation of all the pairwise interaction potentials between cell i and other cells within an effective interaction length; then the gradient of this total potential provides a torque-like force that evolves the orientation of the cell velocity.

Interaction force of the continuum model

In the same way as in Eq. (7.45), the pairwise interaction forces exerted on a continuum density ρ_i ($i \le N$) by the discrete cells of mass m_j ($j > N$) add up to the following macroscopic force and stress:

$$\mathbf{F}_i^{\text{dc}}(\mathbf{x}) + \nabla \cdot \mathbf{T}_i^{\text{dc}}(\mathbf{x}) = -\rho_i(\mathbf{x}) \nabla \left(\sum_{j=N+1}^{M} m_j J_{i,j}(|\mathbf{x} - \mathbf{x}_j|) \right)$$

$$= -\rho_i \nabla \frac{\delta E_{\text{cd}}}{\delta \rho_i}; \tag{7.49}$$

note that here we have replaced the individual mass m_i in Eq. (7.45) with the continuum density ρ_i. Following the same concept, the expression for the macroscopic interaction force and stress exerted on ρ_i ($i \le N$) by all the other continuum densities ρ_j ($j \le N$) is:

$$\mathbf{F}_i^{\text{cc}}(\mathbf{x}) + \nabla \cdot \mathbf{T}_i^{\text{cc}}(\mathbf{x}) = -\rho_i(\mathbf{x}) \nabla \left(\sum_{j=1}^{N} \int J_{i,j}(|\mathbf{x} - \mathbf{y}|) \rho_j(\mathbf{y}) d\mathbf{y} \right)$$

$$= -\rho_i \nabla \frac{\delta E_{\text{cc}}}{\delta \rho_i}, \tag{7.50}$$

where E_{cc} is defined by Eq. (7.41). Because $\delta E_{\text{dd}}/\delta \rho_i = 0$ we may combine Eqs. (7.49), (7.50) to obtain the total interaction force and stress acting on ρ_i ($i \leq N$):

$$\mathbf{F}_i(\mathbf{x}) + \nabla \cdot \mathbf{T}_i(\mathbf{x}) = -\rho_i(\mathbf{x}) \nabla \left(\sum_{j=1}^{M} \int J_{i,j}(|\mathbf{x} - \mathbf{y}|) \rho_j(\mathbf{y}) \, d\mathbf{y} \right)$$
$$= -\rho_i \nabla \frac{\delta E}{\delta \rho_i}. \tag{7.51}$$

Note that this form is consistent with the flux of the generalized Fick's law, as in Eq. (5.19) for the mixture model; indeed, by assuming overdamped evolution we can use Eq. (7.10) to calculate the flux $\mathbf{J}_i$:

$$\mathbf{J}_i = -\frac{1}{\zeta_i}(\mathbf{F}_i + \nabla \cdot \mathbf{T}_i) = -\frac{1}{\zeta_i} \rho_i \nabla \frac{\delta E}{\delta \rho_i}, \tag{7.52}$$

which is essentially the same as Eq. (5.18) of Chapter 5 if ρ_i/ζ_i corresponds to the cell mobility $\bar{M}_i$. For a purely continuum model, $E = E_{\text{cc}}$ and Eq. (7.52) results in the same Darcy velocity as in Eq. (5.21).

7.3.4 Summary of the hybrid modeling framework

Before we present examples of hybrid tumor growth models, let us summarize the general framework of a hybrid continuum–discrete model using the illustration in Figure 7.1. A hybrid continuum–discrete model contains a continuum and a discrete component; each component is described by equations of motion that satisfy the conservation of mass and momentum, as discussed in Sections 7.2.1 and 7.2.2. Momentum conservation is often replaced by an inertialess assumption for dissipative systems, where the state variables quickly relax to a terminal velocity and any extra momentum rapidly damps away, as described in Section 7.2.3. Using statistical mechanics, in Section 7.2.4 the connection between the continuum and discrete mass and momentum is established, from which the coupling mechanism of hybrid models can be derived. The components of the hybrid model are coupled through a mass and momentum exchange mechanism, including continuum-to-discrete (C2D) conversion (Section 7.3.1), discrete-to-continuum (D2C) conversion (Section 7.3.2), and interactions that transfer momentum between the components (Section 7.3.3). Next, we present existing models from the literature to show how a hybrid model can be constructed for particular cases, using the basic elements of the general framework.

7.4 A hybrid model of multicellular tumor spheroids (MCTSs)

The first example is a pioneering work in Kim *et al.* [382]. A model was proposed for the growth of *multicellular tumor spheroids* (MCTSs), for which data are readily available. Spheroids are tumors grown in a controlled *in vitro* environment mimicking the *in vivo* growth of tumors. The tumors generally grow to the scale of a few millimeters,

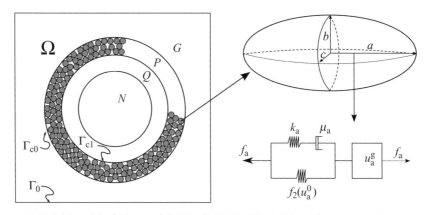

Figure 7.2 Hybrid model of Kim *et al.* [382] of MCTSs. The left-hand figure shows the construction of a model for the growth of multicellular tumor spheroids. The region of necrotic tumor cells is denoted by N, the region of quiescent tumor cells by Q, and the tumor proliferating rim by P. Outside the tumor the space is filled by gel, denoted by G. The N, Q, and G regions are modeled by continuously evolving the interfaces. The tumor cells in the proliferating rim P are modeled discretely using the extended cell-based three-dimensional model of Dallon and Othmer [161]. A schematic of this cell-based three-dimensional model is shown on the right. Reprinted from Kim *et al.* [382], with permission from World Scientific.

estimated to contain on the order of 10^6 cells, a size which is more appropriate for the continuum approach. However, cellular-scale mechanisms such as proliferation and apoptosis may have a great impact on the tumor morphology. Discrete models allow for explicit formulation of the microscopic physics, offering an opportunity to directly manipulate these mechanisms in order to test various hypotheses; in particular, the changes in cell size during proliferation can only be described using a discrete approach. Kim *et al.* [382] adopted a discrete model to describe the cells in the proliferating rim of a tumor, generally 100–200 μm thick, while the rest of the tumor and the surrounding gel are modeled as continuum fields. Following the general framework, we first introduce the continuum and the discrete components of this hybrid model in Section 7.4.1 and then follow with the coupling mechanism in Section 7.4.2. The simulation results and presented in Section 7.4.3, with further discussion in Section 7.4.4.

7.4.1   The continuum and the discrete components

Figure 7.2 shows a schematic of the hybrid model presented in Kim *et al.* [382]. As mentioned earlier, a discrete model is adopted to simulate the individual cells in the proliferating rim P. The rest of the tumor consists of a quiescent region Q and a necrotic region N, which, along with the surrounding gel G, are described by a continuum model.

The continuum model used for regions Q, N, and G is a sharp-interface model with a kinematic description that treats these regions as linear viscoelastic materials:

$$\nabla \cdot \sigma_m = 0 \qquad \text{on } \Omega_m \times (0, T), \tag{7.53}$$

where Ω_m, $m = 0, 1, 2$, denotes the domains of G, Q, and N, respectively; T represents the time when the *in vitro* experiment terminates. The stress tensor σ_m is taken to be of the Kelvin–Voigt form for viscoelastic materials:

$$\sigma_m = \mathcal{C}_m \epsilon + \mathcal{D}_m \frac{d\epsilon}{dt} \qquad \text{on } \Omega_m \times (0, T), \tag{7.54}$$

where $\mathcal{C}_m$ and $\mathcal{D}_m$ are respectively the stiffness tensor and the viscosity tensor and the strain tensor ϵ is defined by

$$\epsilon = \frac{1}{2} \left[\nabla \mathbf{u} + (\nabla \mathbf{u})^T \right], \tag{7.55}$$

$\mathbf{u}$ being the displacement field. A sharp-interface continuum model is derived from the general conservation equations (7.1) and (7.3) by assuming that a sharp interface separates the regions of 100% density and 0% density. The interfaces Γ_{c0} and Γ_{c1} between regions P and G and between regions P and Q, respectively, are evolved by Eqs. (7.53)–(7.55) coupled with the boundary conditions

$$\mathbf{u} = 0 \qquad \text{on} \quad \Gamma_0 \times (0, T), \tag{7.56}$$
$$\sigma_0 \cdot \mathbf{n}_0 = \mathbf{q}_0 \qquad \text{on} \quad \Gamma_{c0} \times (0, T), \tag{7.57}$$
$$\sigma_1 \cdot \mathbf{n}_1 = \mathbf{q}_1 \qquad \text{on} \quad \Gamma_{c1} \times (0, T). \tag{7.58}$$

Here Γ_0 is the fixed boundary of the entire computational domain and $\mathbf{n}_0$, $\mathbf{n}_1$ are respectively the outer normal vectors of regions G and Q on the boundaries Γ_{c0} and Γ_{c1}. The boundary forces $\mathbf{q}_0$ and $\mathbf{q}_1$ are calculated from the interactions between the continuum regions G, Q and the discrete cells in P, which will be described in Section 7.4.2.

The discrete model used to track the cells in P is modified from the cellular slime-mold model of Dallon and Othmer [161]. An illustration of the model is shown on the right in Figure 7.2. In the model each cell is treated as a viscoelastic ellipsoid whose axes and position are regulated by the combination of four forces:

1. the active force $\mathbf{T}_{j,i}$ exerted by the neighboring cells;
2. the reactive force $\mathbf{M}_{j,i}$ due to the reciprocal action in response to $\mathbf{T}_{j,i}$ exerted on the neighboring cells;
3. the dynamic drag $\mu_{i,j}(\mathbf{v}_j - \mathbf{v}_i)$, where $\mathbf{v}_i$ is the velocity of cell i and $\mu_{i,j}$ is a damping coefficient due to the friction between cells i and j;
4. the static friction $\mathbf{S}_{j,i}$ due to cell adhesion through attachment.

By neglecting inertial effects, the equation of motion takes on a kinematic form:

$$\mathbf{F}_i = \sum_{j \in N_i^a} (\mathbf{T}_{j,i} + \mathbf{M}_{j,i}) + \sum_{j \in N_i^d} \mu_{i,j}(\mathbf{v}_j - \mathbf{v}_i) + \sum_{j \in N_i^s} \mathbf{S}_{j,i} = 0, \tag{7.59}$$

where N_i^a, N_i^d, and N_i^s respectively specify the cells interacting with cell i by means of each force. Moreover, the continuum boundaries Γ_{c0} and Γ_{c1} are treated as a substrate, denoted by $j = 0$, and included in Eq. (7.59). With an additional constraint to mechanically conserve the cell volume, the cells can uptake nutrient and grow in size. When the size of a cell reaches a threshold, it divides to create a new cell. The cell growth rate

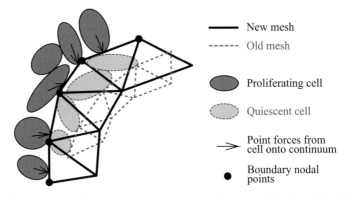

Figure 7.3 Schematic of the discrete-to-continuum conversion in the hybrid model of Kim *et al.* [382]. As the tumor spheroid grows, proliferating cells become quiescent due to lack of oxygen and nutrients and hence are switched from the discrete model to the continuum quiescent field. The sketch shows the re-meshing of the new quiescent region as well as the new nodal points where the interfacial strain and stress are evaluated. Reprinted from Kim *et al.* [382], with permission from World Scientific.

depends on the mechanical stress exerted on the cell, as large stresses and strains may retard growth.

In the hybrid modeling framework, the continuum and discrete models are coupled through the cell–substrate interaction and the continuum–discrete conversion. Next we describe the coupling mechanism.

7.4.2   Continuum–discrete coupling mechanism

The coupling mechanism in the hybrid model of Kim *et al.* [382] is illustrated in Figure 7.3. As discussed in the general framework given in Section 7.3, the continuum and the discrete component may interact and exchange momentum; in addition, mass and momentum may also be directly converted between the components. Since the continuum and the discrete models are applied on separate subdomains the continuum–discrete interaction occurs along the boundaries between such subdomains, more specifically the P–Q interface and the P–G interface. With continuum–discrete interactions the discrete model modifies the momentum of the continuum system through the boundary forces $\mathbf{q}_1$ and $\mathbf{q}_2$ in Eqs. (7.57), (7.58). These forces are calculated by interpolating the forces exerted by the discrete cells at the nearby nodes of a mesh triangulation on which the continuum interfaces are positioned, thus setting up the boundary conditions of the continuum model. This is consistent with the calculation of $\mathbf{F}^{dc}$ in Eq. (7.49), where the forces are written in the form of a potential gradient. Conversely, the viscoelastic continuum boundaries act as substrates in the discrete model and exert forces on the discrete cells which are included in Eq. (7.59) of the discrete model, consistently with the calculation of $\mathbf{F}^{cd}$ in Eq. (7.46).

Continuum–discrete conversion occurs when proliferating regions turn quiescent or vice versa. In typical spheroid experiments, tumors grow larger in size while maintaining

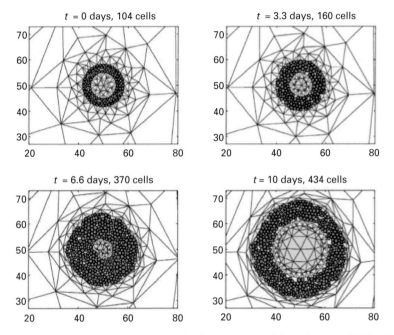

Figure 7.4 A two-dimensional simulation result of the hybrid model in Kim *et al.* [382]. The discrete circular cells represent the proliferating region, while the mesh triangulation of the continuum model is shown in the gel region, the quiescent region, and the necrotic core. In the simulation the tumor grows in an agarose gel. It is found that the viable rim thickness reaches a steady state in 7 days. Reprinted from Kim *et al.* [382], with permission from World Scientific.

a gradient of oxygen and nutrient levels from the periphery into the interior of the tumor. Hence, proliferating regions originally on the periphery turn quiescent as they are advected into the interior, but the reverse tends not to occur. As a result, the conversion mechanism presented in Kim *et al.* [382] consists only of a D2C conversion that turns proliferating cells into the quiescent cell density. If the oxygen and nutrient levels in region P drop below the quiescent threshold, the proliferating cells become quiescent and are absorbed into region Q. The conversion algorithm is illustrated in Figure 7.3. The discrete cells that turn quiescent are removed from the discrete model and converted to the continuum field by updating the $P-Q$ interfacial region. The sharp-interface continuum model in region Q is re-meshed to describe the new interface, and the stress and strain for each mesh node are recalculated accordingly.

7.4.3   Results

Figure 7.4 shows a two-dimensional MCTS simulation using the hybrid model of Kim *et al.* [382]. The results suggest that the viable rim thickness (i.e., the thickness of regions $P + Q$) stabilizes quickly to about 100 μm, while the MCTS keeps growing. The model explicitly formulates the response of cell growth to the stress exerted on the cell, which the authors utilized to study the impact of gel stiffness on the bulk tumor growth. For simplicity, the effect of stress on discrete cell growth in the presented simulations was assumed linear and isotropic, depending only on the magnitude of the stress but not the orientation. It was found that the gel stiffness had no effect on the viable rim thickness, but as predicted, stiff gels hindered the overall growth rate of the entire bulk tumor.

7.4.4   Discussion

The hybrid modeling framework simultaneously allows for the direct modification of microscopic mechanisms and for the observation of macroscopic phenomena. In particular, the discrete component of the hybrid model in Kim *et al.* [382] takes into account the cellular morphology, explicitly describing cell interactions through contact points, which enables more realistic descriptions of cell–cell and cell–ECM adhesion via modeling of the integrins and their impact on cell proliferation. The model currently has no C2D conversion mechanism, as mentioned earlier, because the MCTSs neither shrink nor induce angiogenesis and hence do not turn quiescent regions into proliferating regions. For more general applications, a C2D conversion process may be constructed as the inverse process of the D2C conversion in Figure 7.3 with the initial positions and velocities of the converted cells determined either randomly or by prescribed rules. A potential shortcoming of the model is that its particular D2C conversion algorithm may not conserve mass. The discrete cells are absorbed into the continuum quiescent region when the oxygen and nutrient levels drop beneath the quiescent threshold, regardless of the density of the converted cells; this is inconsistent with the basic assumption of a saturated interior density in the sharp-interface model. The interior density affects the distribution of the interior stress and strain and how they propagate to the boundary; hence, the assumption of a saturated interior density is essential for reducing the problem from the dynamics of the entire bulk mass to simply the evolution of its boundary. This can be improved by adding a density criterion for D2C conversion, so that only densely packed discrete cells are absorbed by the continuum field. However, too strict a criterion may nullify the advantage provided by hybrid modeling and thus some error tolerance may be necessary. If the continuum scale is much larger than the discrete scale, the error may well be negligible.

7.5   A hybrid model of vascularized tumor growth

Another example of a hybrid tumor growth model can be found in Bearer *et al.* [57] and Frieboes *et al.* [228], where vascularized tumor growth is presented. Tumor cells are known to secrete vascular endothelial growth factors (VEGFs) under hypoxic conditions, thus inducing vascular growth; for such phenomena the composite tumor-angiogenesis model introduced in Chapter 5 is employed. In addition, hypoxic tumor cells are observed to down regulate cell–cell and cell–matrix adhesion and upregulate cell motility; this is known as an epithelial–mesenchymal transition (EMT) in the case of tumor cells of epithelial origin [234, 235, 309, 359]. Motility leads individual cells to break away from the bulk tumor and invade the surrounding tissues up gradients of oxygen and nutrient level via chemotaxis, haptotaxis, and other processes; a discrete model component is necessary to describe appropriately such individual cell migration. The model of Bearer *et al.* [57] and Frieboes *et al.* [228], based on a general framework described in Lowengrub *et al.* [425], couples the composite tumor-angiogenesis model to a discrete cell model through the coupling mechanism introduced in Section 7.3.

In particular Bearer *et al.* [57] simulated the invasive characteristics of glioblastomas, which are highly aggressive neoplasms originating in the glial cells in the brain. A glioblastoma can range in size from a few millimeters to several centimeters, corresponding to as many as 10^{11} cells, which requires a continuum approach for correct modeling. Glial cells also exhibit a transition of cell phenotype, switching between collective aggregation behaviors and highly infiltrative cell migration in response to the conditions in the surrounding local microenvironment. A discrete approach would be more appropriate to model the infiltration and movement of individual glial cells. Hence, the hybrid tumor growth model in Frieboes *et al.* [228] can also be applied to glioblastoma, where the phenotype transition is described by the continuum–discrete coupling mechanism of the model.

In Section 7.5.1 we briefly describe the continuum and discrete components adopted for the hybrid vascularized tumor growth model. The hybrid coupling mechanism, consisting of a C2D conversion algorithm, a D2C conversion algorithm, and continuum–discrete interactions, will then be introduced respectively in Sections 7.5.2–7.5.4. Finally, simulation results of this model will be presented in Section 7.5.5, with further discussion in Section 7.5.6.

7.5.1 The continuum and the discrete components

In the hybrid model in Bearer *et al.* [57] and Frieboes *et al.* [228], the majority of the tumor bulk is handled by the Cahn–Hilliard-type continuum multiphase model introduced in Chapter 5. The bulk tumor evolves according to Eq. (5.1), which is essentially the same as the continuity equation (7.1). The tumor subspecies, such as the viable tumor cells and the necrotic tumor cells, are represented by the volume fractions φ_i and the Cahn–Hilliard flux in Eq. (5.27) is adopted to describe cell adhesion, consistently with the kinematic description in Eq. (7.10).

While the tumor bulk is described by the continuum model, the individual cells that have escaped from the tumor bulk are modeled discretely. The discrete component of the hybrid model in Bearer *et al.* [57] and Frieboes *et al.* [228] is a simplified variation of the agent-based model in Chapter 6. In this simplified model the cells are described as point masses that can proliferate, die, or migrate in response to their surrounding microenvironment. Because the distribution of the discrete cells is generally sparse, the cell adhesion forces ($\mathbf{F}_{cca}$ and $\mathbf{F}_{cma}$) are assumed negligible, making the motile locomotive forces $\mathbf{F}_{loc}$, such as chemotaxis and haptotaxis, dominant. A short-range cell–cell repulsion $\mathbf{F}_{ccr}$ is included to prevent the discrete cells from overlapping. Consistently with Eq. (6.30), the overdamped evolution of the discrete cells is governed by the following equation of motion:

$$\frac{d\mathbf{x}_i}{dt} = \chi_{c,i}(1 - n)(1 + k_{c,i}f)\nabla n + \chi_{h,i}(1 - n)\nabla f + \mathbf{v}_{int,i}, \qquad (7.60)$$

where n is the generic cell substrate (oxygen and/or nutrient) concentration, f is the extracellular matrix (ECM) concentration, and $\mathbf{v}_{int,i}$ is the velocity component induced by various cell interactions; both n and f are normalized such that the saturation levels are $n = 1$ and $f = 1$. The first two terms in Eq. (7.60) respectively define the chemotactic and the haptotactic velocities, with the chemotaxis and haptotaxis coefficients denoted by $\chi_{c,i}$ and $\chi_{h,i}$. Both chemotaxis and haptotaxis are upregulated by a lack of cell substrates and thus these two terms are each multiplied by a factor $1 - n$. Moreover, chemotaxis is also facilitated by the presence of the ECM; hence the chemotaxis term should be multiplied by a factor $1 + k_{c,i}f$, where $k_{c,i}$ is a coefficient specifying the effect of the ECM. The interaction-induced velocity $\mathbf{v}_{int,i}$ can be rigorously derived as in Eq. (7.48) of the general hybrid-model formulation. The specific choice for $\mathbf{v}_{int,i}$ adopted by Bearer *et al.* [57] and Frieboes *et al.* [228] will be described in Section 7.5.4 when we discuss momentum exchange via interactions.

In addition to undergoing cell migration, the discrete cells may proliferate by uptaking cell substrates and may die via apoptosis and necrosis. They uptake oxygen and nutrient at the rate

$$\Gamma_u(\mathbf{x}, t) = \sum_i v_{d,i} G_i(\mathbf{x} - \mathbf{x}_i), \qquad (7.61)$$

where $v_{d,i}$ is the *per-volume* uptake rate of cell i and $G_i(\mathbf{x} - \mathbf{x}_i)$ is a spatial distribution function localized around $\mathbf{x}_i$, of which the functional form will be presented later in Eq. (7.66). The cell mitosis rate $\mathcal{R}_M$ is proportional to the concentration of cell substrates:

$$\mathcal{R}_M = \lambda_M n, \qquad (7.62)$$

where λ_M denotes a constant mitosis-rate coefficient. The death rate $\mathcal{R}_D$ is formulated as follows:

$$\mathcal{R}_D = \lambda_A + \lambda_N \mathcal{H}(n_{nec} - n)(n_{nec} - n), \qquad (7.63)$$

where the first term on the right-hand side describes cell apoptosis, with a constant apoptosis-rate coefficient λ_A, and the second term represents cell necrosis with a rate coefficient λ_N; $\mathcal{H}$ is the Heaviside function. Cell necrosis occurs when the cell substrates are lower than n_{nec}, the threshold for necrosis. Note that Eqs. (7.62), (7.63) are consistent with the cell-proliferation and the cell-death mechanisms of the continuum model: see Eq. (5.31) in the case of the continuum model and a simplification of Eqs. (6.10)–(6.13), obtained by neglecting the intracellular dynamics, in the case of the agent-based model.

7.5.2   The C2D conversion process

After our description of the continuum and discrete components of the hybrid vascularized tumor growth model in Section 7.5.1, we now introduce the hybrid-model coupling mechanism, which links the two components. When we discussed the general framework of the coupling mechanism in Section 7.3, we broke it down to three basic elements, C2D conversion, D2C conversion, and momentum exchange through continuum–discrete interactions. In this and the next subsection we will formulate the mass-conserving C2D and D2C conversion processes for the hybrid coupling mechanism. Momentum exchange via interactions will be discussed in Section 7.5.4.

The C2D conversion process converts continuum densities to discrete cells, for which a rate of occurrence is defined, and mass-exchange terms are formulated following the discussion in Section 7.3.1. The rate of occurrence of C2D conversion in the model of Bearer *et al.* [57] and Frieboes *et al.* [228] is determined by the following cell emergence rate:

$$\mathcal{R}^{c \to d}(\mathbf{x}) = \mathcal{C}_{\text{emg}} \mathcal{H}(n_{\text{hyp}} - n(\mathbf{x})) \varphi_{\text{V}}(\mathbf{x}), \tag{7.64}$$

where n_{hyp} is the hypoxia threshold, φ_{V} is the viable-cell volume fraction in the continuum model, and $\mathcal{C}_{\text{emg}}$ is a coefficient defining the frequency of cell emergence. Equation (7.64) states that discrete cells may emerge from the viable-cell species φ_{V} under a hypoxic condition ($n < n_{\text{hyp}}$).

In C2D conversion events, emerging cells are created for the discrete model and, correspondingly, an equivalent amount of mass is subtracted from the continuum model by introducing sink terms into the continuity equation, where $S_{\text{C2D}}^{(k)}(\mathbf{x}, t)$, Eq. (7.17), is expressed as follows:

$$S_{\text{C2D}}^{(k)}(\mathbf{x}, t) = \mathcal{R}_{\text{C2D}}^{(k)}(\mathbf{x}, t) \varphi_{\text{V}}(\mathbf{x}, t), \tag{7.65}$$

which is consistent with the general form in Eq. (7.18) for a C2D conversion event at position $\mathbf{x}_k$ and time t_k. Note that in Eq. (7.65), the cell density ρ in Eq. (7.18) has been replaced by the viable-tumor-cell volume fraction φ_{V}, because the Cahn–Hilliard model described in Chapter 5 uses volume fractions as the continuum variables. Assuming a constant mass per cell volume, as in Chapter 5, the cell density ρ_i and the corresponding cell volume fraction φ_i are related simply by $\rho_i = \rho_0 \varphi_i$, where ρ_0 is a constant cell mass density, introduced for nondimensionalization. Then the general formulas in Sections 7.2 and 7.3 are still applicable. Bearer *et al.* [57] and Frieboes *et al.* [228] adopted the C2D conversion rate $\mathcal{R}_{\text{C2D}}^{(k)}(\mathbf{x}, t)$ in the form of Eq. (7.20). The rate coefficient $\mathcal{C}_k$ is an arbitrary constant that is low enough to retain the stability of the continuum solver of the Cahn–Hilliard equations. The temporal function $\Pi_k(t - t_k)$ is taken to be the rectangular function in Eq. (7.21), where the duration of conversion Δt_k is determined by the "lagged cell-conversion" technique (Eqs. (7.26), (7.27)) presented in Section 7.3.1. Note that because here the continuum variable is the volume fraction φ, the average cell mass $\bar{m}$ in Eq. (7.27) should be replaced by an average cell volume $\bar{V} \equiv \bar{m}/\rho_0$. The localized spatial distribution function $G_k(\mathbf{x} - \mathbf{x}_k)$ is defined as

$$G_k(\mathbf{x} - \mathbf{x}_k) = \frac{1}{2} \left(1 - \tanh \frac{\|\mathbf{x} - \mathbf{x}_k(t)\| - r_c}{\epsilon} \right), \tag{7.66}$$

where r_c is the radius of influence affected by this C2D conversion event and ϵ is an infinitesimal nondimensionalization constant. Note that if we take $\epsilon \to 0$

then $G_k(\mathbf{x} - \mathbf{x}_k) \to \mathcal{H}(r_c - \|\mathbf{x} - \mathbf{x}_k\|)$, defining the domain of conversion as $\Omega_c \equiv \{\mathbf{x} \in \Omega \mid \|\mathbf{x} - \mathbf{x}_i\| \leq r_c\}$. In other words, Eq. (7.66) is a smooth approximation to the Heaviside function that defines Ω_c.

From the subtracted continuum mass in Eq. (7.65), new cells are created for the discrete model. The number $N_c^{(k)}$ of such newly created cells is calculated by Eq. (7.22) with $\bar{m}$ replaced by $\bar{V}$ corresponding to the use of volume fractions as the continuum variables. Since the discrete model is a kinematic model assuming overdamped evolution, it is not necessary to transfer the momentum of the subtracted mass from the continuum model to the discrete model. As discussed in Section 7.3.1, the transferred momentum results in impulsive forces and the effects quickly dissipate in the kinematic assumption. The velocities of the new cells are directly evaluated by Eq. (7.60) of the discrete model.

7.5.3   The D2C conversion process

In Eq. (7.60) of the discrete model, the motile locomotive forces such as chemotaxis and haptotaxis are the dominant mechanisms driving the migration of the discrete cells. Chemotaxis causes the cells to move up the gradient of oxygen and nutrients, resulting in cell aggregation in better vascularized regions; here the cells downregulate cell mobility and upregulate cell adhesion, forming a satellite tumor colony. The tumor cells within such colonies behave more collectively than individually; hence, the evolution of these satellite tumor colonies should be described by the continuum model. The process that converts the discrete cells to the continuum fields is the D2C conversion introduced in Section 7.3.2 of the general hybrid modeling framework.

In the D2C conversion there is a criterion to determine whether a conversion takes place; if so, the mass of the converted cells is transferred from the discrete model to the continuum model via the mass-exchange mechanism presented in Section 7.3.2. For a discrete cell k at position $\mathbf{x}_k$ and time t_k, the D2C conversion criterion in the hybrid model of Bearer *et al.* [57] and Frieboes *et al.* [228] is whether the local population around cell k exceeds a certain threshold, i.e. whether

$$N_k = \sum_{j \neq k} G_k(\mathbf{x} - \mathbf{x}_k)\,\delta_{\mathbf{x},\mathbf{x}_j} \geq N^{\text{D2C}} \tag{7.67}$$

holds. Here N_k is the cell count around $\mathbf{x}_k$, N^{D2C} is the threshold, $\delta_{\mathbf{x},\mathbf{x}_j}$ is the Kronecker delta function, and $G_k(\mathbf{x} - \mathbf{x}_k)$ is the localized spatial distribution defined in Eq. (7.66) to specify the vicinity of cell k. If Eq. (7.67) is satisfied, the cell population around cell k is high enough for cell k to change its phenotype and exhibit more collective behavior. In consequence we convert cell k to the continuum volume fraction field.

Once a cell has been converted to the continuum volume fraction, the mass of the cell is injected into the continuum model by introducing a source term of the type (7.28) into the continuity equation of the continuum model. To preserve the stability of the continuum solver, the "lagged cell-conversion" technique is again adopted, and Eq. (7.34) is used to express the source term $S_{\text{D2C}}^{(k)}(\mathbf{x}, t)$ for the conversion of cell k. Furthermore, Bearer *et al.* [57] and Frieboes *et al.* [228] formulate the conversion rate

$\mathcal{R}^{(k)}_{\text{D2C}}(\mathbf{x}, t)$ as

$$\mathcal{R}^{(k)}_{\mathcal{D} \in \mathcal{C}}(\mathbf{x}, t) = \mathcal{C}_{\text{D2C}} \left(1 - \varphi_{\text{T}}(\mathbf{x}, t)\right) \Pi(t - t_k), \tag{7.68}$$

where $\mathcal{C}_{\text{D2C}}$ is a constant conversion-rate coefficient, φ_{T} is the total tumor-cell volume fraction, and $\Pi(t - t_k)$ is the rectangular temporal function defined in Eq. (7.21) for C2D conversion. The factor $1 - \varphi_{\text{T}}$ prevents the unphysical situation of $\varphi_{\text{T}} > 1$ and reflects that a discrete cell needs to squeeze into the continuum field by pushing the surrounding mass aside (see the end of the chapter for further discussion). Like C2D conversion, the duration of the conversion $\Delta t_k = t - t_k$ in $\Pi(t - t_k)$ is determined dynamically using Eqs. (7.36), (7.37) of the lagged cell-conversion technique.

7.5.4 Momentum exchange through continuum–discrete interactions

As discussed in Section 7.3.3 in the context of the general hybrid modeling framework, the continuum and discrete components may exchange momentum via interactions. The interaction forces exerted on the continuum and discrete components can respectively be derived from Eqs. (7.51) and (7.47). Here we describe the formulations of the interaction terms chosen in Bearer $et\ al.$ [57] and Frieboes $et\ al.$ [228].

Assuming overdamped evolution, the interaction forces on a discrete cell result in the interaction velocity $\mathbf{v}_{\text{int},i}$ in Eq. (7.60), which is specifically formulated as

$$\mathbf{v}_{\text{int},i} = \eta_i \mathbf{p}(\mathbf{x}_i) + \frac{1}{\zeta_i} \nabla_{\mathbf{x}_i} \left(\sum_{j \neq i} J^{\text{ccr}}_{i,j}(|\mathbf{x}_i - \mathbf{x}_j|) \right). \tag{7.69}$$

The first term on the right-hand side of Eq. (7.69) is a simplified version of the continuum–discrete interaction $\mathbf{F}^{cd}_i$ in Eq. (7.46), while the second term is consistent with the discrete–discrete interaction $\mathbf{F}^{dd}_i$ in Eq. (7.45). The continuum–discrete interaction term describes discrete cells being passively dragged along by the local continuum momentum; η_i represents the coefficient of inertia of cell i. By assuming negligible cell–cell and cell–ECM adhesion ($\mathbf{F}_{\text{cca}} \approx 0$ and $\mathbf{F}_{\text{cma}} \approx 0$), owing to the sparse distribution of the discrete cells, the discrete–discrete interaction consists of only a cell–cell repulsion $\mathbf{F}_{\text{ccr}}$ characterized by the repulsive potential

$$J^{\text{ccr}}_{i,j}(x) \equiv \begin{cases} \gamma_i \left(2\ell_{\text{r}} - \frac{\ell_{\text{r}}^2}{x} - x \right) & \text{if } x \leq \ell_{\text{r}}, \\ 0 & \text{otherwise,} \end{cases} \tag{7.70}$$

where ℓ_{r} is the repulsive length, γ_i specifies the repulsive strength, and we recall that ζ_i in Eq. (7.69) is the same damping coefficient as in Eq. (7.48). This form of the potential $J^{\text{ccr}}_{i,j}(x)$ in (7.70) implies that a repulsive force is exerted on a pair of cells only if their distance is smaller than ℓ_{r}. Note that the constant $2\ell_{\text{r}}$ in Eq. (7.70) makes $J^{\text{ccr}}_{i,j}(x)$ continuous at $x = \ell_{\text{r}}$ but has no effect on the evaluation of $\mathbf{v}_{\text{int},i}$ after the gradient has been taken. Also note that Eq. (7.70) is a hard-core repulsion, i.e., the magnitude of the repulsion increases indefinitely as the cell distance $|\mathbf{x}_i - \mathbf{x}_j| \to 0$, which is necessary in the point-mass discrete model to prevent cell overlap. In contrast, the repulsion using the functional form presented in Eq. (6.5) is a soft-core repulsion, which is an adequate approximation for a discrete model with nonzero cell size in a relatively low cell-speed regime. However, a hard-core repulsion is more numerically challenging, requiring an adaptive time-step method for its accurate computation, and it may impose a severe restriction on the numerical time-step size.

The interaction force acting on the continuum component consists of a continuum–continuum interaction term and a continuum–discrete interaction term, as in Eqs. (7.49), (7.50). As discussed in Section 7.3, the continuum–continuum interaction leads to the same adhesion flux as in Eq. (5.18) for the Cahn–Hilliard model. However, while the discrete cells are passively dragged along by the continuum density, the interaction force exerted by the discrete cells on the continuum density is simply assumed to be negligible. Hence there is no additional continuum–discrete interaction term to be added to the continuum model, and this is why the Cahn–Hilliard flux in Eq. (5.18) remains intact. This assumption is valid when the continuum scale is significantly larger than the discrete scale. For general cases, a more realistic treatment will be discussed later, in Section 7.5.6.

7.5.5 Results

Figure 7.5 shows the evolution of a vascularized tumor predicted by the hybrid continuum–discrete model in Bearer *et al.* [57] and Frieboes *et al.* [228]. Discrete cells, represented by small dots, are released from hypoxic perinecrotic regions of the continuum tumor density. The cells migrate up the oxygen and nutrient gradient via chemotaxis and thus move away from the tumor bulk. The cell substrates are supplied via diffusion from the far-field boundary as well as from the newly formed vasculature, depicted by the dark curves. As the discrete cells evolve, they degrade and remodel the extracellular matrix by laying down fibronectin macromolecule networks. The cells respond to the remodeled microenvironment by forming Indian-file-like strands of palisading cells from low-oxygen perinecrotic regions to high-oxygen environments. As the cells reach oxygen-releasing vessels, they begin to aggregate by upregulating their proliferation and downregulating their motility. When the local cell density exceeds a certain threshold, the hybrid model converts the discrete cells back to a continuum bulk, forming tumor microsatellites around the neovasculature. The proliferating cells produce a mechanical pressure, which in turn shuts off the existing blood vessels, and the vasculature may regress as a result. This phenomenon can be seen around some of the newly formed continuum clusters. The resulting morphology compares well with pathology data for human brain tumor specimens (e.g., [57, 229]): for example, one can see the growth of tumor cell clusters "cuffing" the conducting blood vessels and also infiltrating cell strands moving away from the bulk tumor into the healthy tissue in regions where neovascularization has not yet occurred.

7.5.6 Discussion

The hybrid model of Bearer *et al.* [57] and Frieboes *et al.* [228] utilizes the advantages of both the continuum and the discrete modeling approaches to overcome the limitations

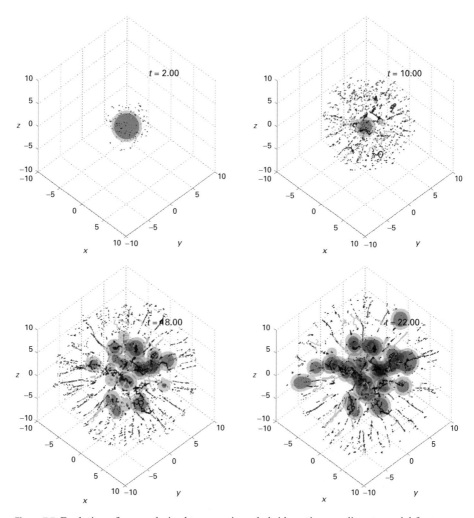

Figure 7.5 Evolution of a vascularized tumor using a hybrid continuum–discrete model for tumor cells (nondimensionalized time units). Discrete cells (dots) are released from hypoxic regions of the continuous tumor regions (surfaces). Conversely, discrete cells are converted back to continuum volume fractions when their local density is sufficiently large. Vessel sprouts are shown as light-color lines, which may anastomose and form looped vessels (dark-color lines), releasing oxygen and nutrients. Note that the infiltrating discrete cells form a palisading pattern as they invade the surrounding healthy tissue, as is observed to occur *in vivo* in regions of lower neo-vascularization [57].

of either a stand-alone continuum model or a stand-alone discrete model. It efficiently and realistically describes the phenomenon of single-cell migration along with bulk tumor growth. The model can be improved by constructing continuum–discrete interaction terms, following the derivation in Section 7.3 of the general hybrid modeling framework. Recall that, in the continuum–discrete interaction of the hybrid model, the discrete cells are dragged along by the continuum field but the reciprocal action of the discrete cells on the continuum field is ignored. As an approximation of cell–cell adhesion, this simplification is appropriate when the size of the discrete cells is negligible at the continuum scale and the discrete cells distribute very sparsely. In a more rigorous treatment of cell–cell adhesion, an adhesion potential $J_{i,j}^{cca}$ is introduced, and this combines with the repulsive potential $J_{i,j}^{ccr}$ in Eq. (7.70) to obtain the total interaction potential $J_{i,j} = J_{i,j}^{cca} + J_{i,j}^{ccr}$. Then the interaction energy E in Eq. (7.38) can be constructed, and more realistic interaction mechanisms can be derived, using Eq. (7.46) for the discrete model and Eq. (7.49) for the continuum model.

A rigorous formulation of the interaction terms may help to resolve a potential shortcoming of the model algorithm, which is that, in the C2D and the D2C conversions, a lagged cell conversion scheme is adopted and so the converted mass is "in limbo" while waiting either for new discrete cells to be created or to be completely absorbed into the continuum field. Such in-limbo mass may introduce errors in the hybrid model simulations. In particular, the factor $1 - \varphi_T$ in Eq. (7.68) for the D2C conversion can be very close to zero, significantly prolonging the duration of lagged cell conversion and magnifying the errors caused by in-limbo mass. Like the simplified interaction formula, this problem is less of a concern if the scale of the continuum model is much larger than that of the discrete model. For such cases the in-limbo mass is negligible on the continuum scale and, as a result, the conversions generally complete very quickly. For problems that have relatively larger discrete cells, the converted mass may make a notable impact on the continuum model if a considerable amount of mass is removed from or added to the continuum model. To ease the impact, a prolonged conversion process is required but this comes at the expense of accuracy. This problem can be resolved by accounting fully for the continuum–discrete interaction. If the discrete cells are allowed to interact with the continuum density, the converted mass does not suddenly disappear from or appear in the continuum model and this reduces the necessity for a prolonged conversion process. In particular, if continuum mass is repelled from a discrete cell then the factor $1 - \varphi_T$ is not close to zero in the presence of discrete cells and does not delay the conversion process as much as it does in the current algorithm.

In summary, the hybrid model of Bearer *et al.* [57] and Frieboes *et al.* [228] utilizes a continuum mixture model to simulate efficiently the bulk tumor growth but at the same time individual discrete cells shed from the tumor bulk are precisely described by a discrete point-mass model. The proliferation and the migration of individual cells are modeled by a C2D conversion process and a D2C conversion process, each consisting of a transition criterion and a mass-exchange algorithm. The discrete cells and the continuum densities co-exist in space and exchange momentum through interactions. In the current model, a simplification is made to the interaction terms and the conversion processes by assuming that the discrete mass is negligible at the continuum scale. For more general cases the interaction terms should be derived more rigorously, as in Section 7.3.3, to prevent errors due to the conversion algorithms.

8 Numerical schemes[1]

With Y.-L. Chuang, F. Jin, and S. M. Wise

In this chapter, we provide readers with a numerical framework for solving the multiphase mixture model described in Chapter 5. It is highly challenging to develop an accurate and efficient numerical method to solve such a model, which consists of high-order (fourth-order) nonlinear equations and is characterized by the interaction of small-scale features (e.g. diffuse tumor–host boundaries) with larger-scale features (e.g. the tumor morphology). Here we describe a finite-difference nonlinear multigrid method that can be used to solve the system accurately and efficiently.

The primary component of the model is a Cahn–Hilliard-type equation for the total tumor volume fraction; this is a fourth-order nonlinear advection–diffusion equation. The solutions are characterized by nearly constant states, with complex morphologies, separated by evolving narrow transition layers that describe the diffuse interfaces between the tumor and host tissues. "Slaved" to the Cahn–Hilliard (CH) equation are the equations for the remaining tumor-cell volume fractions, the cell-substrate concentrations (e.g., the oxygen and nutrient levels), and the mechanical pressure that determines cell velocity. The coupling of phenomena across widely varying length scales is a hallmark of multiphase problems. The use of a uniform mesh in such cases leads to either infeasibly large computational problems or significant limitations on the scales of problems that can be simulated. We overcome this difficulty by adopting multigrid methods with adaptive spatial refinement that uses block-structured Cartesian meshes. This balances the needs of localized fine spatial resolution and computational efficiency. We utilize the *full approximation storage* (FAS) nonlinear multigrid method, which enables the efficient solution of nonlinear problems by performing only local linearizations.

While adaptive mesh-refinement schemes have been developed in a finite element framework [259, 363, 559], in this chapter we adopt a finite difference discretization generalized from Wise *et al.* [697]. Finite difference discretizations are advantageous for the multigrid method because the refined mesh patches are Cartesian, thus not requiring any changes to the finite difference stencil as the mesh grids are refined. In contrast, a finite element method requires the modification of the discretization (e.g. the stiffness tensor) for different meshes. We note that other multilevel multigrid algorithms have been developed as part of the CHOMBO [140, 455] and the BEARCLAW [471] software packages. In this chapter, we modify the FAS method so as to fit it within the framework

---

[1] This chapter is partly based on work in preparation by Wise *et al.* [696] and describes work in Wise *et al.* [695] and Kim *et al.* [380].

of the block-structured *multilevel adaptive technique* (MLAT) developed by Brandt (see the books [75] and [662]) and by Wise *et al.* [697].

In Section 8.1 we review the multiphase mixture model presented in Chapter 5 and develop finite difference discretization in Section 8.2. The block-structured adaptive mesh is presented in Section 8.3 along with the multilevel mesh hierarchy. Finally, the MLAT–FAS method is introduced in Section 8.4.

8.1   Review of the multiphase mixture model

Let us recapitulate the nondimensionalized multispecies phase-field tumor growth model given in Chapter 5. The basic components of this model include Cahn–Hilliard-type (CH-type) equations that describe the growth of various tumor-cell species and diffusion–reaction equations representing the time evolution of the environmental substrate concentrations that affect tumor growth. By putting together Eqs. (5.13), (5.27), and (5.28) we may write the tumor growth equations as follows:

$$\frac{\partial \varphi_T}{\partial t} + \nabla \cdot (\mathbf{u}_S \varphi_T) = S_T - M \nabla \cdot (\varphi_T \nabla \mu), \tag{8.1}$$

$$\frac{\partial \varphi_p}{\partial t} + \nabla \cdot (\mathbf{u}_S \varphi_p) = S_p - M \nabla \cdot (\varphi_p \nabla \mu), \quad p = 1, \dots, N_T, \tag{8.2}$$

$$\mu = \mu(\varphi_T) = f'(\varphi_T) - \epsilon^2 \nabla^2 \varphi_T, \tag{8.3}$$

where φ_p is the volume fraction of tumor subspecies p and φ_T is the total volume fraction of the tumor cells; S_p and S_T represent the net sources of, respectively, the tumor subspecies and the entire tumor. Note that, in this generalized form, φ_p includes the dead-cell species φ_D, the viable-cell species φ_V, and any mutated-cell species φ_M present in the system. Furthermore, since $\varphi_T = \sum_{p=1}^{N_T} \varphi_p$, where N_T denotes the total number of tumor subspecies, we only need to solve for $N_T - 1$ subspecies since and the last is directly constrained by the others, i.e., $\varphi_{N_T} = \varphi_T - \sum_{p=1}^{N_T-1} \varphi_p$. The cell mobility M characterizes the response of the tumor cells to the cell-adhesion potential μ. Here we assume that μ depends only on φ_T, which simplifies the numerical solver presented in Section 8.4.3, where the numerical treatment of more complicated forms of μ is also discussed. In Eq. (8.3), the function f is defined by Eqs. (5.24)–(5.26) and ϵ is the interfacial thickness parameter, a numerical treatment for which will be discussed further in Section 8.2.4. The dimensionless mass-averaged cell velocity $\mathbf{u}_S$ is given by Eq. (5.30):

$$\mathbf{u}_S = -k(\varphi_T, \varphi_p)\left(\nabla \bar{p} - \frac{\bar{\gamma}}{\epsilon} \mu \nabla \varphi_T\right). \tag{8.4}$$

Here we use $\bar{p}$ to represent the pressure, so distinguishing it from the tumor sub-species index p, and $\bar{\gamma}$ is a dimensionless measure of the adhesion force.

The tumor growth equations Eqs. (8.1)–(8.3) are complemented by equations for the microenvironmental variables that regulate the source terms S_T and S_p and the mass-averaged cell velocity $\mathbf{u}_S$ (evaluated by Eq. (8.4)). The microenvironmental variables, such as the nutrient and VEGF concentrations, are generally described by the reaction–diffusion equations

$$\nabla \cdot (D_q \nabla c_q) + \Gamma_q = \begin{cases} 0 & \text{steady-state or fast time scales,} \\ \dfrac{\partial c_q}{\partial t} & \text{slow time scale,} \end{cases} \tag{8.5}$$

where c_q represents some measure of the microenvironmental variable q, with $q = 1, \ldots, N_c$, and N_c is the total number of these variables. Some c_q evolve on a much faster time scale than the tumor growth time and can reasonably be approximated by their steady-state solution, which typically depends on the instantaneous tumor morphology (hence they are termed quasi-steady). Other c_q evolve on a slower time scale and the time derivative $\partial c_q/\partial t$ should be taken into consideration. The corresponding diffusion coefficients D_q can be constant or have a functional dependence on the cell volume fractions, for instance. The reaction terms Γ_q comprise the transfer and the uptake or decay of the microenvironmental variable q. For example, in the nutrient equation (5.35),

$$\Gamma_q = T_{\mathrm{C}} - \sum_{\substack{p=1 \\ p \neq D}}^{N_c-1} v_p^u \varphi_p.$$

and, in the VEGF concentration equation (3.36),

$$\Gamma_q = -\bar{\lambda}_{\mathrm{decay}}^c c - \bar{\lambda}_{\mathrm{binding}}^c c \mathbf{1}_{\mathrm{sprout\ tips}} + \lambda_{\mathrm{prod}}^c.$$

In addition, even the pressure equation derived by taking the divergence of Eq. (5.30),

$$\nabla \cdot \left(k(\varphi_{\mathrm{T}}, \varphi_p)\nabla \bar{p}\right) + S_{\mathrm{T}} - \nabla \cdot \left(k(\varphi_{\mathrm{T}}, \varphi_p)\frac{\bar{\gamma}}{\epsilon}\mu\nabla\varphi_{\mathrm{T}}\right) = 0, \tag{8.6}$$

can be written in the form of Eq. (8.5) by setting $D_q = k(\varphi_{\mathrm{T}}, \varphi_p)$ and $\Gamma_q = S_T - \nabla \cdot [(\bar{\gamma}/\epsilon)k\mu\nabla\varphi_{\mathrm{T}}]$.

Equation (8.1) is a Cahn–Hilliard-type equation of fourth order, where the fourth-order derivative of φ_{T} is given by the interfacial term $\epsilon^2\nabla^2\varphi_{\mathrm{T}}$ in Eq. (8.3). The presence of the fourth-order term makes numerical integration of the equations a challenge. The time-step size s of a general explicit method is constrained by $s \propto h^4$, where h is the spatial grid size. To remove this restriction we use a Crank–Nicholson-like implicit time-integration method, which results in nonlinear equations at the implicit time level. Then a multilevel nonlinear FAS multigrid method is adopted to solve this discrete system. With the initial and boundary conditions given, Eqs. (8.1)–(8.5) can be numerically integrated with respect to space and time. The multigrid numerical solver presented here can easily be adapted to various types of boundary condition; thus, we do not restrict ourselves to a particular type in this chapter. Some common types of boundary conditions are discussed in Section 8.2.3. In the following sections we first discretize the governing equations (8.1)–(8.5) in space and time, transforming the partial differential equations into an algebraic problem. Next we present a nonlinear multigrid method to solve this algebraic problem. The multigrid scheme is introduced first on a uniform mesh and then extended to an adaptive-mesh version that allows more efficient computation for complex tumor morphology. To simplify the mathematical expressions in this chapter, we will illustrate the numerical formulations in two dimensions. The three-dimensional application is merely a straightforward extension of the two-dimensional version. For one-dimensional problems the multigrid method is equally applicable; however, multigrid methods in one dimension usually reduce to other well-known one-dimensional optimal methods and thus are not needed as such [662].

8.2   Uniform mesh discretization

In this section, we introduce a cell-centered finite difference scheme for spatial discretization [380], which can conveniently be extended to a block-structured adaptive mesh later, in Section 8.3, for the adaptive multigrid method [695]. Time discretization is obtained using a Crank–Nicolson algorithm for most terms, except $f'(\varphi_T)$ in Eq. (8.3) for the adhesion potential. This term is treated differently to obtain numerical stability, making the scheme deviate from the traditional Crank–Nicholson algorithm. However, a fully implicit method is used for the equations that solve for steady-state solutions, also for better numerical stability [477]. The discretization is first-order accurate in time and second-order accurate in space and can easily be extended to be formally second-order accurate in time as well.

Let us assume that the computational domain Ω is a rectangular region $(x_{min}, x_{max}) \times (y_{min}, y_{max})$. A typical spatial discretization using $N_x \times N_y$ uniform mesh grids is as follows:

$$x_i = x_{min} + \left(i - \tfrac{1}{2}\right)h,$$ (8.7)

$$y_j = y_{min} + \left(j - \tfrac{1}{2}\right)h,$$ (8.8)

where i and j are integers or half-integers and $h > 0$ is the grid spacing, defined by $h = (x_{max} - x_{min})/N_x = (y_{max} - y_{min})/N_y$. The mesh grid cell $(x_{i-1/2}, x_{i+1/2}) \times (y_{j-1/2}, y_{j+1/2})$ is denoted $I_{i,j}$, where $0 \le i \le N_x + 1$ and $0 \le j \le N_y + 1$ with i and j integers. Note that the cells $I_{0,j}$, $I_{N_x+1,j}$, $I_{i,0}$, and I_{i,N_y+1} are actually outside the computational domain $(x_{min}, x_{max}) \times (y_{min}, y_{max})$. These are called *ghost cells* and are necessary for implementing the boundary conditions. The model variables $(\varphi_p(x_i, y_j), c_q(x_i, y_j))$ are discretized as $(\varphi_{p,(i,j)}^n, c_{q,(i,j)}^n)$, where the subscripts (i, j) indicate that the values are defined at the center of the grid cell $I_{i,j}$ and the superscript n denotes the nth time step. Using s as the time-step size, Eqs. (8.1)–(8.3) and (8.5) are discretized using the following central difference form [697, 696]:

$$\varphi_{T,(i,j)}^n - \varphi_{T,(i,j)}^{n-1} = s\nabla_d \cdot \left[M\left(\frac{\varphi_{T,(i,j)}^n + \varphi_{T,(i,j)}^{n-1}}{2} \right) \nabla_d \mu_{(i,j)}^{n-1/2} \right]$$

$$+ \frac{s}{2}\left(S_{T,(i,j)}^n + S_{T,(i,j)}^{n-1} \right)$$

$$- \frac{s}{2}\left\{ \nabla_d \cdot \left[\left(\mathbf{u}_{S,(i,j)}^n \varphi_{T,(i,j)}^n \right) + \left(\mathbf{u}_{S,(i,j)}^{n-1} \varphi_{T,(i,j)}^{n-1} \right) \right] \right\},$$ (8.9)

$$\varphi_{p,(i,j)}^{n} - \varphi_{p,(i,j)}^{n-1} = s\nabla_{\mathrm{d}} \cdot \left[M\left(\frac{\varphi_{p,(i,j)}^{n} + \varphi_{p,(i,j)}^{n-1}}{2} \right) \nabla_{\mathrm{d}}\mu_{(i,j)}^{n-1/2} \right]$$

$$+ \frac{s}{2}\left(S_{p,(i,j)}^{n} + S_{p,(i,j)}^{n-1} \right)$$

$$- \frac{s}{2}\left\{ \nabla_{\mathrm{d}} \cdot \left[\left(\mathbf{u}_{\mathrm{S},(i,j)}^{n}\varphi_{p,(i,j)}^{n} \right) + \left(\mathbf{u}_{\mathrm{S},(i,j)}^{n-1}\varphi_{p,(i,j)}^{n-1} \right) \right] \right\}, \quad (8.10)$$

$$\mu_{(i,j)}^{n-1/2} = \frac{1}{2}\left[f_{\mathrm{c}}'\left(\varphi_{\mathrm{T},(i,j)}^{n} \right) + f_{\mathrm{c}}'\left(\varphi_{\mathrm{T},(i,j)}^{n-1} \right) \right] - f_{\mathrm{e}}'\left(\varphi_{\mathrm{T},(i,j)}^{n-1} \right)$$

$$- \frac{\epsilon^2}{2}\left(\nabla_{\mathrm{d}}^2\varphi_{\mathrm{T},(i,j)}^{n} + \nabla_{\mathrm{d}}^2\varphi_{\mathrm{T},(i,j)}^{n-1} \right), \quad (8.11)$$

$$c_{q,(i,j)}^{n} - c_{q,(i,j)}^{n-1} = s\nabla_{\mathrm{d}} \cdot \left[\frac{D_q}{2}\nabla_{\mathrm{d}}\left(c_{q,(i,j)}^{n} + c_{q,(i,j)}^{n-1} \right) \right]$$

$$+ \frac{s}{2}\left(\Gamma_{q,(i,j)}^{n} + \Gamma_{q,(i,j)}^{n-1} \right). \quad (8.12)$$

For the steady-state or fast-time-scale version of Eq. (8.12) we adopt a fully implicit method, thus enhancing the numerical stability of the solver:

$$0 = \nabla_{\mathrm{d}} \cdot \left(D_q \nabla_{\mathrm{d}}c_{q,(i,j)}^{n} \right) + \Gamma_{q,(i,j)}^{n}. \quad (8.13)$$

Note that instead of directly substituting μ into Eqs. (8.9) and (8.10), we discretize Eq. (8.3) separately as in Eq. (8.11) to simplify the mathematical expression and that f_{c} and f_{e} are convex functions exemplified in Eq. (5.25) and (5.26); the two components of f are treated differently for numerical stability. Furthermore, the discretization of the f_{e} term in Eq. (8.11) is first-order accurate in time; to make it second-order, we may use an Adams–Bashforth discretization, replacing $f_{\mathrm{e}}'(\varphi_{\mathrm{T},(i,j)}^{n-1})$ with $f_{\mathrm{e}}'(3\varphi_{\mathrm{T},(i,j)}^{n}/2 - \varphi_{\mathrm{T},(i,j)}^{n-1}/2)$. Finally, the discrete gradient operator ∇_{d} will be defined explicitly in Section 8.2.1. Note that in both Eqs. (8.12) and (8.13) the diffusion coefficient D_q is assumed constant; a functional D_q can be discretized following the derivation in Section 8.2.1.

8.2.1   The discrete gradient operator ∇_{d}

In the discretized equations (8.10)–(8.13) the gradients and the divergences are expressed in terms of a discrete operator, ∇_{d}. Here, we explicitly define this operator. Let us consider its appearance first in a gradient and then in a divergence. The advection terms need special treatment to avoid oscillatory numerical errors at the tumor–host interface. We will discuss advection in Section 8.2.2.

Assuming that $c = c(x, y)$ is a scalar function, its discretized gradient, the vector $\nabla_{\mathrm{d}}c_{(x,y)}$, is defined as follows:

$$\nabla_{\mathrm{d}}c_{(i,j)} = \left(\frac{c_{(i+1/2,j)} - c_{(i-1/2,j)}}{h}, \frac{c_{(i,j+1/2)} - c_{(i,j-1/2)}}{h} \right). \quad (8.14)$$

The subscript (i, j) indicates that the gradient is defined at the cell center of $I_{i,j}$, while $(i + 1/2, j)$, $(i - 1/2, j)$, $(i, j + 1/2)$, and $(i, j - 1/2)$ are the values at the right, left, top, and bottom cell-edge centers. Since we are using a central difference scheme, the values are only explicitly defined at the cell centers. To obtain the value at a cell-edge center, we may use linear interpolation; for example, we set $c_{(i+1/2,j)} \equiv (c_{(i+1,j)} + c_{(i,j)})/2$. Conversely, if the values are instead defined on the cell edges, we may also obtain the cell-center values by interpolating the cell-edge center values.

The definition of ∇_d in a divergence is similar. Given a two-dimensional vector $\mathbf{u}(x, y) = (u(x, y), v(x, y))$, its discrete divergence is expressed as:

$$\nabla_d \cdot \mathbf{u}_{(i,j)} = \frac{u_{(i+1/2,j)} - u_{(i-1/2,j)}}{h} + \frac{v_{(i,j+1/2)} - v_{(i,j-1/2)}}{h}, \tag{8.15}$$

where the subscripts have the same meaning as in Eq. (8.14). With both the discrete gradient and the discrete divergence defined, we may define and derive the discrete Laplacian Δ_d:

$$\Delta_d c_{(i,j)} = \nabla_d \cdot \nabla_d c_{(i,j)} = \frac{c_{(i+1,j)} + c_{(i-1,j)} + c_{(i,j+1)} + c_{(i,j-1)} - 4c_{(i,j)}}{h^2}, \tag{8.16}$$

which is useful for expressing the diffusion terms if the diffusion coefficient is a constant.

8.2.2 Treatment of the advection terms

Various numerical schemes have been proposed to discretize the advective flux, which plays an essential role in many PDE systems and especially in the development of discontinuities or shock waves. Classical discretization methods, such as central difference approximations, have the disadvantage of causing unphysical oscillations across discontinuities (or near discontinuities) known as the Gibbs phenomenon [277]. Upwind schemes discretize the advective flux by capturing the upwind direction, to which the flux is heading, to integrate the PDE numerically. To suppress the Gibbs phenomenon, Harten *et al.* proposed an essentially nonoscillatory (ENO) scheme based on the Godunov upwind scheme, which achieves an accuracy of arbitrarily high order [304, 611, 612]. Efforts to improve the accuracy and efficiency of the ENO scheme have led to the derivation of the weighted essentially nonoscillatory (WENO) scheme [349, 420]. Since solutions to the multiphase mixture model (Eqs. (8.1)–(8.5)) may have sharp gradients at the interfaces of the tumor-cell species, we shall adopt an accurate shock-capturing scheme, such as the ENO or the WENO scheme, to discretize the convective flux. In this section, we first use a second-order-accurate upwinding ENO scheme to express the advection terms in Eq. (8.10), from which a third-order-accurate WENO scheme can be derived to replace the ENO scheme.

Let us start by writing down the discretized advection term explicitly using the definition of the discrete divergence in Eq. (8.15):

$$\nabla_d \cdot \left(\mathbf{u}_{(i,j)} \varphi_{(i,j)} \right) = \nabla_d \cdot (\mathbf{u}\varphi)_{(i,j)}$$

$$= \frac{(u\varphi)_{(i+1/2,j)} - (u\varphi)_{(i-1/2,j)}}{h} + \frac{(v\varphi)_{(i,j+1/2)} - (v\varphi)_{(i,j-1/2)}}{h}, \tag{8.17}$$

where $\mathbf{u} = (u, v)$. The key idea of ENO schemes is to use the stencil that provides the smoothest approximation of the numerical fluxes. This involves several logical statements. Here, we use the evaluation of $(u\varphi)_{i+1/2,j}$ in Eq. (8.17) to exemplify the procedure of the second-order ENO scheme:

$$k = \begin{cases} i & \text{if } u_{(i+1/2,j)} \geq 0, \\ i+1 & \text{otherwise,} \end{cases} \tag{8.18}$$

$$f^{(1)}_{(i+1/2,j)} = u_{(k,j)}\varphi_{(k,j)}, \tag{8.19}$$

$$C_a = \frac{u_{(k,j)}\varphi_{(k,j)} - u_{(k-1,j)}\varphi_{(k-1,j)}}{h}, \tag{8.20}$$

$$C_b = \frac{u_{(k+1,j)}\varphi_{(k+1,j)} - u_{(k,j)}\varphi_{(k,j)}}{h}, \tag{8.21}$$

$$C_c = \begin{cases} C_a & \text{if } |C_a| \leq |C_b|, \\ C_b & \text{otherwise,} \end{cases} \tag{8.22}$$

$$(u\varphi)_{(i+1/2,j)} = f^{(2)}_{(i+1/2,j)} = f^{(1)}_{(i+1/2,j)} + \tfrac{1}{2}hC_c\left[1 - 2(k-1)\right]. \tag{8.23}$$

The first logical statement (Eq. (8.18)) selects the upwinding direction while the second (Eq. (8.22)) is used to find the smoothest approximation.

Further, to derive the WENO formula let us assume that $k = i$ is the upwinding direction in the ENO scheme. For the case $k = i + 1$, the procedure is similar. For $k = i$, we may rewrite Eqs. (8.19)–(8.23) as

$$(u\varphi)_{(i+1/2,j)} = f^{(2)}_{(i+1/2,j)} = \begin{cases} -\tfrac{1}{2}(u\varphi)_{(i-1,j)} + \tfrac{3}{2}(u\varphi)_{(i,j)} & \text{if } C_a < C_b, \\ \tfrac{1}{2}(u\varphi)_{(i,j)} + \tfrac{1}{2}(u\varphi)_{(i+1,j)} & \text{otherwise,} \end{cases}$$

$$= q_\ell\left((u\varphi)_{(i+\ell-1,j)}, (u\varphi)_{(i+\ell,j)}\right), \tag{8.24}$$

where

$$q_\ell(g_0, g_1) = \sum_{m=0}^{1} a_{\ell,m} g_m \tag{8.25}$$

with $a_{0,0} = -\tfrac{1}{2}$, $a_{0,1} = \tfrac{3}{2}$, $a_{1,0} = a_{1,1} = \tfrac{1}{2}$, and $\ell = 0$ or 1. The ENO scheme (Eqs. (8.19)–(8.23)) selects the smoothest q_ℓ from the candidates (q_0 and q_1 in our case), and this is the key idea in circumventing discontinuities and suppressing the Gibbs phenomenon. However, in a smooth region, we can use all the candidates to obtain a linear combination of $(u\varphi)_{(i-1,j)}$, $(u\varphi)_{(i,j)}$, and $(u\varphi)_{(i+1,j)}$. With an appropriate weight on each stencil we may approximate $(u\varphi)_{(i+1/2,j)}$ to third-order accuracy, in contrast with the second-order accuracy of the ENO scheme above. Weighted essentially nonoscillatory schemes provide algorithms to weight the stencils in order (i) to achieve a higher-order accuracy using lower-order ENO stencils and (ii) to exclude discontinuous stencils by giving them essentially zero weight.

Following the derivation in [349, 420], we may give a third-order-accurate WENO formula for the flux $(u\varphi)_{(i+1/2,j)}$:

$$(u\varphi)_{(i+1/2,j)} = \hat{f}^{(3)}_{(i+1/2,j)} = \sum_{\ell=0}^{1} \omega_\ell q_\ell, \tag{8.26}$$

where q_ℓ is defined in Eq. (8.25). The weight coefficient ω_ℓ is as follows:

$$\omega_\ell = \frac{\alpha_\ell}{\alpha_0 + \alpha_1}, \tag{8.27}$$

where

$$\alpha_\ell = \frac{\gamma_\ell}{(\epsilon + IS_\ell)^p} \tag{8.28}$$

for $\ell = 0, 1$. The values of γ_ℓ are respectively $\gamma_0 = \frac{1}{3}$ and $\gamma_1 = \frac{2}{3}$, which constitute the *optimal weighting*; $\epsilon = 10^{-6}$ is a small number that prevents the denominator from becoming zero; the empirical power value $p = 2$ is sufficiently high to distinguish continuous stencils from discontinuous stencils for the flux approximation. The smoothness of the flux function on the stencils are measured by IS_ℓ:

$$IS_\ell = \left((u\varphi)_{(i+\ell,j)} - (u\varphi)_{(i+\ell-1,j)}\right)^2, \tag{8.29}$$

the 2-norm of the first undivided difference. A discontinuity results in a larger IS_ℓ measurement and, hence, a lower-weight ω_ℓ, which essentially excludes discontinuous stencils and suppresses Gibbs phenomena.

Recall that Eqs. (8.24)–(8.29) were derived by assuming that $k = i$ is the upwinding direction. For the case where $k = i + 1$ is the upwinding direction, the WENO formula can be obtained similarly, by replacing $(u\varphi)_{(i-1,j)}$ by $(u\varphi)_{(i+2,j)}$, $(u\varphi)_{(i,j)}$ by $(u\varphi)_{(i+1,j)}$, and $(u\varphi)_{(i+1,j)}$ by $(u\varphi)_{(i,j)}$. To simplify the mathematical notation, from now on we write for, the flux $(u\varphi)_{(i+1/2,j)}$ along the x-direction

$$(u\varphi)_{(i+1/2,j)} = W_{(i+1/2,j)}(u\varphi) \tag{8.30}$$

and, similarly, for the flux $(v\varphi)_{(i,j+1/2)}$ along the y-direction

$$(v\varphi)_{(i,j+1/2)} = W_{(i,j+1/2)}(v\varphi), \tag{8.31}$$

where $W(.)$ represents the upwind WENO flux reconstruction given above.

Readers may notice that, for the cell next to the boundaries of the computational domain, the evaluation of the WENO fluxes involves the cells immediately outside the domain, the *ghost cells*. The values in the ghost cells are not updated by the model equations; rather, they are defined by the boundary conditions. In Section 8.2.3 we discuss the discretization of the boundary conditions, which determines the values in the ghost cells.

8.2.3   Discretization of the boundary conditions

The formulation of boundary conditions depends on the problem; different variables and different physical circumstances may give rise to different boundary conditions. As mentioned earlier, the multigrid method can readily be used with various types of boundary condition. In this sub section, we describe the discretization of two common types of boundary condition, Dirichlet and Neumann. A third type, Robin boundary conditions, is a linear combination of the other two.

Dirichlet boundary conditions explicitly specify the values of the variables on the boundaries of the computational domain:

$$c = \eta \quad \text{on} \quad \partial\Omega, \tag{8.32}$$

where c is the model variable, η is the boundary value, and $\partial\Omega$ denotes the boundaries of the computational domain. Such boundary conditions may be used for the nutrient concentration, which can diffuse into the system from afar; thus, we may assume that the nutrient concentration on the boundaries equals a far-field nutrient level provided that the tumor, as a nutrient sink, remains reasonably far away from the boundaries. Another example is the pressure, which requires the definition of a reference point; hence, a convenient reference point for computation is setting the pressure level to be zero on the boundaries. For cell-centered variables $c_{i,j}^n$, the discrete form of the Dirichlet boundary conditions can be expressed as

$$\frac{1}{2}\left(c_{0,j}^n + c_{1,j}^n\right) = \eta_{\text{L}}^n, \tag{8.33}$$

$$\frac{1}{2}\left(c_{N_x,j}^n + c_{N_y+1,j}^n\right) = \eta_{\text{R}}^n, \tag{8.34}$$

$$\frac{1}{2}\left(c_{i,0}^n + c_{i,1}^n\right) = \eta_{\text{B}}^n, \tag{8.35}$$

$$\frac{1}{2}\left(c_{i,N_y}^n + c_{i,N_y+1}^n\right) = \eta_{\text{T}}^n, \tag{8.36}$$

where η_{L}^n, η_{R}^n, η_{B}^n, and η_{T}^n respectively represent the left, right, bottom, and top boundary conditions at the nth time step. We interpolate the cell-center values to cell edges, where the results are matched according to the boundary conditions; from this the values of the ghost points $i, j = 0, i = N_x + 1, j = N_y + 1$ are determined.

Neumann boundary conditions define the first-order derivatives of the variables across the boundaries:

$$\frac{\partial c}{\partial \mathbf{n}} \equiv \nabla c \cdot \mathbf{n} = \eta \quad \text{on} \quad \partial\Omega, \tag{8.37}$$

where $\mathbf{n}$ is a normal vector pointing outward on the boundary $\partial\Omega$, and here η represents the boundary value of the normal derivative. In many cases the normal derivative of a variable is equivalent to its flux; hence, homogeneous Neumann boundary conditions, for which $\eta = 0$ in Eq. (8.37), are also known as no-flux boundary conditions. In our tumor-growth model, if the computational domain boundaries are reasonably far from the growing tumor then we may assume that there is no flux of tumor-cell volume fractions across the computational boundaries and adopt no-flux boundary conditions for the tumor-cell volume fractions. Assuming a rectangular domain with cell-center variables $c_{i,j}^n$, Neumann boundary conditions can typically be discretized as

$$\frac{1}{h}\left(c_{1,j}^n - c_{0,j}^n\right) = \eta_{\text{L}}^n, \tag{8.38}$$

$$\frac{1}{h}\left(c_{N+1,j}^n - c_{N,j}^n\right) = \eta_{\text{R}}^n, \tag{8.39}$$

$$\frac{1}{h}\left(c_{i,1}^n - c_{i,0}^n\right) = \eta_{\text{B}}^n, \tag{8.40}$$

$$\frac{1}{h}\left(c_{i,N+1}^n - c_{i,N}^n\right) = \eta_{\text{T}}^n, \tag{8.41}$$

which determines the ghost points.

Robin boundary conditions are typically defined as follows:

$$Ac + B\frac{\partial c}{\partial \mathbf{n}} = \eta \qquad \text{on} \qquad \partial\Omega, \tag{8.42}$$

which is a linear combination of Eqs. (8.32) and (8.37). Hence, it is straightforward to formulate the corresponding discrete Robin boundary conditions using linear combinations of Eqs. (8.33)–(8.36) and Eqs. (8.38)–(8.41).

At every time step $c_{1,j}^n$, $c_{N_x,j}^n$, $c_{i,1}^n$, and c_{i,N_y}^n, located inside the computational domain, are updated by solving the model equations for the ghost points given above.

8.2.4   Mesh discretization across interfaces

For a numerical method accurately to capture the evolution of a moving boundary in a phase-field model, a large enough number (five to six) of grid cells must span the interfacial region, where the thickness is controlled by the interfacial thickness parameter ϵ and the double-well potential f in Eq. (8.3). As illustrated in the upper panel of Figure 8.1, an interface sharper than the mesh resolution becomes "invisible" when it is traveling between two grid points, in which region no grid points can describe its movement. However, an interface that diffuses across more grid cells can be described more accurately, as shown in the lower panel of Figure 8.1. Physically, the interfacial thickness of the phase-field model can be interpreted as the region of mixed tumor and host cells. However, this physical mixing region may sometimes be negligible on the spatial scale of bulk tumor growth and thus too small for reasonable mesh discretization [482].

8.2.5   Treatment of the source terms

The specific formulations of the source terms S_{T}, S_p, and Γ_q in Eqs. (8.1)–(8.5) may vary for different model constitutions. In this subsection, we use the source terms of the single-viable-species model in Eqs. (5.31), (5.32) to illustrate the treatment of some common difficulties regarding the source terms of multiphase mixture models.

Owing to the diffuse nature of multiphase mixture models, a common phenomenon is that some tumor cells may leak from the bulk tumor. These may then start to grow and form satellite tumors proximal to the primary tumor. While this phenomenon may be

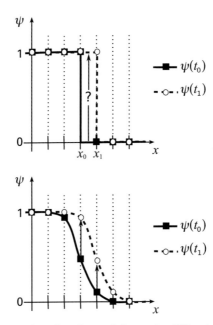

Figure 8.1 Evolution of (top) a sharp interface and (bottom) a diffuse interface with respect to mesh discretization in one dimension. The solid lines represent the solution for the phase variable ψ at time t_0; the solid squares denote the values at the grid points. The broken lines represent the solution for ψ at t_1, where $t_1 > t_0$; the open circles denote the values on the grid points. The upper panel shows that between t_0 and t_1 the sharp interface has "disappeared" between two mesh grid points, and it is unclear when the value at the grid point x_1 should make the jump from 0 to 1. In contrast, the lower panel shows that the mesh grid points are able to capture the evolution of the moving diffuse interface, provided that there are enough grid cells across the interfacial area.

physical under some circumstances, in fact it is more appropriate to describe the leaked cells using a discrete model, thus forming a hybrid continuum–discrete model as already discussed in Chapter 7. Here, for a pure continuum model we suppress such cell growth by introducing a cutoff function in the source terms describing cell proliferation. We rewrite Eqs. (5.31), (5.32) for the numerical solver as

$$S_p = \bar{\lambda}_{\mathrm{M},p} \frac{n}{\bar{n}_\infty} \varphi_p G\left(\varphi_{\mathrm{T}}\right) - \bar{\lambda}_{\mathrm{A},p} \varphi_p - \bar{\lambda}_{\mathrm{N},p} \mathcal{H}(\bar{n}_{\mathrm{N},i} - n)\varphi_p,$$
$$N_{\mathrm{T}} \geq p > 0 \quad \text{and} \quad p \neq D \tag{8.43}$$

$$S_{\mathrm{D}} = \sum_{\substack{p=1 \\ p \neq D}}^{N_{\mathrm{T}}} \left(\bar{\lambda}_{\mathrm{A},p} + \bar{\lambda}_{\mathrm{N},p} \mathcal{H}(\bar{n}_{\mathrm{N},p} - n) \right) \varphi_p - \bar{\lambda}_{\mathrm{L}} \varphi_{\mathrm{D}}, \tag{8.44}$$

where n is the nutrient concentration, solved as one of the c_q from Eq. (8.5), and, for the total tumor volume fraction,

$$S_{\mathrm{T}} = \sum_{\substack{p=1 \\ p \neq D}}^{N_{\mathrm{T}}} S_p + S_{\mathrm{D}} = \sum_{\substack{p=1 \\ p \neq D}}^{N_{\mathrm{T}}} \bar{\lambda}_{\mathrm{M},p} \frac{n}{\bar{n}_\infty} \varphi_p G\left(\varphi_{\mathrm{T}}\right) - \bar{\lambda}_{\mathrm{L}} \varphi_{\mathrm{D}}, \tag{8.45}$$

where we multiply the cutoff function $G(\varphi)$ by the proliferation terms, and $\mathcal{H}(\varphi)$ is the Heaviside step function. Note that φ_p here does not include the volume fraction of the host cells; hence, N_T is equivalent to $N - 1$ in Chapter 5. We define $G\varphi$ as follows:

$$
G(\varphi) = \begin{cases} 1 & \text{if } \dfrac{3\epsilon_\varphi}{2} \le \varphi, \\[2mm] \dfrac{\varphi}{\epsilon_\varphi} - \dfrac{1}{2} & \text{if } \dfrac{\epsilon_\varphi}{2} \le \varphi < \dfrac{3\epsilon_\varphi}{2}, \\[2mm] 0 & \text{if } \varphi \le \dfrac{\epsilon_\varphi}{2}, \end{cases} \tag{8.46}
$$

where ϵ_φ is a small number setting a minimum volume fraction above which tumor cells may proliferate. This cutoff function prevents the growth of a small perturbation $\varphi < \epsilon_\varphi/2$, avoiding the accumulation of mass after a long simulation. The Heaviside function is commonly used in source terms, since some phenomena, such as necrosis in Eqs. (8.43), (8.44), occur with respect to a certain threshold. The exact Heaviside function satisfies $\mathcal{H}(x) = 1$ when $x > 0$ and $\mathcal{H}(x) = 0$ when $x < 0$. However, the discontinuity at $x = 0$ may cause problems for a numerical solver. Therefore, we approximate it by the smooth function,

$$
\mathcal{H}(x) = \frac{1}{2} + \frac{1}{2} \tanh\left(\frac{x}{\sqrt{2}\epsilon}\right). \tag{8.47}
$$

Thus far we have discretized the tumor growth model equations (8.9)–(8.12) on a uniform mesh with grid size h. To solve the model using a multigrid method, a multilevel mesh hierarchy needs to be constructed. This is presented in the next section.

8.3   Multigrid mesh hierarchy and block-structured adaptive mesh

In Section 8.2 we discretized the model equations using a cell-center finite difference scheme on a single uniform Cartesian mesh. To implement a multigrid method, multiple levels of coarse-to-fine meshes are constructed, on which we formulate the discretized equations (8.10)–(8.13) accordingly. As mentioned earlier, the advantage of adopting a finite difference discretization for a multigrid method is that the stencil does not change with the mesh size. Using this advantage the multilevel mesh structure can easily be made adaptive, and a *block-structured adaptive mesh* can be constructed. Block-structured adaptive-mesh discretization dynamically generates local refined-mesh patches to address localized small-scale phenomena, which characterize many multiscale problems.

Figure 8.2 shows the multilevel mesh structure, as well as the construction, of the two-dimensional block-structured adaptive mesh which will be described in more detail in Section 8.3.1. The values on the dynamically generated refined-mesh grids need to be initialized by transferring information from other existing mesh grids; we discuss this in Section 8.3.2. Thus in this section we are setting up the

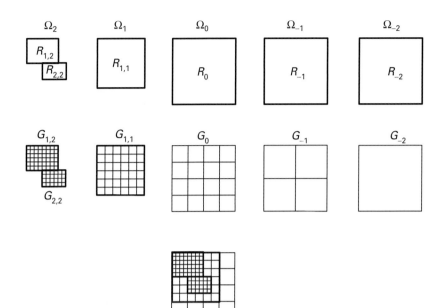

Figure 8.2 A block-structured locally refined adaptive multigrid mesh. The various meshes are denoted by Ω_L, where L is the mesh level. We call Ω_0 the root level, above which (i.e., for $L > 0$) the meshes do not cover the entire computational domain and only some local block regions are refined. The levels $L > 0$ are called refinement levels. Each refinement-level mesh may consist of multiple block-refinement regions, called *patches* and denoted by $R_{k,L}$. Reprinted from *J. Comput. Phys.*, 226, Wise *et al.* [695], p. 424, © 2007, with permission from Elsevier.

adaptive multigrid mesh structure for the MLAT–FAS multigrid method presented in Section 8.4.

8.3.1   Block-structured adaptive-mesh refinement

As in Section 8.2, we consider a rectangular computational domain

$$\Omega = (x_{min}, x_{max}) \times (y_{min}, y_{max}).$$

Figure 8.2 illustrates the multigrid discretization of the computational domain Ω; multiple levels of meshes are constructed, with the coarsest level on the right and the finest on the left. We denote the domain of mesh level L by Ω_L, where $L_{min} \le L \le L_{max}$; for the example in Figure 8.2, $L_{min} = -2$ and $L_{max} = 2$. The $L = 0$ mesh level is called the *root mesh* and is the finest uniform mesh covering the whole domain. Furthermore, patches of block-structured mesh grids can be constructed on Ω. Each patch occupies a rectangular subdomain of Ω, denoted by $R_{k,L}$, indicating that it is the kth patch on Ω_L, as shown in the top row of Figure 8.2. The patch $R_{k,L}$ is discretized into an $N_x^{(k,L)} \times N_y^{(k,L)}$ grid $G_{k,L}$ with grid spacing h_L, as depicted in the middle row of Figure 8.2. Patches at the same mesh level do not overlap, i.e., $R_{k_1,L} \cap R_{k_2,L} = \emptyset$ for $k_1 \ne k_2$. Later we will introduce ghost cells for each patch in order to set the boundary conditions across patches; note that ghost cells are not in the interior of a patch and hence are allowed to overlap. Furthermore, let us explicitly define Ω_L:

$$\Omega_L \equiv \bigcup_{k=1}^{k_{L,\max}} R_{k,L} \subseteq \Omega. \tag{8.48}$$

For $L \leq 0$ the grid mesh structure is essentially rectangular and uniform and consists of only one patch, which covers the entire computational domain: $R_{1,L} = \Omega_L = \Omega$. To simplify the notation we will write the patches $R_{1,L}$ as R_L for $L \leq 0$ and similarly the grids $G_{1,L}$ as G_L. For $L > 0$, the patches $R_{k,L}$ represent localized mesh refinement, and there can be multiple patches at one mesh level. For example, in Figure 8.2 we have $R_{1,2}$ and $R_{2,2}$ on Ω_2. Consequently, we call Ω_0 the *root level*, on which mesh-refined patches are constructed; the corresponding grid G_0 is called the *root grid*. The bottom row of Figure 8.2 shows the composite mesh for this block-structured adaptive-mesh discretization.

The advantage of block-structured adaptive mesh is that each grid $G_{k,L}$ corresponding to the patch $R_{k,L}$ is a rectangular uniform mesh grid. The FAS method, originally derived for a uniform-mesh discretization, can be applied directly to patch grids without reformulation of the approximating stencils. Hence, we may easily make the FAS method adaptive within the block-structured MLAT framework (Section 8.4). Note that by using the root level as the finest mesh level, i.e., $L_{\max} = 0$, we obtain a uniform multigrid mesh structure. Hence the uniform-mesh FAS multigrid method can simply be deduced from the MLAT–FAS multigrid scheme. Also note that Figure 8.2 implies that the mesh grid sizes of mesh $L - 1$ and mesh L are related by

$$h_{L-1} = \frac{h_L}{2}, \tag{8.49}$$

which is called the *standard coarsening* [662]. A general multigrid method is not restricted to this choice of mesh coarsening or refinement. Readers may refer to Sections 2.3.1 and A.7.1 of [662] for alternative options for coarsening rules.

To maintain systematically the hierarchy of the meshes during refinement, the meshes must be nested. Each level-L grid $G_{k_1,L}$ with grid spacing h_L completely covers a subgrid $\hat{G}_{k_1,L} \subset G_{k_2,L-1}$ with grid spacing h_{L-1} at level $L - 1$. In the hierarchy structure, $G_{k_1,L}$ is a "child" of $G_{k_2,L-1}$ while the latter is the "parent" of the former. This hierarchy structure also implies that the level-L domains satisfy

$$\Omega_L \subseteq \Omega_{L-1} \quad \text{for } L_{\min} + 1 \leq L \leq L_{\max}. \tag{8.50}$$

Note that this is more general and less restrictive than the "definitive" nesting requirement for a mesh generated in CHOMBO [140]. Using the standard coarsening rule (8.49) the size of the subgrid $\hat{G}_{k_1,L}$ is $(N_x^{(k_1,L)}/2) \times (N_y^{(k_1,L)}/2)$, where $N_d^{k_1,L}$ $(d = x, y)$ are the dimensions of $G_{k_1,L}$.

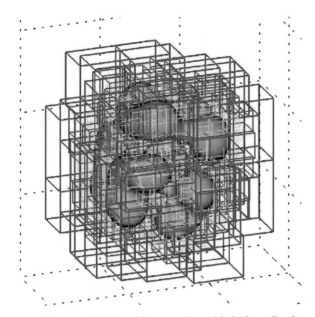

Figure 8.3 A snapshot of the $\varphi_T = 0.5$ isosurface, together with the bounding boxes of a three-dimensional block-structured adaptive mesh. The variable φ_T represents the volume fraction of the total tumor cells. The root level ($L = 0$) of the adaptive mesh is a 32^3 mesh. There are three refinement levels above it; each level of refinement has half the grid spacing of the one on the level below. Therefore, the finest level ($L = 3$) has the equivalent resolution of a 256^3 mesh. Reprinted from Wise *et al.* [696].

The construction of a multilevel refined mesh begins at the root-level grid, G_0. The following pseudocode outlines the process.

BLOCK-STRUCTURED MESH-REFINING ALGORITHM

Loop: **for** $L = 0$ **until** $L_{\max} - 1$

- tag grids to be refined on Ω_L,
- cover the tagged grids using multiple nonoverlapping rectangular patches $R_{k_{L+1}, L+1} \subset \Omega_L$,
- construct the refined-mesh grids $G_{k_{L+1}, L+1}$ on $R_{k_{L+1}, L+1}$,
- $\Omega_{L+1} \equiv \bigcup_{k_{L+1}=1}^{k_{L+1,\max}} R_{k_L, L}$

end for Loop

There exist various *a priori* and *a posteriori* criteria to tag grids for mesh refinement, which we will discuss later in Section 8.4.5 in the context of the MLAT–FAS method. Figure 8.3 shows an example of the block-structured adaptive Cartesian mesh in three dimensions; the rectangular boxes represent refined-mesh patches at various levels. We next discuss the initialization of the values on the dynamically generated refined-mesh patches.

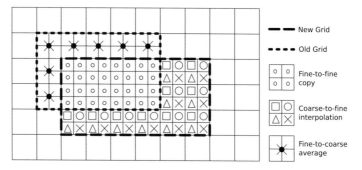

Figure 8.4 Example of dynamic mesh refinement: transferring data from an old level-1 grid and the root-level grid to a newly constructed level-1 grid. In the region where the new and old level-1 meshes overlap, the values of the discretized variables are directly copied from the old mesh to the new one. In the region where the new mesh finds no corresponding old mesh grid, the values on the new mesh grid are obtained by interpolating the values on the coarser root-level mesh. In the region where the old level-1 mesh finds no corresponding new mesh grid, the values on the old mesh are used to correct the values on the coarser root-level mesh. Reprinted from *J. Comput. Phys.*, vol. 226, Wise *et al.* [695], p. 426, © 2007, with permission from Elsevier.

8.3.2   Initialization of the refined meshes

During the course of simulation, meshes finer than the root level are dynamically constructed and destroyed. The values on an existing mesh need to be transferred to newly generated meshes, thus initializing the new meshes. Moreover, the smoothing procedure of the FAS scheme requires the construction of ghost cells for each patch. While the ghost cells of the meshes covering the entire computational domain Ω are set up by the physical boundary conditions (Section 8.2.3), the refined patches in the interior of Ω need to obtain information from existing mesh grids.

First, we discuss the details of how to transfer the data to initiate the values on a newly created patch for the dynamical mesh refinement scheme. Figure 8.4 shows an example of dynamic mesh refinement; in this example, the coarser mesh is the root-level mesh, and the two finer meshes are at level-1. The finer mesh enclosed by the broken line is newly constructed while the finer mesh enclosed by the dotted line is an old mesh, which was constructed at the previous time step and will be destroyed after the data have been transferred to the new mesh. Note that the overlaps comprise "fine-to-fine copy" regions, "coarse-to-fine interpolation" regions, and "fine-to-coarse average" regions. The "fine-to-fine copy" region indicates where the new level-1 patch overlaps the old level-1 patch. The algorithm is straightforward: copy the data from the old grid cells to the corresponding grid cells on the new mesh. "Coarse-to-fine interpolation" is applied to regions on the new mesh that do not overlap the same-level old mesh. For such regions in Figure 8.4, the values on the new level-1 mesh grid are obtained by interpolating the values on the coarser-mesh grid. The "fine-to-coarse average" is used for regions on the old level-1 mesh that do not overlap the new level-1 mesh. For such regions, the values on the old level-1 mesh are averaged to correct the values on the corresponding level-0 grid, given that the values on the finer-level grids are more accurate than those on

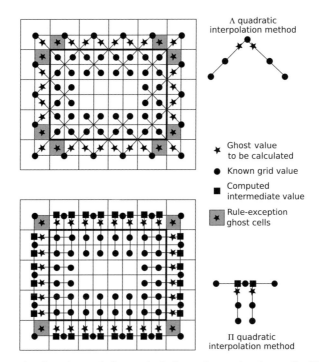

Figure 8.5 Schematic of two interpolation methods for setting up the ghost cells. The Λ-method (top) interpolates the values on the ghost cells of the finer mesh using two finer grid values and one coarser grid value. However, there are eight rule-exception ghost cells on a rectangular mesh for this method, owing to the lack of information. The Π-method (bottom) first calculates intermediate values using the values on the coarser grid; then the ghost-cell values are interpolated from these intermediate values and the finer-grid values. For this method, there exists four rule-exception ghost cells on each rectangular mesh. Reprinted from *J. Comput. Phys.*, vol. 226, Wise *et al.* [695], p. 427, © 2007, with permission from Elsevier.

the coarser-level grid. The coarse-to-fine interpolation operators and the fine-to-coarse average operators are respectively the same as the interpolation and restriction operators of the FAS multigrid method and will be introduced in Section 8.4.4.

After the values on all patches of a refinement mesh level are initialized, we set up the values at the corresponding ghost cells by transferring the data from the existing meshes. Since a coarser-level mesh completely covers a finer one, we may set the ghost cell values of the patches on a finer mesh by interpolating the values on the coarser-level mesh as well as on the patch itself. Illustrated in Figure 8.5 are two quadratic interpolation rules, Λ- and Π-methods. The Λ-method, which uses two points on the patch and one point from the coarser mesh, is efficient to compute and generally more accurate; however, there exist eight rule-exception ghost cells for each rectangular patch owing to the lack of data points. The Π-method has only four rule-exception ghost cells. It first uses three points on the coarser mesh to compute an intermediate value; then the intermediate value is used with another two points within the patch to interpolate the value at a ghost cell. Thus, the Π-method requires twice as much computation as the Λ-method. Although neither the Λ-method nor the Π-method is conservative, we

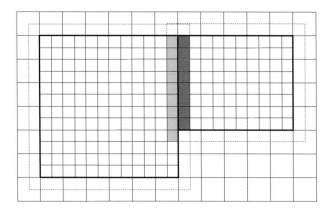

Figure 8.6 Transferring data between adjacent patches to set up the ghost cells of each patch. The ghost cells of one refined patch may overlap the interior cells of adjacent patches. Data on the overlapped interior cells are directly copied to the corresponding ghost cells. Reprinted from *J. Comput. Phys.*, vol. 226, Wise *et al.* [695], p. 427, © 2007, with permission from Elsevier.

found that the mass is maintained to a high degree of accuracy [695]. Other ghost-cell interpolation methods can be found in [84].

Frequently a refined patch is located adjacent to another patch of the same level, and so its ghost cells overlap the grid cells of the adjacent patch, as shown in Figure 8.6. Given that finer grids have higher accuracy in a multigrid scheme, in such a case we copy the data directly from the grid cells of the adjacent patch to the corresponding ghost cells.

The following pseudocode summarizes the process of constructing and initializing the adaptive meshes for the adaptive-mesh refinement (AMR) framework.

AMR PROCESS

Set the mesh grids from the previous time step to $G^{\text{old}}_{k'_L, L}$,
tag the mesh grid cells on Ω_0 for mesh refinement,
Loop: **for** $L = 1$ **until** $L_{\max}$

- construct $G_{k_L, L}$ to cover all the tagged grid cells on Ω_{L-1},
- initialize all $G_{k_L, L}$ by interpolating the data on Ω_{L-1},
- copy data from $G^{\text{old}}_{k'_L, L}$ to the overlapping grid cells on $G_{k_L, L}$,
- restrict the nonoverlapping $G^{\text{old}}_{k'_L, L}$ grid cells to the grid cells on Ω_{L-1},
- set the ghost cells for all $G_{k_L, L}$ by interpolation,
- copy the data if the ghost cells overlap adjacent $G_{k_L, L}$ grid cells,
- tag the mesh grid cells on Ω_L for mesh refinement.

end for Loop
destroy $G^{\text{old}}_{k'_L, L}$.

In Section 8.4 we present the MLAT–FAS scheme, which fits the AMR structure described above.

8.4   Multigrid method on adaptive meshes: the MLAT–FAS scheme

The modified Crank–Nicholson method used for time discretization yields the nonlinear implicit equations (8.9)–(8.12) for the discrete form of the model. There are two approaches to solving such a nonlinear implicit system using multigrid methods. The first uses a global linearization algorithm on the nonlinear system, and so a linear multigrid method can be applied; however, the disadvantage of this technique is that the globally evaluated Jacobian matrix has to be stored in the memory when the numerical solver sweeps through the mesh grids, resulting in a high computational demand. The second approach, which we follow here, uses local linearization within the multigrid method to solve the nonlinear system. We will present a nonlinear *full approximation storage* (FAS) multigrid method within the *multilevel adaptive technique* (MLAT) framework using the block-structured adaptive multilevel multigrid mesh previously constructed in Section 8.3 [75, 379, 380, 662, 695, 696]. The FAS method was first applied to Cahn–Hilliard-type equations by Kim, Kang, and Lowengrub [379, 380]. It was then modified within the MLAT framework to solve strongly anisotropic Cahn–Hilliard equations by Wise *et al.* [695]; the MLAT–FAS scheme was later applied to a CH-type tumor growth model by Wise *et al.* [696, 697].

To introduce the nonlinear MLAT–FAS method for solving the tumor growth model on a block-structured adaptive mesh, in Section 8.4.1 we rewrite the discrete nonlinear implicit equations (8.9)–(8.12), separating the operator terms and the source terms. The operator terms are subject to local linearization in the smoothing step of the nonlinear FAS method. In Section 8.4.2 the overall procedure of the MLAT–FAS multigrid method, known as V-cycles, will be presented, and this is followed in Sections 8.4.3 and 8.4.4 by a detailed description of each V-cycle component. With the MLAT–FAS components defined, we will discuss the criteria for mesh refinement in Section 8.4.5.

8.4.1   Operators and sources

Before introducing the FAS method, let us simplify the mathematical notation and define

$$\psi^{\star}_{(i,j)} = \left(\varphi^{\star}_{T,(i,j)}, \varphi^{\star}_{p,(i,j)}, \mu^{\star-1/2}_{(i,j)}, c^{\star}_{q,(i,j)}, \right), \tag{8.51}$$

where the superscript star is a place-holder for any index. Setting $\psi^{\star} \equiv \{\psi^{\star}_{(i,j)} | \forall i, j\}$, we can write Eqs. (8.9)–(8.12) in the generalized form

$$\mathbf{N}^{n}_{(i,j)}\left(\psi^{n}, \psi^{n-1}\right) = \mathbf{F}^{n}_{(i,j)}\left(\psi^{n}, \psi^{n-1}\right), \tag{8.52}$$

where we have split the equations into source terms, represented by $\mathbf{F}$, and operator terms, represented by $\mathbf{N}$. The components of $\mathbf{F}^{n}_{(i,j)}$ and $\mathbf{N}^{n}_{(i,j)}$ are explicitly defined as follows:

$$
\begin{aligned}
F_{(i,j)}^{(\mathrm{T}),n}\left(\psi^{n}, \psi^{n-1}\right) &= \varphi_{\mathrm{T},(i,j)}^{n-1} + \frac{s\,M}{2}\nabla_{\mathrm{d}}\cdot\left(\varphi_{\mathrm{T},(i,j)}^{n-1}\nabla_{\mathrm{d}}\mu_{(i,j)}^{n-1/2}\right) \\
&\quad + \frac{s}{2}\left(S_{\mathrm{T},(i,j)}^{n} + S_{\mathrm{T},(i,j)}^{n-1}\right) \\
&\quad - \frac{s}{2}\left[\nabla_{\mathrm{d}}\cdot\left(\mathbf{u}_{\mathrm{S},(i,j)}^{n}\varphi_{\mathrm{T},(i,j)}^{n}\right) + \nabla_{\mathrm{d}}\cdot\left(\mathbf{u}_{\mathrm{S},(i,j)}^{n-1}\varphi_{\mathrm{T},(i,j)}^{n-1}\right)\right],
\end{aligned}
\tag{8.53}
$$

$$
\begin{aligned}
F_{(i,j)}^{(p),n}\left(\psi^{n}, \psi^{n-1}\right) &= \varphi_{p,(i,j)}^{n-1} + s\,\nabla_{\mathrm{d}}\cdot\left[M\left(\frac{\varphi_{p,(i,j)}^{n} + \varphi_{p,(i,j)}^{n-1}}{2}\right)\nabla_{\mathrm{d}}\mu_{(i,j)}^{n-1/2}\right] \\
&\quad + \frac{s}{2}\left(S_{p,(i,j)}^{n} + S_{p,(i,j)}^{n-1}\right) \\
&\quad - \frac{s}{2}\left[\nabla_{\mathrm{d}}\cdot\left(\mathbf{u}_{\mathrm{S},(i,j)}^{n}\varphi_{p,(i,j)}^{n}\right) + \nabla_{\mathrm{d}}\cdot\left(\mathbf{u}_{\mathrm{S},(i,j)}^{n-1}\varphi_{p,(i,j)}^{n-1}\right)\right],
\end{aligned}
\tag{8.54}
$$

$$
F_{(i,j)}^{(\mu),n}\left(\psi^{n}, \psi^{n-1}\right) = \frac{1}{2}f_{\mathrm{c}}'\left(\varphi_{\mathrm{T},(i,j)}^{n-1}\right) - f_{\mathrm{e}}'\left(\varphi_{\mathrm{T},(i,j)}^{n-1}\right) - \frac{\epsilon^{2}}{2}\nabla_{\mathrm{d}}^{2}\varphi_{\mathrm{T},(i,j)}^{n-1},
\tag{8.55}
$$

$$
F_{(i,j)}^{(q),n}\left(\psi^{n}, \psi^{n-1}\right) = c_{q,(i,j)}^{n} + \frac{s}{2}\nabla_{\mathrm{d}}\cdot\left(D_{q}\nabla_{\mathrm{d}}c_{q,(i,j)}^{n-1}\right) + \frac{s}{2}\Gamma_{q,(i,j)}^{n-1},
\tag{8.56}
$$

and

$$
N_{(i,j)}^{(\mathrm{T}),n}\left(\psi^{n}, \psi^{n-1}\right) = \varphi_{\mathrm{T},(i,j)}^{n} - \frac{s\,M}{2}\nabla_{\mathrm{d}}\cdot\left(\varphi_{\mathrm{T},(i,j)}^{n}\nabla_{\mathrm{d}}\mu_{(i,j)}^{n-1/2}\right),
\tag{8.57}
$$

$$
N_{(i,j)}^{(p),n}\left(\psi^{n}, \psi^{n-1}\right) = \varphi_{p,(i,j)}^{n},
\tag{8.58}
$$

$$
N_{(i,j)}^{(\mu),n}\left(\psi^{n}, \psi^{n-1}\right) = \mu_{(i,j)}^{n-1/2} - \frac{1}{2}f_{\mathrm{c}}'\left(\varphi_{\mathrm{T},(i,j)}^{n}\right) + \frac{\epsilon^{2}}{2}\nabla_{\mathrm{d}}^{2}\varphi_{\mathrm{T},(i,j)}^{n},
\tag{8.59}
$$

$$
N_{(i,j)}^{(q),n}\left(\psi^{n}, \psi^{n-1}\right) = c_{q,(i,j)}^{n} - \frac{s}{2}\nabla_{\mathrm{d}}\cdot\left(D_{q}\nabla_{\mathrm{d}}c_{q,(i,j)}^{n}\right) - \frac{s}{2}\Gamma_{q,(i,j)}^{n},
\tag{8.60}
$$

where $\mathbf{F} = \left(F^{(\mathrm{T})}, F^{(p)}, F^{(\mu)}, F^{(q)}\right)$ and $\mathbf{N} = \left(N^{(\mathrm{T})}, N^{(p)}, N^{(\mu)}, N^{(q)}\right)$. Equation (8.52) is an algebraic equation in which ψ^{n} is the unknown variable to be solved while the previous-time-step solution ψ^{n-1} is known. In principle, $\mathbf{N}$ contains the unknown variables while all terms in $\mathbf{F}$ are known. Hence, terms containing only ψ^{n-1} should go to $\mathbf{F}$, while those containing only ψ^{n} are placed in $\mathbf{N}$; the assignment of cross terms (containing both ψ^{n} and ψ^{n-1}), however, is rather arbitrary. The ψ^{n} terms assigned to $\mathbf{F}$ are actually lagged in the multigrid V-cycle, which will be detailed in Section 8.4.2. In other words, there is no unique way to set up the components in Eq. (8.52). Solvability, stability, accuracy, convergence, and efficiency may all affect the decision about how the assignment is made. For example, including more cross terms in the operator may improve the accuracy but make the solution difficult to converge; conversely, assigning more cross terms to the source decouples the equations, making them easier to solve, but it may reduce the stability and the efficiency of the solver.

For simplicity of presentation, from now on we will use the abstract Eq. (8.52) to represent the discretized tumor growth model (8.9)–(8.12). Next we introduce a nonlinear

MLAT–FAS multigrid method to solve Eq. (8.52) on a block-structured adaptive mesh [75, 379, 380, 662, 695, 696]. A multigrid method is based on the idea that a smoothing procedure quickly damps high-frequency noise and coarse-grid corrections accurately correct low-frequency errors [662]. For a nonlinear problem such as our tumor growth model, a local linearization algorithm is used to evaluate locally the Jacobian matrix of the system and to construct the smoothing operator. Then the smoothing step and the coarse-grid correction of a multigrid method can be directly applied to a nonlinear system.

Thus a multigrid method consists of smoothing procedures, fine-to-coarse grid restrictions, and coarse-to-fine grid interpolations, all of which are combined in an iterative cycle, termed the V-cycle because its operations move down and up the multilevel mesh structure. In Section 8.4.2 we present an overview of FAS V-cycles, followed by more detailed descriptions of each component.

8.4.2 The MLAT–FAS V-cycle method

The MLAT–FAS multigrid method is carried out on multiple levels of uniform mesh patches, as shown in Figure 8.2. Let us denote the coarsest, $2^0 \times 2^0$, mesh level by $L = L_{\min}$ and the finest, $2^{\bar{L}} \times 2^{\bar{L}}$, mesh level by $L = L_{\max}$, where $\bar{L} = L_{\max} - L_{\min}$. As in Figure 8.2 the domain of the Lth-level mesh is Ω_L, and we will use the notation ψ_L^n to represent the discretized variable ψ^n on Ω_L, where n represents the nth time step. During FAS V-cycle iterations, an additional superscript m is added to indicate the mth cycle in the process, and ψ^n is written as $\psi_L^{n,m}$. In the process of solving Eq. (8.52) let us express the result of one FAS V-cycle iteration as follows [380]:

$$\psi_L^{n,m+1} = \mathbf{MLATFAScycle}\left(L, \psi_L^{n-1}, \psi_L^{n,m}, \mathbf{N}_L, \mathbf{F}_L, \nu\right), \qquad (8.61)$$

where ν specifies the number of relaxation sweeps of the pre- and post-smoothing procedure, which will be introduced in Section 8.4.3; $\mathbf{N}_L$ and $\mathbf{F}_L$ are the operator and source functions of Eq. (8.52), the subscript L indicating that these functions depend on h_L, the grid spacing of Ω_L. Because in Eq. (8.61) and the following equations the FAS iteration index m is only associated with the current time step n, not the previous time step $n - 1$, it is convenient to drop the time index for such cases; thus, from now on, $\varphi^m \equiv \varphi^{n,m}$ unless specified otherwise. Next, let us define an MLAT–FAS V-cycle [380, 695, 696]. The V-cycle starts with the finest mesh level $L = L_{\max}$, using the following steps to update $\psi_L^{n,m+1}$ for each mesh level $L_{\max} \geq L \geq L_{\min}$.

1. **Pre-smoothing**
   Compute the smoothed variable $\bar{\psi}_L^m$ by applying ν smoothing steps to ψ_L^m:

$$\bar{\psi}_L^m = \mathbf{SMOOTH}^\nu\left(\psi_L^{n-1}, \psi_L^m, \mathbf{N}_L, \mathbf{F}_L\right). \qquad (8.62)$$

   We defer the description of the **SMOOTH** relaxation operator to Section 8.4.3.

2. **Defect**
   Compute the residual of the equations on Ω_L after the smoothing step:

$$\bar{d}_L^m = \mathbf{F}_L\left(\bar{\psi}_L^m, \psi_L^{n-1}\right) - \mathbf{N}_L\left(\bar{\psi}_L^m, \psi_L^{n-1}\right). \qquad (8.63)$$

3. **Restriction**

   Project $(\bar{d}_L^m, \bar{\psi}_L^m)$ from Ω_L to the coarser mesh Ω_{L-1}:

$$\bar{d}_{L-1}^m = \mathbf{I}_L^{L-1}\bar{d}_L^m \qquad \text{on } \Omega_{L-1} \cap \Omega_L, \tag{8.64}$$

$$\bar{\psi}_{L-1}^m = \begin{cases} \mathbf{I}_L^{L-1}\bar{\psi}_L^m & \text{on } \Omega_{L-1} \cap \Omega_L, \\ \psi_{L-1}^m & \text{on } \Omega_{L-1} \setminus \Omega_L. \end{cases} \tag{8.65}$$

   For $\bar{d}_{L-1}^m$ on $\Omega_{L-1} \setminus \Omega_L$, we apply Eq. (8.63) of step 2 to the $(L-1)$th level. The restriction operator $\mathbf{I}_L^{L-1}$ will be discussed in Section 8.4.4.

4. **Computing the source term**

   Compute the source term on the coarser mesh Ω_{L-1}:

$$F_{L-1}^m = \begin{cases} \bar{d}_{L-1}^m + \mathbf{N}_{L-1}\left(\bar{\psi}_{L-1}^m, \psi_{L-1}^{n-1}\right) & \text{on } \Omega_{L-1} \cap \Omega_L, \\ \mathbf{F}_{L-1}\left(\bar{\psi}_{L-1}^m, \psi_{L-1}^{n-1}\right) & \text{on } \Omega_{L-1} \setminus \Omega_L. \end{cases} \tag{8.66}$$

5. **Coarse-grid solution**

   Solve

$$\mathbf{N}_{L-1}\left(\hat{\psi}_{L-1}^m, \psi_{L-1}^{n-1}\right) = F_{L-1}^m \tag{8.67}$$

   for $\hat{\psi}_{L-1}^m$:

   { If $L > L_{\min} + 1$, proceed to the next level by solving

$$\hat{\psi}_{L-1}^m = \mathbf{MLATFAScycle}\left(L-1, \psi_{L-1}^{n-1}, \bar{\psi}_{L-1}^m, \mathbf{N}_{L-1}, F_{L-1}^m, \nu\right). \tag{8.68}$$

   { If $L = L_{\min} + 1$, the mesh Ω_{L-1} reduces to the coarsest mesh $\Omega_{L_{\min}}$; thus, we can explicitly invert the matrix to obtain the solution $\hat{\psi}_{L_{\min}}^m$. It will become clear in Section 8.4.3 that solving for $\hat{\psi}_{L_{\min}}^m$ is essentially the same as the smoothing procedure:

$$\hat{\psi}_{L_{\min}}^m = \mathbf{SMOOTH}^\nu\left(\psi_{L_{\min}}^{n-1}, \bar{\psi}_{L_{\min}}^m, \mathbf{N}_{L_{\min}}, F_{L_{\min}}^m\right). \tag{8.69}$$

   Note that in contrast with a linear multigrid method, which solves for the approximated defects at this step, the FAS method solves directly for the full approximated solution of $\hat{\psi}_{L-1}^m$ for nonlinear problems, which is the origin of the name "full approximation storage."

6. **Coarse-grid correction (CGC)**

   On $\Omega_{L-1} \cap \Omega_L$, compute the corrections of the variables:

$$\hat{\theta}_{L-1}^m = \hat{\psi}_{L-1}^m - \bar{\psi}_{L-1}^m. \tag{8.70}$$

   On $\Omega_{L-1} \setminus \Omega_L$, set the solution:

$$\psi_{L-1}^{m+1} = \hat{\psi}_{L-1}^m. \tag{8.71}$$

7. **Interpolation**

   Project the coarse-grid correction onto the finer grid Ω_L:

$$\hat{\theta}_L^m = \mathbf{I}_{L-1}^L \hat{\theta}_{L-1}^m. \tag{8.72}$$

   The interpolation operator $\mathbf{I}_{L-1}^L$ is also discussed in Section 8.4.4.

8. **Correction**

   Use the interpolated coarse-grid corrections to approximate the solutions on Ω_L:

   $$\psi_L^{m,\text{after CGC}} = \bar{\psi}_L^m + \hat{\theta}_L^m. \tag{8.73}$$

9. **Post-smoothing**

   Apply ν steps of the smoothing procedure to $\psi_L^{m,\text{after CGC}}$ to obtain the solution of one MLAT–FAS cycle:

   $$\psi_L^{m+1} = \textbf{SMOOTH}^\nu\left(\psi_L^{n-1}, \psi_L^{m,\text{after CGC}}, \textbf{N}_L, \textbf{F}_L\right). \tag{8.74}$$

The above steps complete one MLAT–FAS V-cycle.

To advance the solution ψ^{n-1} at the $(n-1)$th time step to ψ^n at the nth time step, we iterate through multiple MLAT–FAS V-cycles until at the finest level a small-residual-defect criterion is satisfied. The following pseudocode illustrates the essence of the algorithm.

MLAT–FAS V-CYCLE ALGORITHM

set $\psi_L^{n,m=0} = \psi_L^{n-1}, \forall L$
Vcycle loop: **for** $m = 0$ **until** $m_{\max} - 1$
   $\psi_L^{n,m+1} = \textbf{MLATFAScycle}\left(L, \psi_L^{n-1}, \psi_L^m, \textbf{N}_L, \textbf{F}_L, \nu\right)$
   if $\left\|\textbf{F}_{L_{\max}}\left(\psi_{L_{\max}}^{n,m+1}, \psi_{L_{\max}}^{n-1}\right) - \textbf{N}_{L_{\max}}\left(\psi_{L_{\max}}^{n,m+1}, \psi_{L_{\max}}^{n-1}\right)\right\| < tol$ **exit** loop
end for Vcycle loop
set $\psi_L^n = \psi_L^{n,m+1}, \forall L$

Recall that $L_{\max}$ is the finest mesh and that we check whether the residual $\|\textbf{N} - \textbf{F}\|$ is below a tolerance level tol. Next, in Section 8.4.3 we discuss smoothing procedures further and in Section 8.4.4 the restriction and interpolation operators.

8.4.3 Smoothing procedures

The smoothing procedures are essentially iterative methods toward solving an implicit problem such as Eq. (8.52). Typically, iterative methods rapidly damp high-frequency errors within just a few iterations, but they do not converge to the solution fast enough to be efficient solvers by themselves owing to the slow rate at which low-frequency errors are decreased. Hence, the reason for using an iterative method here is not to solve the model equations directly but to quickly kill the high-frequency errors; the low-frequency errors are more efficiently dealt with using the coarse-grid correction (CGC) step of the MLAT–FAS V-cycle described in Section 8.4.2.

Within the MLAT–FAS V-cycle, we attempt to solve the nonlinear equation (8.52) in the following form:

$$\textbf{N}_{L,(i,j)}^n\left(\psi^{n,m}, \psi^{n-1}\right) = \textbf{F}_{L,(i,j)}^n\left(\psi^{n,m-1}, \psi^{n-1}\right), \tag{8.75}$$

recalling that $\psi^{n,m}$ denotes the solution at the nth time step and the mth V-cycle. The conventional Jacobi relaxation method solves Eq. (8.75) by iteration:

$$\psi^{n,m,\ell} = \psi^{n,m,\ell-1} + \omega J_{\mathbf{N}_L^n}^{-1}\left(\psi^{n,m,\ell-1}, \psi^{n-1}\right) \tag{8.76}$$
$$\times \left[\mathbf{F}_L^n\left(\psi^{n,m-1}, \psi^{n-1}\right) - \mathbf{N}_L^n\left(\psi^{n,m,\ell-1}, \psi^{n-1}\right)\right],$$

where ℓ denotes the ℓth iterative step and ω is a relaxation parameter. If the Jacobian $J_{\mathbf{N}_L^n}(\psi^{n,m}, \psi^{n-1})$ of $\mathbf{N}_L^n(\psi^{n,m}, \psi^{n-1})$ is globally calculated with respect to $\psi^{n,m}$ on the entire computational domain, Eq. (8.76) is a *global linearization* of the nonlinear equation (8.75), for which linear multigrid methods may be applied. The FAS nonlinear multigrid solver approximates the full solution of a nonlinear problem by employing only *local linearization*, for which the Jacobian $J_{\mathbf{N}_L^n}(\psi^{n,m}, \psi^{n-1})$ of the nonlinear operator $\mathbf{N}_L^n(\psi^{n,m}, \psi^{n-1})$ is evaluated locally for each grid point. Assuming that we are working in two dimensions, let us rewrite the dependence of $\mathbf{N}_L^n(\psi^{n,m}, \psi^{n-1})$ as $\mathbf{N}_{L,(i,j)}^n(\psi_{(i_1,j_1)}^{n,m,\ell-1}, \psi_{(i_2,j_2)}^{n,m,\ell}, \psi^{n-1})$ for each grid point (i, j), where (i_1, j_1) denotes the grid points to be evaluated using the values of $\psi^{n,m,\ell-1}$ from the previous iterative step and (i_2, j_2) denotes the grid points to be evaluated by the values of $\psi^{n,m,\ell}$ in the current iterative step. For local linearization, the Jacobian is calculated with respect to $\psi_{(i_2,j_2)}^{n,m,\ell}$ and rewritten as $J_{\mathbf{N}_{L,(i,j)}^n}(\psi_{(i_1,j_1)}^{n,m,\ell-1}, \psi_{(i_2,j_2)}^{n,m,\ell}, \psi^{n-1})$ for a grid point (i, j). Then the relaxation method uses the following iterative steps to relax Eq. (8.75) for the solution $\psi_{(i,j)}^{n,m}$ on each local grid point (i, j):

$$\psi_{(i,j)}^{n,m,0} = \psi_{(i,j)}^{n,m-1}, \tag{8.77}$$

loop: **for** $\ell = 1$ **until** ν

$$\psi_{(i,j)}^{n,m,\ell} = \psi_{(i,j)}^{n,m,\ell-1} + \omega J_{\mathbf{N}_{L,(i,j)}^n}^{-1}\left(\psi_{(i_1,j_1)}^{n,m,\ell-1}, \psi_{(i_2,j_2)}^{n,m,\ell}, \psi^{n-1}\right)$$
$$\times \left[\mathbf{F}_{L,(i,j)}^n\left(\psi^{n,m-1}, \psi^{n-1}\right) - \mathbf{N}_{L,(i,j)}^n\left(\psi_{(i_1,j_1)}^{n,m,\ell-1}, \psi_{(i_2,j_2)}^{n,m,\ell}, \psi^{n-1}\right)\right],$$

end   **for** loop   (8.78)

$$\psi_{(i,j)}^{n,m} = \psi_{(i,j)}^{n,m,\nu}, \tag{8.79}$$

where ν is the number of total iterative steps. The conventional Jacobi method for local linearization at a grid point (i, j) is straightforward, with $(i_2, j_2) = (i, j)$ and the other grid points assigned to (i_1, j_1) using the lagged values from the previous iterative step. However, the conventional method provides a relatively slow smoothing effect and can be improved by adopting an alternative relaxation scheme. Here, we adopt Gauss–Seidel (GS) relaxation as the smoothing operator of the multigrid solver. Note that to solve Eq. (8.75) numerically on a spatial grid, we evaluate Eq. (8.78) by sweeping through the discretized grid points one after another at each iterative step. The GS relaxation method utilizes the values of the already updated grid points, i.e., $\psi^{n,m,\ell}$, at the current iterative step. Therefore, while the conventional Jacobi method is independent of the grid point ordering, it would appear that the GS method depends on which grid points are swept first. The most straightforward ordering is the lexicographic ordering, where we sweep the grid points $I_{(i,j)}$ from $i = 0$ to $i = N_x$ and from $j = 0$ to $j = N_y$, increasing i and j

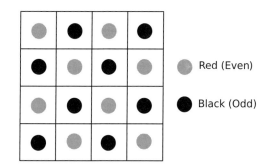

Figure 8.7 The illustration of red–black ordering. The gray and the black dots represent the cell-center values of the red and the black cells respectively. Using the red–black ordering in a mesh-grid relaxation method, we sweep all the red cells first and then the black cells. The red cells are also called even cells because the cell index add up to even numbers; likewise, the black cells are also called odd cells. Hence, red–black ordering is also known as the odd–even ordering.

by 1 each time. Thus, the rule to implement the GS relaxation method with lexicographic ordering (GS–LEX) can be concisely written as

$$
\begin{cases} i_1 > i \\ j_1 > j \end{cases} \quad \text{and} \quad \begin{cases} i_2 \leq i, \\ j_2 \leq j. \end{cases} \tag{8.80}
$$

However, analysis shows that the smoothing factor of GS relaxation can be improved further by adopting red–black ordering (GS–RB) [662]. Red–black ordering is also known as odd–even ordering, which is self-explanatory, and is shown in Figure 8.7; all the red (odd) grids are swept first and then all the black (even) grids; or vice versa. A potential disadvantage of the Gauss–Seidel method is that the results may be biased due to the dependence on grid ordering. A strategy to reduce this shortcoming is to rotate the sweep direction; for red–black ordering we may even alternate the order in which the red and black grid groups are swept, further reducing bias due to the grid ordering. In this respect, red–black ordering is more favorable than lexicographic ordering; an analysis shows that red–black smoothing has a higher degree of independence [662].

The choice of the relaxation parameter ω that enhances the rate of high-frequency error smoothing depends on the relaxation method as well as on the grid ordering. For the GS–RB method $\omega = 1$ gives a satisfactory smoothing factor, but the smoothing effect can generally be improved by using $\omega > 1$ [662]; in practice, we use the empirical value $\omega = 1.2$ for the tumor growth model. Readers may refer to Sections 2.1 and 4.3–4.5 in [662] for more in-depth discussions of these relaxation schemes.

The rule for GS–RB implementation cannot be expressed concisely like Eq. (8.80) of the GS–LEX scheme; thus, from now on, we will write the equations in the GS–LEX form whenever explicit expressions are necessary. In a programming code, the GS–LEX implementation can easily be modified to obtain the GS–RB algorithm by changing the order of mesh grid sweeping loops. Next, we formulate explicitly the GS–LEX smoothing algorithm for the tumor growth model by applying the source equations (8.53)–(8.56) and the operator equations (8.57)–(8.60) to the relaxation method in

Eqs. (8.77)–(8.79). The operator terms $N_{L,(i,j)}^n$ involve the discrete gradient operator ∇_d introduced in Section 8.2.1, which utilizes values at the cell-edge centers. Hence, for the nth time step, mth V-cycle step, and ℓth relaxation step, let us define the cell-edge-center variables using the discretized values at the cell centers:

$$\varphi_{(i+1/2,j)}^{n,m,\ell} = \tfrac{1}{2}\left(\varphi_{(i+1,j)}^{n,m,\ell-1} + \varphi_{(i,j)}^{n,m,\ell-1}\right), \qquad \varphi_{(i-1/2,j)}^{n,m,\ell} = \tfrac{1}{2}\left(\varphi_{(i,j)}^{n,m,\ell-1} + \varphi_{(i-1,j)}^{n,m,\ell}\right),$$

$$\varphi_{(i,j+1/2)}^{n,m,\ell} = \tfrac{1}{2}\left(\varphi_{(i,j+1)}^{n,m,\ell-1} + \varphi_{(i,j)}^{n,m,\ell-1}\right), \qquad \varphi_{(i,j-1/2)}^{n,m,\ell} = \tfrac{1}{2}\left(\varphi_{(i,j)}^{n,m,\ell-1} + \varphi_{(i,j-1)}^{n,m,\ell}\right),$$

where φ can be either φ_T, φ_p, or c_q. Note that, in GS–LEX relaxation, $\varphi_{(i+1,j)}$ and $\varphi_{(i,j+1)}$ are from the previous iterative step and $\varphi_{(i-1,j)}$ and $\varphi_{(i,j-1)}$ are at the current iterative step. Similarly, we define the cell-edge-center diffusion coefficient D_q by

$$D_{q,(i+1/2,j)}^{n,m-1} = \tfrac{1}{2}\left[D_q\left(\varphi_{T/p,(i+1,j)}^{n,m-1}, c_{q,(i+1,j)}^{n,m-1}\right) + D_q\left(\varphi_{T/p,(i,j)}^{n,m-1}, c_{q,(i,j)}^{n,m-1}\right)\right],$$

$$D_{q,(i-1/2,j)}^{n,m-1} = \tfrac{1}{2}\left[D_q\left(\varphi_{T/p,(i,j)}^{n,m-1}, c_{q,(i,j)}^{n,m-1}\right) + D_q\left(\varphi_{T/p,(i-1,j)}^{n,m-1}, c_{q,(i-1,j)}^{n,m-1}\right)\right],$$

$$D_{q,(i,j+1/2)}^{n,m-1} = \tfrac{1}{2}\left[D_q\left(\varphi_{T/p,(i,j+1)}^{n,m-1}, c_{q,(i,j+1)}^{n,m-1}\right) + D_q\left(\varphi_{T/p,(i,j)}^{n,m-1}, c_{q,(i,j)}^{n,m-1}\right)\right],$$

$$D_{q,(i,j-1/2)}^{n,m-1} = \tfrac{1}{2}\left[D_q\left(\varphi_{T/p,(i,j)}^{n,m-1}, c_{q,(i,j)}^{n,m-1}\right) + D_q\left(\varphi_{T/p,(i,j-1)}^{n,m-1}, c_{q,(i,j-1)}^{n,m-1}\right)\right].$$

The diffusion coefficient D_q for substrate species q may also depend on the variables φ_T, φ_p, and c_q; to simplify the numerical calculations, we use the results obtained from the previous V-cycle step $m-1$ to evaluate D_q. The operator equations (8.57)–(8.60) can be rewritten for the GS–LEX method as:

$$\begin{aligned}
N_{L,(i,j)}^{(T),n,m,\ell} = \varphi_{T,(i,j)}^{n,m,\ell} - \frac{\tfrac{1}{2}sM}{h_L^2}&\left[\left(\varphi_{T,(i+1/2,j)}^{n,m,\ell}\mu_{(i+1,j)}^{n-1/2,m,\ell-1}\right.\right.\\
&+ \varphi_{T,(i-1/2,j)}^{n,m,\ell}\mu_{(i-1,j)}^{n-1/2,m,\ell} + \varphi_{T,(i,j+1/2)}^{n,m,\ell}\mu_{(i,j+1)}^{n-1/2,m,\ell-1}\\
&+ \left.\varphi_{T,(i,j-1/2)}^{n,m,\ell}\mu_{(i,j-1)}^{n-1/2,m,\ell}\right) - \left(\varphi_{T,(i+1/2,j)}^{n,m,\ell}\right.\\
&+ \left.\left.\varphi_{T,(i-1/2,j)}^{n,m,\ell} + \varphi_{T,(i,j+1/2)}^{n,m,\ell} + \varphi_{T,(i,j-1/2)}^{n,m,\ell}\right)\mu_{(i,j)}^{n-1/2,m,\ell}\right],
\end{aligned} \tag{8.81}$$

$$N_{L,(i,j)}^{(p),n,m,\ell} = \varphi_{p,(i,j)}^{n,m,\ell}, \tag{8.82}$$

$$\begin{aligned}
N_{L,(i,j)}^{(\mu),n,m,\ell} = {}&\mu_{(i,j)}^{n-1/2,m,\ell} - \tfrac{1}{2}f_c'\left(\varphi_{T,(i,j)}^{n,m,\ell}\right)\\
&+ \frac{\tfrac{1}{2}\epsilon^2}{h_L^2}\left(\varphi_{T,(i+1,j)}^{n,m,\ell-1} + \varphi_{T,(i-1,j)}^{n,m,\ell} + \varphi_{T,(i,j+1)}^{n,m,\ell-1} + \varphi_{T,(i,j-1)}^{n,m,\ell} - 4\varphi_{T,(i,j)}^{n,m,\ell}\right),
\end{aligned} \tag{8.83}$$

$$\begin{aligned}
N_{L,(i,j)}^{(q),n,m,\ell} = {}&c_{q,(i,j)}^{n,m,\ell} - \tfrac{1}{2}s\Gamma_{q,(i,j)}^{n,m,\ell} - \frac{\tfrac{1}{2}s}{h_L^2}\left[\left(D_{q,(i+1/2,j)}^{n,m-1}c_{q,(i+1,j)}^{n,m,\ell-1}\right.\right.\\
&+ D_{q,(i-1/2,j)}^{n,m-1}c_{q,(i-1,j)}^{n,m,\ell} + D_{q,(i,j+1/2)}^{n,m-1}c_{q,(i,j+1)}^{n,m,\ell-1}\\
&+ \left.D_{q,(i,j-1/2)}^{n,m-1}c_{q,(i,j-1)}^{n,m,\ell}\right) + \left(D_{q,(i+1/2,j)}^{n,m-1} + D_{q,(i-1/2,j)}^{n,m-1}\right.\\
&+ \left.\left.D_{q,(i,j+1/2)}^{n,m-1} + D_{q,(i,j-1/2)}^{n,m-1}\right)c_{q,(i,j)}^{n,m,\ell}\right],
\end{aligned} \tag{8.84}$$

in which h_L denotes the grid spacing at mesh level L. Here we assume that $h_{L,x} = h_{L,y} = h_L$ to avoid further complicated mathematical expressions; it is not difficult to derive the appropriate formula for $h_{L,x} \neq h_{L,y}$. The unknowns to be solved in Eqs. (8.81)–(8.84) are $\varphi_{T,(i,j)}^{n,m,\ell}$, $\varphi_{p,(i,j)}^{n,m,\ell}$, $\mu_{(i,j)}^{n-1/2,m,\ell}$, and $c_{q,(i,j)}^{n,m,\ell}$. Hence we can see that only Eqs. (8.81) and (8.83) need to be solved together, while the equation for the tumor subspecies (8.82) and the equation for the microenvironmental variables (8.84) are decoupled. This allows us to solve separately Eqs. (8.81) and (8.83) as 2×2 systems for φ_T and μ; then the decoupled "slave" equations (8.82) and (8.84), driven by the solutions of φ_T and μ, can be solved independently. This simplification results from the assumption that the adhesion potential μ depends only on φ_T. The Jacobians of Eqs. (8.81)–(8.84) are derived as follows, for $x = T, \mu$:

$$
J_{L,(i,j)}^{(x),n,m,\ell} = \begin{bmatrix} 1 & \begin{array}{c} \varphi_{T,(i+1/2,j)}^{n,m,\ell} + \varphi_{T,(i-1/2,j)}^{n,m,\ell} \\ + \varphi_{T,(i,j+1/2)}^{n,m,\ell} + \varphi_{T,(i,j-1/2)}^{n,m,\ell} \end{array} \\ -\frac{1}{2} \left[f_c'' \left(\varphi_{T,(i,j)}^{n,m,\ell-1} \right) + \frac{4\epsilon^2}{h_L^2} \right] & 1 \end{bmatrix}.
$$
(8.85)

For p, q we have

$$
J_{L,(i,j)}^{(p),n,m,\ell} = 1,
$$
(8.86)

$$
J_{L,(i,j)}^{(q),n,m,\ell} = 1 - \frac{s}{2} \left[\frac{\partial \Gamma_{q,(i,j)}^{n,m,\ell-1}}{\partial c_q} + \frac{1}{h_L^2} \left(D_{q,(i+1/2,j)}^{n,m-1} + D_{q,(i-1/2,j)}^{n,m-1} \right. \right.
$$

$$
\left. \left. + D_{q,(i,j+1/2)}^{n,m-1} + D_{q,(i,j-1/2)}^{n,m-1} \right) \right].
$$
(8.87)

Here Eq. (8.85) is the Jacobian of the coupled equations (8.81) and (8.83). For a more general adhesion potential μ, which depends also on the tumor subspecies φ_p, we need to solve a fully coupled system. Alternatively, we may try lagging the φ_p variables for the evaluation of μ and obtain a decoupled system similar to Eqs. (8.81) and (8.84).

The smoother is primarily responsible for damping the high-frequency noise. Because of its good smoothing properties, the GS–RB method does not need very many iterations to be effective in the multigrid numerical solver. In practice, we have found that two iterative steps (i.e., $\nu = 2$ in Eq. (8.61)) is sufficient for our tumor growth model. As mentioned earlier, the low-frequency noise is more efficiently dealt with by the coarse grid correction (CGC) process, which essentially corrects the defects in the coarse grids and interpolates the results back to the fine grids. In the next subsection we discuss two key operators of the CGC process, the restriction operator and the interpolation operator; the former operator projects discretized quantities from a finer mesh grid to a coarser one, while the latter does the opposite.

8.4.4 Restriction and interpolation

The construction of the restriction and the interpolation operators depends on the grid coarsening rules. For standard coarsening we may simply use bilinear interpolation

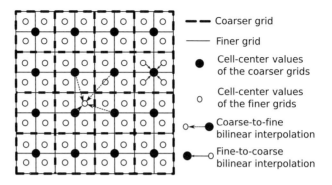

Figure 8.8 Bilinear interpolation between a coarser and a finer mesh grid. The values are defined at the cell centers on each mesh. The displacements from a to-be-interpolated coarser-grid point to the neighboring finer-grid points are $(\pm h/2, \pm h/2)$, where h is the mesh grid size of the finer mesh. Conversely, the displacements from a to-be-interpolated finer grid point to its neighboring coarser grid points depend on where the finer-grid point is with respect to the coarser-mesh grids. Using the point indicated in the figure as an example, the displacement vectors to the four neighboring coarser grid points are respectively $(-h/2, 3h/2)$, $(-h/2, -h/2)$, $(3h/2, 3h/2)$, and $(3h/2, -h/2)$.

as our coarse-to-fine interpolation operator, also known as the prolongation operator (trilinear interpolation in the case of three dimensions). Recall that the values of the model variables ψ are defined at the cell centers, as shown in Figure 8.8; hence, the bilinear interpolation is performed as follows:

$$\psi_{L+1} := \mathbf{I}_L^{L+1}\psi_L, \tag{8.88}$$

where

$$
\psi_{L+1,(i_2,j_2)}
$$
$$
= \begin{cases}
\frac{1}{16}\big(9\psi_{L,(i_1,j_1)} + 3\psi_{L,(i_1+1,j_1)} \\
\quad + 3\psi_{L,(i_1,j_1+1)} + \psi_{L,(i_1+1,j_1+1)}\big) & \text{for } i_2 = 2i_1, \ j_2 = 2j_1, \\[1.2em]
\frac{1}{16}\big(3\psi_{L,(i_1,j_1)} + 9\psi_{L,(i_1+1,j_1)} \\
\quad + \psi_{L,(i_1,j_1+1)} + 3\psi_{L,(i_1+1,j_1+1)}\big) & \text{for } i_2 = 2i_1 + 1, \ j_2 = 2j_1, \\[1.2em]
\frac{1}{16}\big(3\psi_{L,(i_1,j_1)} + \psi_{L,(i_1+1,j_1)} \\
\quad + 9\psi_{L,(i_1,j_1+1)} + 3\psi_{L,(i_1+1,j_1+1)}\big) & \text{for } i_2 = 2i_1, \ j_2 = 2j_1 + 1, \\[1.2em]
\frac{1}{16}\big(\psi_{L,(i_1,j_1)} + 3\psi_{L,(i_1+1,j_1)} \\
\quad + 3\psi_{L,(i_1,j_1+1)} + 9\psi_{L,(i_1+1,j_1+1)}\big) & \text{for } i_2 = 2i_1 + 1, \ j_2 = 2j_1 + 1.
\end{cases}
$$
$$\tag{8.89}$$

Here $0 \leq i_1, j_1 \leq N_L + 1$ are the grid indices on mesh L and $0 \leq i_2, j_2 \leq 2N_L + 1$ are the grid indices on mesh $L + 1$. (See Section 2.8.4 of [662] for further discussion.) Similarly, the restriction operator is constructed as follows:

$$\psi_{L-1} := \mathbf{I}_L^{L-1}\psi_L, \tag{8.90}$$

where

$$\psi_{L-1,(i,j)} = \tfrac{1}{4}\left(\psi_{L,(2i-1,2j-1)} + \psi_{L,(2i-1,2j)} + \psi_{L,(2i,2j-1)} + \psi_{L,(2i,2j)}\right), \quad (8.91)$$

and $1 \le i, j \le \tfrac{1}{2}N_L$. Note that information on the finer-level mesh is not fully preserved when it is restricted to a coarser-level mesh and, hence, cannot be recovered by interpolation. In other words, the restriction operator is not invertible, i.e., $\mathbf{I}_L^{L-1} \ne (\mathbf{I}_{L-1}^L)^{-1}$.

Recall that in Section 8.3.2 we referred to the present section for "coarse-to-fine interpolation" and the "fine-to-coarse average;" we use the interpolation operator $\mathbf{I}_{L-1}^L$ for the former and the restriction operator $\mathbf{I}_L^{L-1}$ for the latter. However, $\mathbf{I}_{L-1}^L$ and $\mathbf{I}_L^{L-1}$ do not update the values in the ghost cells. After the interior grid cells have been updated by $\mathbf{I}_{L-1}^L$ and $\mathbf{I}_L^{L-1}$, we recalculate the values in the ghost cells using the algorithm in Section 8.2.3 for ghost cells outside the computational domain Ω; for the ghost cells of an interior patch located inside Ω we use the algorithm in Section 8.3.2.

In the next subsection we discuss the mesh refinement criteria of the block-structured adaptive mesh. The discussion was postponed from Section 8.3 because some elements introduced in the present section for the MLAT–FAS multigrid method are utilized to determine where the locally refined patches should be constructed.

8.4.5 Criteria for mesh refinement

As mentioned in Section 8.3.1, there are several *a priori* and *a posteriori* criteria for tagging adaptive mesh grids for mesh refinement. We will adopt and mix two such tests, the undivided-gradient test and the relative-truncation-error test, which will be described in this subsection.

Typically, interfacial regions characterized by large gradients G have a significant impact on the overall dynamics owing to these sharp and rapid changes in physical properties. Hence, it is desirable to have a fine mesh resolution around the large-gradient regions. The *undivided-gradient test* is a convenient criterion for tagging such regions:

$$G_{k,L}^{\text{tag},1} = \left\{ I_{(i,j)} \in G_{k,L} \left| \sqrt{\left(\varphi_{(i+1,j)} - \varphi_{(i-1,j)}\right)^2 + \left(\varphi_{(i,j+1)} - \varphi_{(i,j-1)}\right)^2} > \delta_L \right. \right\}.$$

$$(8.92)$$

Here, $I_{(i,j)}$ denotes the grid cells and φ is the model variable whose gradient needs to be calculated precisely. The testing criterion is defined by a critical value δ_L which depends on the level-specific grid spacing h_L.

Furthermore, accurately discretizing the equations may also require enhanced local spatial accuracy. The operators may oscillate, introducing large truncation errors. Such regions can be tagged by the *relative-truncation-error test*, which compares the restricted finer-level operators and the coarser-level operators for truncation errors and tags the fine-level mesh grids where the truncation error is large. We use $\mathbf{N}_{(i_1,j_1),L}$ to represent the values of the operator $\mathbf{N}$ on the Lth-level mesh grids and $\mathbf{N}_{(i_2,j_2),L-1}$ for the operator values on the $(L-1)$th-level mesh grids. The subscripts (i_1, j_1) and (i_2, j_2) respectively indicate the grid cells $I_{(i_1,j_1),L}$ on Ω_L and $I_{(i_2,j_2),L-1}$ on Ω_{L-1}. For the standard coarsening $h_{L-1} = 2h_L$, we may set the grid indexing in such a way that $I_{(2i_2-1,2j_2-1),L}, I_{(2i_2-1,2j_2),L}$,

$I_{(2i_2,2j_2-1),L}$, $I_{(2i_2,2j_2),L}$ on Ω_L are the four grid points adjacent to $I_{(i_2,j_2),L-1}$ on Ω_{L-1}. The restriction operator $\mathbf{I}_L^{L-1}$ in Eq. (8.90) projects the values of these four grid points on Ω_L to the grid point $I_{(i_2,j_2),L-1}$ on Ω_{L-1}. Then the relative-truncation-error test tags the mesh grids for refinement as follows:

$$G_{k,L}^{\text{tag},2} = \big\{ I_{(2i_2-1,2j_2-1),L}, \, I_{(2i_2-1,2j_2),L}, \, I_{(2i_2,2j_2-1),L}, \, I_{(2i_2,2j_2),L}$$

$$\in G_{k,L} \big\| \, (\mathbf{I}_L^{L-1} \mathbf{N}_L)_{(i_2,j_2)} - \mathbf{N}_{(i_2,j_2),L-1} \big\| > \delta_L \big\}. \tag{8.93}$$

The critical value δ_L, in this test also, depends on the level-specific grid spacing h_L. Readers may refer to Sections 5.3.7 and 9.4.1 in [662] for further discussion about the relative-truncation-error test.

After the sets $G_{k,L}^{\text{tag},1}$ and $G_{k,L}^{\text{tag},2}$ have been constructed, conventionally we expand the tagged region by a process called *buffering* [140]. The buffering process defines, for each level L, a buffered set B_L which includes the tagged sets $G_{k,L}^{\text{tag},1}$ and $G_{k,L}^{\text{tag},2}$ as well as $n_b \in \{0, 1, 2, \ldots\}$ layers of grid cells around them. The grid cells in B_L are then selected for mesh refinement. To do this the grid cells are grouped into rectangular patches, using the clustering algorithm given in Berger and Rigoutsos [65], and the refined grids $G_{k,L+1}$ are constructed on these patches.

Part II

Applications

9 Continuum tumor modeling: a multidisciplinary approach[1]

With H. B. Frieboes

9.1 Introduction

In this chapter we describe recent multidisciplinary efforts that combine the models of cancer described in this book with laboratory experiments to model tumor growth and invasion. We focus on morphology, invasion, and anti-angiogenic therapy. The work has been accomplished in stages that progressively incorporate the complexity of the tumor environment: (i) the modeling of avascular tumors *in vitro* and *in silico* to assess the stages of tumor growth; (ii) the modeling of vascularized tumors *in silico* to assess the angiogenic switch and vascular growth; (iii) the modeling of vascularized tumors *in vivo* and *in silico* to assess tumor progression in the body.

9.1.1 Avascular growth

Even though *in vitro* conditions offer tumor cells unlimited space for expansion and no interaction with the host environment [299, 588, 647], the study of tumor growth and invasion using a three-dimensional multicellular tumor spheroid *in vitro* model has provided novel insights. A spheroid can be viewed as a network of individual cells, each with its own proliferation potential. Growth of the spheroid is the outcome of the balance between individual cell proliferation and internal cohesive forces. Variations in spheroid growth rates and in the extent of central necrosis have been attributed to fluctuations in the oxygen and nutrient concentrations [224, 226], and tumor growth characteristics *in vitro* and *in vivo* have been extensively modeled [5, 172, 210, 456, 577]. Spheroids may have the complexity of self-organized dynamical systems, which are regulated by both environmental and internal noise [139, 693].

Although it is usually vascular tumors that exhibit irregular shapes and complex morphology, there have been reports of irregular avascular tumors exhibiting complex growth characteristics [77, 104, 356, 480]. This may carry important implications for vascularized tumors *in vivo*. *In vitro* tumors that deviate from the spheroidal shape by expressing branched or chained structures have been observed (e.g., [356]). An

---

[1] This chapter is based partly on the following articles: *Clin. Cancer Res.*, Cristini *et al.*, vol. 11, pp. 6772–6779 (© 2005 American Association for Cancer Research); *Cancer Res.*, Frieboes *et al.*, vol. 66, pp. 1597–1604 (© 2006 American Association for Cancer Research); and *Cancer Res.*, Bearer *et al.*, vol. 69, pp. 4493–4501 (© 2009 American Association for Cancer Research).

expanding tumor may exert both mechanical pressure and traction on its microenvironment [288]. The tumor mass may compress the bulk matrix radially outward, but the tips of invasive cells pull the surrounding matrix inward. A feedback mechanism may link volumetric growth and invasive expansion [168]. Mechanical forces influence tumor shape and may be related to tumor proliferation and invasive growth. Cellular traction may cause extracellular matrix alignment, in which matrix filaments form tracks that promote cellular elongation and directed migration, leading in some cases to the formation of multicellular tubular structures [671] and intratumor cellular swirls. To explain the occurrence of invasive branches in brain tumors, a scenario has been proposed [168] in which cells follow each other because of reduced mechanical resistance, enhanced haptotactic gradient, and increased chemical attraction, all as part of a self-organizing adaptive biosystem. The emergence of multicellular clusters in networks formed by migrating cells may occur as well, and this may represent a guiding influence on the invasion dynamics [168].

In Section 9.2.1 we study in detail tumor invasion in an avascular environment. The main hypothesis [145] is that the tumor shape, while being affected by random cellular proliferation and adhesion and by complex mechanical interactions at the tumor viable rim, may result deterministically from the competition between individual cell proliferation and internal cohesive forces in the presence of microenvironmental substrate gradients [145]. It is well known that spatial gradients of nutrient, oxygen, and growth factors are formed in tumors and are affected by the three-dimensional heterogeneous arrangement of the cells and the extracellular matrix [3, 111, 114, 212] as well as by the vasculature *in vivo* [339, 340]. In turn, intratumoral regions of hypoxia and acidosis may generate spatially heterogeneous tumor-cell proliferation and migration [145]. In an avascular environment, the need of cells to maximize their exposure to the culture medium to allow for optimal substrate uptake may lead to shape instability and invasive tumor morphology. When the substrate is abundant, cell survival may be better served by increased cell–cell contact, which generates compact morphologies; in contrast, in a low-substrate medium, cells strive to minimize cell contact while maximizing exposure to the medium. Under these conditions, invasive morphologies manifest themselves with the development of low-wavenumber deterministic fluctuations in cell positions at the tumor viable rim as cells proliferate and regulate adhesion on the basis of the local diffusion gradients; this leads to the recursive formation and growth of a finite number of sub-tumors, which eventually may break off from the parent tumor (Figure 9.2). The growth of low-wavenumber modes is characteristic of diffusion-driven instabilities in materials and biomaterials (see [148] and references therein).

Frieboes *et al.* [230] applied a version of the single-phase model presented below in Section 9.2.1 to quantify this instability. They showed that the model can be fitted accurately to *in vitro* experimental data to determine the model parameters (see Figure 4.11) and that multidimensional computer simulations based on these input parameters predict tumor cell spatial arrangement and tumor morphologies that closely resemble those observed *in vitro* (Figure 9.1). Tumor morphologies may thus be quantified, with the long-term goal of optimizing parameters for therapy application to minimize growth and invasion. The implication is that in an *in vivo* environment, diffusion-driven tumor-shape instability could perhaps be suppressed by enforcing a more spatially homogeneous nutrient and oxygen supply (see Section 9.2.2 below), because normoxic conditions act both by decreasing gradients and by increasing cell adhesion, consequently supporting the mechanical forces that maintain a well-defined tumor morphology [145].

9.1.2   Vascularized tumors

The tumor microvasculature is typically highly disorganized [305, 341], resulting in considerable spatial and temporal heterogeneity in the delivery of oxygen and nutrients and the removal of metabolites [339, 340]. This results in variable regions of acute and chronic hypoxia and of acidosis in most tumors *in vivo* [303, 317, 670]. Clinical studies have demonstrated that hypoxia correlates with a poor clinical outcome and increased risk of metastasis, independently of the therapeutic treatment [82, 83, 315, 316, 645]. Hypoxia may select for cells that are more resistant to apoptosis [289, 712]; it can induce angiogenic regulators [218, 303, 613] and directly increase tumor-cell invasiveness by causing increased production of autocrine motility factor, increased expression of tumor urokinase plasminogen activator receptor and a protease receptor, increased production of Cathepsin B [572] and upregulation of hepatocyte growth factor (HGF) [108, 539, 571, 710, 711]. Pleiotropic effects include cell proliferation, motility, differentiation, and survival, as shown in the collagen invasion assay [474], in which cells were observed to form branched structures and invade a three-dimensional collagen gel. Strong cell–cell and cell–matrix adhesion forces generated by cell-adhesion molecules such as E-cadherins and integrins can attenuate such potentially invasive morphologies [569].

The critical role of angiogenesis in promoting tumor growth and invasion has been well demonstrated. However, the results of clinical trials using various drugs to suppress angiogenesis have been mixed. Although some tumor regression can be observed following therapy, the length of survival remains unchanged [66, 72, 391]. Furthermore, it has been observed experimentally that antiangiogenic treatment can exacerbate hypoxic effects [631] and cause tumor-mass fragmentation, cancer-cell migration, and tissue invasion [514, 596]. Systemic treatment with vascular endothelial growth factor receptor-2 (VEGFR-2) antibody was shown to inhibit glioblastoma angiogenesis in mice and to lead to decreased tumor volume but increased tumor invasiveness along the host microvasculature [392]. It was demonstrated [528] that hypoxia, instead of inhibiting tumor growth, induced Met tyrosine kinase, which increased sensitivity to HGF. In [398] it was noticed that there was a remarkable increase in the number and total area of small satellite tumors clustered around the primary mass in mice treated with VEGFR-2 antibody. These satellites usually contained central vessel cores, and often tumor cells had migrated along pre-existent blood vessels over long distances to reach the surface and spread in the subarachnoid space. A modification in the tumor-cell invasion pattern at the tumor–normal-brain-parenchyma interface was reported [58]. Tumors from animals belonging to the control group and those under antiangiogenic treatment showed similar invasion patterns, consisting of an undefined interface with trails of invading tumor cells and distant tumor satellites. However, tumors treated against both angiogenesis and invasion showed a well-defined tumor–parenchyma interface, without trails of invading cells or tumor satellites, and a clear decrease in peripheral-vessel recruitment. A similar formation of satellite glioblastoma tumors in rats after anti-VEGF antibody treatment was observed [576]. A correlation of hypoxic status *in vivo* with increased metastatic spread has also been observed in D-12 melanoma cells and KHT-C fibrosarcoma cells [170, 571].

In Section 9.2.2 we review in detail the computational and experimental work of Cristini *et al.* [145], who proposed the hypothesis that the morphological stability of tumor tissue *in vivo* may require a reliable, constant, level of cell-substrate concentrations through a robust vascular network. Morphologic instability may occur, during solid-tumor growth and during response to treatment, as a result of oxygen, glucose, acid and drug concentration gradients which drive spatially heterogeneous cell proliferation, migration, and death and which reduce cell adhesion and other mechanical forces in hypoxic and acidotic regions owing to the disruption of cell–cell and cell–matrix interactions (see Figures 9.3–9.5). These conditions result in invasive fingering and branching and even the fragmentation and migration of cell clusters into the surrounding tissue because of differential proliferation along the gradients (the reader may refer to recent mathematical and computational analyses [147, 149, 416, 437, 620, 697, 722]).

9.1.3   Clinical observations

In a clinical setting, the variables currently used for prognosis include epidemiological information, tumor type, and molecular characterization, as well as clinical parameters such as tumor size and the presence of nodal and extranodal metastasis (tumor, node, metastasis (TNM) staging) [57]. The clinical predictive value from histological epidemiology remains limited because it is cumbersome and time consuming. However, a quantitative histopathologic analysis is often subjective to the pathologist. Even when different histopathologic features in a tumor can be adequately described, the relative impact of each feature on patient outcome and tumor progression remains uncertain.

Bearer *et al.* [57] used a version of the multiphase mixture model presented in Chapter 5 to demonstrate that molecular phenomena regulating cell proliferation, migration, and adhesion forces (including those associated with genetic evolution from lower- to higher-grade tumors) generate, in a predictable and quantifiable way, heterogeneous proliferation and oxygen and nutrient demand (and the suppression of apoptosis) across the three-dimensional tumor mass, determining its morphology. The model is thus used to hypothesize phenomenological functional relationships linking genetic and phenotypic effects, the microenvironment, and tissue-scale growth and morphology. We will describe these results in Section 9.2.3.

9.2   Application to brain cancer

9.2.1   Avascular growth

Tissue architecture and cellular environment are believed to be dominant determinants of tumor shape [201, 587, 594], yet the intracellular and extracellular factors that promote a particular tumor to adopt a specific morphology are not well understood. Frieboes *et al.* [230] varied the glucose and serum concentrations in *in vitro* experiments to vary the cell adhesion and rate of cell proliferation (corresponding to model parameters G and A in the model, as described in Section 4.4). According to the theory of the model, the heterogeneous access to oxygen, nutrients, and growth factors at the cellular scale owing to spatial gradients of these substances established by diffusion leads to differential proliferation within the viable rim of *in vitro* tumor spheroids: groups of cells proliferate faster in regions of higher substance concentrations. Under these conditions fluctuations in shape are predicted, as cells proliferate heterogeneously at the viable rim and lead, for relatively weak cell adhesion (high values of G), to morphological instability manifesting itself in the development and growth of sub-spheroidal structures. These sub-spheroids, consisting of groups of predominantly viable cells, may even break off from the "mother" spheroid and grow into separate tumor spheroids. This process repeats itself on the sub-tumors, leading to recursive sub-spheroid growth as the main mechanism of tumor spheroid morphogenesis. In contrast, for low values of the parameter G, surface perturbations do not grow and the overall spheroidal tumor shape remains stable. In the experiments, unstable morphologies *in vitro* seemed to exhibit mostly tumor-surface perturbations characterized by low wavenumbers (e.g., 3 or 4) at the onset of instability, which is consistent with the diffusional instability theory [149]. Frieboes *et al.* [230] were able to observe in the experiments both stable and unstable spheroids (Chapter 4), according to the theory, when the parameter G was varied.

Further, Frieboes *et al.* [230] performed computer simulations of glioblastoma spheroid growth in two spatial dimensions, in order to solve in the nonlinear regime (where there are large shape deformations) a version of the single-phase model described in Chapter 3. A compact spheroid morphology was achieved for values of G within the stable region (as defined in Figure 4.11).

For an unstable case, with large G ($G > 0.9$), snapshots of the evolution of a tumor spheroid predicted from a computer simulation are shown in the upper part of Figure 9.1. The outer boundary tracks the surface of the spheroid; the inner boundary encloses regions of hypoxia, where necrotic cells are also present. A thin rim (with thickness about equal to the oxygen diffusion length, 100 μm) of viable and actively proliferating cells is predicted, surrounding a large hypoxic core. Irregularities arising on the spheroid surface introduce low-wavenumber perturbations, as seen in the snapshots A, B from the simulation and in the photographs C from the *in vitro* experiments. The perturbations grow (owing to the large G value), leading to formation of sub-spheroidal structures that eventually separate from the mother spheroid (see for comparison the photographs C). Clusters of spheroids are thus formed and this allows the tumor mass to grow to a larger size and over a much larger region than would have been possible had the spheroid maintained a compact shape and had instability not occurred, as predicted in [145]. If the spheroid remains compact, nutrient- and oxygen-diffusion limitations to mass growth cause the spheroid to reach a final stable size. The bottom right snapshot from the simulation shows that the instability repeats itself on the sub-spheroids, as was observed in the *in vitro* experiments.

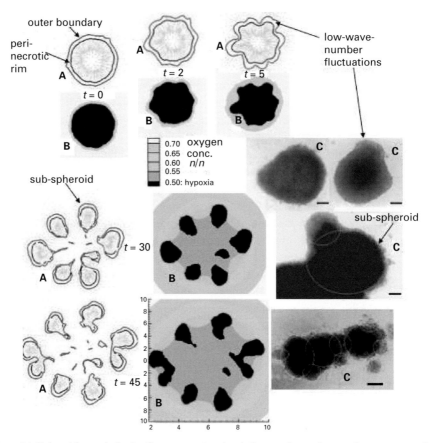

Figure 9.1 Spheroid morphologies from computer simulations and experiments. Low-wavenumber instabilities arise on the spheroid surfaces, eventually leading to the development and separation of sub-spheroids (lower center and left snapshots). The simulation snapshots (with length rescaled by the diffusion length of oxygen, time rescaled by the mitosis time of 1 day) show the outer boundary and inner perinecrotic rims (A) and the local levels of diffusing substances (B) such as oxygen or glucose. The photographs C show glioblastoma spheroids growing in culture; the sub-spheroids are highlighted in the middle and bottom photographs. Bar, 130 μm. Reprinted from Frieboes *et al.* [230], with permission from the American Association for Cancer Research.

Thus the computational results and experimental *in vitro* observations of morphology are in agreement, supporting the hypothesis that invasive tumor morphologies observed *in vitro* are the result of a shape instability driven by diffusion gradients in the tumor microenvironment. This morphological instability develops as a result of differential cell proliferation induced by underlying diffusion gradients across the tumor mass, where cells that are exposed to higher levels of nutrients and oxygen are favored for proliferation. Perturbations of cell positions on the surface of a tumor or of the spatial distribution of these nutrients and oxygen trigger the instability. This differential growth becomes more pronounced as time proceeds, since cells at the leading edge of a shape perturbation (the "bulb" of a cell, e.g., as in Figure 9.1) have a higher exposure to

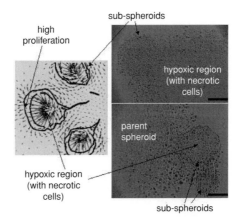

Figure 9.2 Left, cell displacement velocities (arrows) during sub-spheroid formation (zoom from simulation of Figure 9.1 at $t = 30$). Right (top and bottom), histologic cross sections showing cell arrangements in sub-spheroids grown *in vitro*. In the photographs, sub-spheroids can be seen forming from parent spheroids. As a result of a shape instability on the parent spheroid, swirling patterns of cell arrangements can be identified surrounding the necrotic cells in the sub-spheroids both in the simulations and in the experiments. Bar, 130 μm. Reprinted from Frieboes *et al.* [230], with permission from the American Association for Cancer Research.

nutrient, thereby gaining a proliferative advantage, while cells at the receding edge become increasingly disadvantaged. In this figure, the concentration gradients of the substrate that is most important to proliferation (oxygen) are shown as predicted from the simulation. These gradients drive the instability, and proliferation of the sub-spheroids follows the gradients. Morphological instability thus provides an effective means to invasion because it allows tumor cells better access to oxygen and nutrients in the surrounding environment.

In Figure 9.2, a detail of sub-spheroid formation is shown from the simulation and experiment of Figure 9.1. In the simulation snapshot, mass fluxes (associated with the tumor growth velocity) are depicted by arrows. Outside the tumor these are associated with the flow of fluid in the culture dish driven by proliferation and motion of cells. In the tumor, the arrows represent cell mass fluxes. As cells proliferate, they are advected (pushed) by the rising pressure. The cell mass in the unstable regions (i.e., the proliferating rim, acting as a source of cell mass) protrudes owing to the outward movement of cells at the leading edge and undergoes involution in the area of "pinching" that is distal from this growth (the hypoxic and necrotic regions, acting as a sink of cell mass); this leads to eventual tumor breakup and the separation of sub-spheroids from the parent spheroid. The resulting swirling patterns are typical of motion in the presence of a source and a sink in fluids and materials. Swirling arrangements of viable cells are also roughly identifiable in the photographs and are commonly observed experimentally with fibroblast-like cells (e.g., glial cells). Note that the flux of cells into the hypoxic regions predicted by the simulation is exaggerated in magnitude as an artifact of the model used, which considers a uniform cell-mass density. In reality, the cell-mass density in

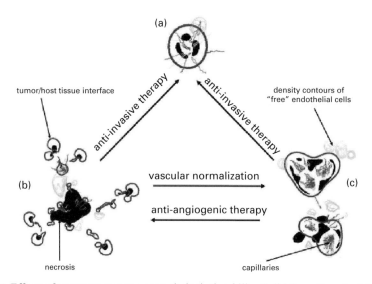

Figure 9.3 Effects of treatment on tumor morphological stability. Solid lines, the calculated tumor boundary; black areas necrosis. The neovasculature forms from "free" endothelial cells (light gray) that chemotax after sprouting from pre-existing vessels (not shown) in the outer tissue toward the angiogenic regulators in the perinecrotic regions. Reprinted from Cristini *et al.* [145], with permission from the American Association for Cancer Research.

the hypoxic regions is much lower than in the viable rim because of necrosis; hence, the real flux of cells is also lower.

9.2.2 Vascularized tumors

The single-phase tumor model from Chapter 3 was used to create two-dimensional computer simulations of vascularized tumors under varying conditions [145]. The results shown in Figure 9.3 can be interpreted as a "phase diagram" for cancer that illustrates the possible morphology states obtained by varying two model parameters, one associated with the cell adhesion forces (the parameter G; see Section 3.2) and the other with the microvascular density (see Section 3.3).

Cell survival and proliferation are dependent on substrate availability and, thus, vascular density. Cellular adhesion is dependent on a variety of membrane proteins such as E-cadherins and integrins that maintain the cell position through contact with other cells, the basement membrane, and the extracellular matrix. However, there is also an interrelationship of vascular density with cell adhesion, since the hypoxia and extracellular acidosis associated with a reduction in vascular density diminish cell adhesion through acid-induced ECM degradation and loss of gap junctions and through the other mechanisms reviewed in Section 9.1. In other words, a reduced vascular density will favor diminished cell adhesion although mutations in intracellular pathways may result in reduced adhesion even in the presence of normal extracellular concentrations of oxygen and acid. The solid contours in Figure 9.3 show the calculated interface between

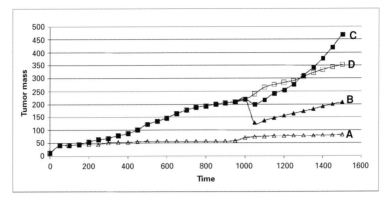

Figure 9.4 The progress of the two-dimensional tumor mass for the simulations shown in Figure 9.3: A, high cell adhesion; B, low cell adhesion and antiangiogenic therapy starting at $t = 1000$; C, low cell adhesion and vasculature normalization starting at $t = 1000$; D, the same as B but without therapy. The tissue density is assumed to be constant and equal to 1 [722], and thus the mass is nondimensionalized by the square of the diffusion length. Reprinted from Cristini *et al.* [145], with permission from the American Association for Cancer Research.

tumoral and nontumoral tissues, and black regions denote necrotic areas. Immature capillaries newly formed during angiogenesis are also shown. Some capillaries have formed loops through anastomosis and conduct blood, whereas others have not. The faint contours show the density of the "free" endothelial cells migrating chemotactically after sprouting from pre-existing vessels (not shown) in the outer tissue towards the source of angiogenic regulators in the tumor.

In all cases gradients of the oxygen level and cell nutrient level are established, the levels being higher in the outer tissue away from the tumor and lower at the lesion location owing to the higher cellularity and uptake therein. Frame (a) in Figure 9.3 corresponds to a sufficiently high value of the cell adhesion parameter that the simulated tumor growth is morphologically stable. Owing to the heterogeneity of the nutrient and oxygen distribution following diffusion from pre-existing vessels in the outer tissue and uptake by tumor cells, the simulated tumor has formed necrotic regions where the nutrient concentration is very low. At this point the tumor has reached a diffusion-limited equilibrium, in which the rate of proliferation in the viable outer layers of the cells balances the rate of cell mass destruction in the necrotic regions. Angiogenic regulators (not shown) emanate from the penumbral region of hypoxic but viable cells adjacent to the necrotic region and diffuse radially outward, reaching pre-existing vessels and triggering angiogenesis. Note that penetration of the tumor by new capillary sprouts has occurred. Even after angiogenesis, the tumor maintains a compact morphology because of high cell adhesion. This result is quantified by the curves labeled "A" in Figures 9.4 and 9.5. Tumor mass growth is very slow and the shape factor is at a minimum, corresponding to a nearly spherical tumor. Herein the shape factor is defined as the perimeter-to-area ratio for the simulated two-dimensional tumors, in order to quantify its fragmentation and invasiveness.

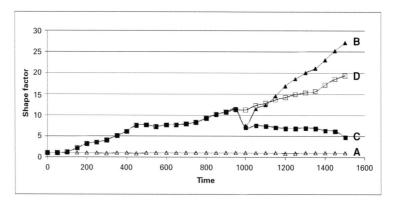

Figure 9.5 The progress of the simulated two-dimensional tumor shape factor for the simulations shown in Figure 9.3. A, high cell adhesion; B, low cell adhesion and antiangiogenic therapy starting at $t = 1000$; C, low cell adhesion and vasculature normalization starting at $t = 1000$; D, the same as B but without therapy. The perimeter is the total length of the tumor–tissue interfaces for each frame in Figure 9.3, and the area is the total mass of each lesion. Reprinted from Cristini *et al.* [145], with permission from the American Association for Cancer Research.

Frame (b) in Figure 9.3 depicts a simulated lesion corresponding to a smaller value of the cell-adhesion parameter. In this case, as the tumor progresses it experiences a diffusional instability [101, 149, 722] that leads to a morphology very different from that in the previous simulation at the same simulation time. The heterogeneous nutrient and oxygen distribution that leads to nonuniform cell proliferation and migration (chemotaxis would enhance this behavior) is responsible for the diffusional instability leading to the proliferation and spread of tumor-cell clusters. In other words, the tumor morphology is shaped by nutrient and oxygen levels that dictate where proliferation and spread will preferentially occur. For the parameters used here, cell adhesion is reduced owing to the presence of hypoxia and acidosis and, as a result, it is insufficient to keep proliferating cells together. The tumor lesion breaks up into fragments that move away from the original location, following the nutrient and oxygen concentration gradients. This motion has been analyzed using computer simulations [722] and involves the accumulation of cell mass in a fragment's leading edge due to the higher oxygen and nutrient concentration therein and to cell death or lesser proliferation in the fragment's trailing edge. Regions where the instability first develops continue to grow at a higher rate than the rest of the tumor mass, leading to bulbed shapes and to the separation of fragments or clusters of cells that grow into regions of higher nutrient and oxygen concentration. Fingering invasive structures predicted by these simulations are consistent with those from the *in vitro* experiments reported in [230, 528]. Note that there is evidence that actual cancer-cell migration is different from that for model cells such as fibroblasts, in that tumor cells develop migrating cell clusters [232, 235] rather than migrating alone. The morphological instability described through these simulations may provide the underlying mechanism to explain this difference.

In vitro experiments (Figure 9.6) on a human glioblastoma multiforme cell line [230], showed that spheroids first grow to a diffusion-limited millimeter size. When

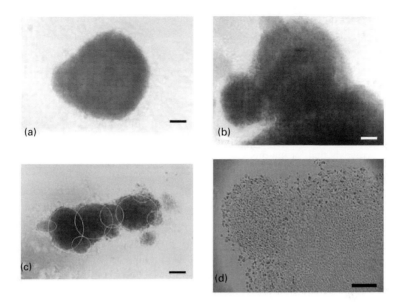

Figure 9.6 Morphology instabilities observed *in vitro* in human glioblastoma spheroids. The individual spheroid subcomponents are highlighted in (c). The spheroid cross section in (d) shows that at the time of separation subcomponents are mainly composed of viable cells. Bar, 130 μm. Reprinted from [145], Cristini *et al.* (2005), with permission from the American Association for Cancer Research.

proliferation rates are low, cell adhesion and possibly other mechanical forces exerted through the extracellular matrix are sufficient to maintain compact spheroid shapes. When proliferation rates are high (at high serum or glucose concentrations), the spheroids are observed to become unstable, assuming dimpled shapes as predicted by the simulations. The photographs in Figure 9.6(a), (b) reveal that sub-spheroids sometimes emanated from the main tumor, separating as a "bubble" of cells or as spheroid fragments. This process could repeat itself on sub-spheroids, leading to recursive sub-spheroidal growth and separation as the main mechanism of spheroid morphogenesis and invasion of the surrounding environment. Figure 9.6(c) shows a colony of successive generations of spheroids from one original "mother" spheroid (the white lines are guides to the eye to identify individual sub-spheroids). Figure 9.6(d) is a histological cross section of a spheroid in the process of splitting from a sub-spheroid. The main spheroid has a viable rim of cells and a central necrotic area. The sub-spheroid has only proliferating cells (no necrosis) and may further grow, after separating, to develop its own necrotic core. The morphological instability observed in these simulations and experiments is characterized by growing low-wavenumber modes, that is, a few "bumps" (typically three or four) develop on the original spherical tumor surface and evolve into separate spheroids. Low-wavenumber unstable modes are typical of diffusion-driven morphological instabilities [149]. These instabilities are regulated by a model parameter [149] that is proportional to the ratio describing the relative strengths of the cell proliferation driving instability and the cell adhesion resisting it. Thus increasing proliferation or decreasing cell adhesion has the equivalent effect of favoring instability. Note that for high values of this parameter (e.g., very low local-cell adhesion) even high wavenumber modes become unstable [149]. This implies that if a single cell mutates and becomes highly aggressive it can leave the tissue mass, migrating and proliferating radially away from the main tumor.

Anti-angiogenic therapy has been initiated on the tumor in frame (c) of Figure 9.3 (at dimensionless time $t = 1000$) and is simulated by decreasing the rate of transfer of nutrients and oxygen from the neovasculature. The curve labeled B in Figure 9.4 reveals that, at the onset of antiangiogenic therapy, the tumor mass initially shrinks by about one-half, and then later recovers through renewed proliferation. Hypoxia caused by therapy leads to enhanced tumor fragmentation, as documented *in vivo* in [58] and elsewhere. Interestingly, frame (b) of Figure 9.3 also shows that some tumor-cell clusters tend to co-opt the vasculature in order to maximize nutrient uptake, as documented previously [392, 398, 576]. Tumor fragmentation is quantified in this simulation by the large shape-factor increase for B illustrated in Figure 9.5. In contrast, the curves labeled D in Figures 9.4 and 9.5 correspond to evolution of this same lesion without therapy. When comparing D (no therapy) with B (therapy), it is important to note that without therapy the tumor mass, shown in Figure 9.4 is higher than after therapy, which is the expected result. However, after therapy the shape factor is about 50% higher than before therapy, reflecting increased scattering and invasiveness in response to hypoxia. Frame (c) in Figure 9.3 corresponds to low cell adhesion, as in (b) but with a more uniform and efficient distribution of newly formed vessels. The simulation predicts that this "vascular normalization" leads to reduced oxygen and cell nutrient gradients and hence to the re-clustering of cells into bigger fragments with a more compact morphology. This result could be achieved by pruning immature and inefficient blood vessels, leading to a more normal vasculature of vessels reduced in diameter, density, and permeability [342, 344].

This effect is quantified in curve C of Figure 9.5, from a simulation. The shape factor and thus the lesion's invasiveness are minimized. Note that the more rapid tumor growth predicted after vascular normalization (see curve C of Figure 9.4) is an artifact of the oversimplified implementation of vascular normalization therapy in the model; this implementation involved increasing the rate of transfer from the vasculature and also increasing the local vessel density. The vasculature responds by not only producing a more homogeneous nutrient profile but also a higher level of nutrient. In addition to this, in reality tumor mass growth with compact, nearly spherical, shapes would be mechanically constrained by the surrounding tissue.

9.2.3   Clinical observations

The work presented so far predicts that the morphologic instability of a tumor mass, i.e., a morphology resulting in "roughness" or harmonic content [149, 416] of the tumor margin, may provide a powerful tissue-invasion mechanism since it allows tumor cells to escape the growth limitations imposed by diffusion (even *in vitro*, [230, 528]) and invade the host independently of the extent of angiogenesis [145, 230]. Experiments with various glioma models *in vivo* [58, 392, 398, 576] have supported these findings. For example, recently published images of rat glioblastoma *in vivo* [441] show that, while the bulk tumor is perfused by blood, infiltrative cell clusters are much less perfused or not at all. These may be universal considerations that apply to tumor invasion across different tissue types [166, 235].

Changes in tumor cells at the biochemical and genetic levels are also implicated in tumor progression. Mutations in genes that regulate cell cycle and adhesion result in the unrestrained proliferation, invasion, and accumulation of further genetic damage characteristic of high-grade disease [63, 443, 466]. In particular, in glioblastoma the receptor for the epidermal growth factor oncogene (EGFR) is frequently overexpressed, amplified, or mutated [395] and promotes mitosis [63] and tumor progression *in vivo* [498] and inhibits apoptosis [488]. See for example [43, 632, 718] for recent mathematical models on the effect of EGFR gene expression on tumor growth patterns and [31] for the phenomenological modeling of multiple mutations. However, the tumor suppressor genes TP53 and retinoblastoma (Rb) downregulate cell division [63] and, secondarily, affect oxygen and/or nutrient consumption, while the phosphatase and tensin homolog gene (PTEN) controls angiogenesis, migration, and invasiveness [665]. These genes are inactivated in most malignant brain tumors [332].

In this section we describe an application of the mixture model of Chapter 5 to study glioblastoma growth. Figure 9.7 shows the onset of diffusion-driven morphologic instability [145, 149, 230, 416] from simulations of an avascular condition that are similar to the work presented in Section 9.2.1 for a single-phase model. As with the latter model, perturbations that arise in the spatial arrangement of cells at the periphery of the spheroids in culture (Figure 9.7(a)) are consistently replicated (Figure 9.7(b)). Once this shape asymmetry is created, local-cell substrate gradients (Figure 9.7(d)) cause spatially heterogeneous cell proliferation and migration (Figure 9.7(c)), as cells that are exposed to more substrates proliferate more. Mechanical forces, e.g., cell–cell and cell–matrix adhesion, which are in general stabilizing [149, 230, 235], are in this case not strong enough to prevent morphologic instability.

When the instability persists, it can lead to the proliferative growth of bud-like clusters or "bumps" of cells (Figures 9.7 and 9.8), as described in the preceding sections. *In vitro*, these may eventually detach as sub-spheroids from the parent spheroid [230], analogously to microsatellites *in vivo*, and also may represent the initial stage of the growth of cell chains, strands, or detached clusters [235, 584] observed *in vitro* and *in vivo* (Figures 9.7(b), (d)). Simulations reveal that when chemotaxis or haptotaxis is dominant, e.g., if mitosis is downregulated, protrusions begin as high-frequency perturbations (linear stability predicts the growth of short wavelengths [147, 149]) on the tumor surface and then develop into cell chains and strands [235], Figure 9.7(a). When proliferation is the prevailing pro-invasion mechanism, the buds grow into round fingers, Figure 9.7(b), which may detach as clusters [230] (linear stability predicts the growth of long-wavelength perturbations [145, 149, 416]). This is clearly seen in Figure 9.8(c), where cells acquire a hypoxia-induced migratory phenotype. These simulations are supported by experimental observations under hypoxic conditions (Figure 9.8(b), (d)) [528, 576].

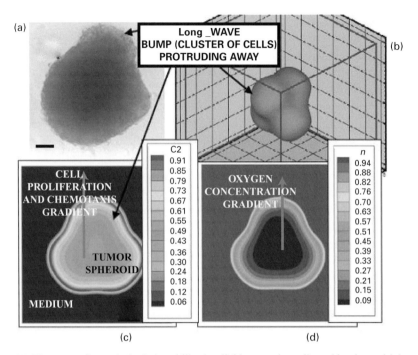

Figure 9.7 The onset of morphologic instability in glioblastoma is replicated by the multiphase model. The collective motion of cells shown is due to heterogeneous proliferation (and possibly chemotaxis) up concentration gradients and under weak cell adhesion. (a) *In vitro* glioblastoma spheroid (bar, 130 μm) and (b)–(d) computer simulations. Adapted from Frieboes *et al.* [230]; reprinted with permission from the American Association for Cancer Research.
(b) Representation of observed tumor shape; (c) Local mass fraction C2 of viable tumor cells.
(d) Oxygen concentration n ($n = 1$ in the medium). Note the hypoxic core in (d) and the correspondingly low cell viability levels (c) within, and the high viability at the outer rim.
(b)–(d) Reprinted from Bearer *et al.* [57], with permission from the American Association for Cancer Research.

As necrosis and substrate gradients develop within the tumor, proliferation may be downregulated and motility upregulated. While various substrate components (e.g., oxygen, glucose, growth factors, metabolites) may be determinants of these phenotypic changes, the present work focuses on oxygen because of its well-established role in regulating cell proliferation and motility (e.g., [528]). Figure 9.9 shows a migratory or hypoxic growth stage, where palisading cells are seen from simulations (a), (b) and from histology (c), (d).

In the simulations, which used the hybrid continuum–discrete approach described in Chapter 7, individual cells originate in the peri–necrotic region (the center) and undergo the phenotypic changes described above. In particular, the cells downregulate proliferation and migrate via chemotaxis up oxygen gradients and via haptotaxis up bound chemokines in the extracellular matrix, which the cells also degrade and remodel. The resulting morphology (Figure 9.9(a), (b)) compares well with the pathology data (Figure 9.9(c), (d)) (see also [229]). As strands of cells move away from the perinecrotic

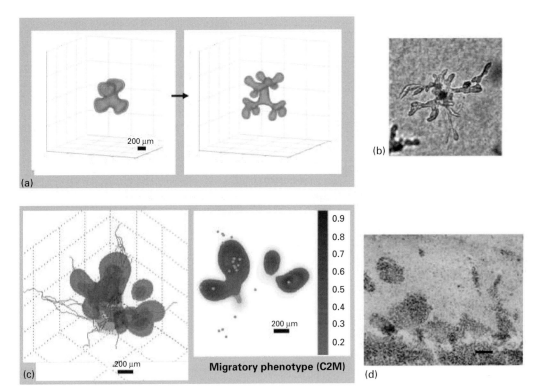

Figure 9.8 Variability and persistence of morphologic patterns predicted by mathematical modeling simulating heterogeneity *in vitro* ((a); refs. [230, 437]) and *in vivo* ((c); refs. [229, 697]). In both cases, only the higher-grade clone 2 is simulated (with no mutations), although the clone is allowed to acquire a hypoxia-induced migratory phenotype [57]. (a) Proliferation is downregulated and the clone migrates up the oxygen gradients towards the far-field boundary (the computational box is not shown; the arrow indicates the time direction). (c) Migratory phenotype of tumor clone 2 (C2M, light gray) and the less motile clone 2 (C2, dark gray). The proliferation of both clones is regulated by the oxygen levels. The darkest interior regions of the three-dimensional graph denote necrosis. The two-dimensional horizontal slice in the right shows the distribution of C2M; the small circles indicate the cross sections of mature, blood-conducting, vessels. Morphologic instability occurs in both simulations because this clone's cell adhesion is low, resulting in cell strands in (a), and fingers and detached clusters in (c). The results of the simulations are supported by experimental observations revealing morphologic instability after hypoxia has been induced in spheroids *in vitro* ((b): reprinted from *Cancer Cell*, vol. 3, Pennacchietti *et al.* [528], p. 354, © 2003 with permission from Elsevier) and in xenografts *in vivo* ((d): bar, 80 μm; reprinted from *Neoplasia*, vol. 2, Rubenstein *et al.* [576], p. 311, © 2000, with permission from Neoplasia). (a), (c): reprinted from Bearer *et al.* [57], with permission from the American Association for Cancer Research.

regions and into the brain stroma, they may trigger angiogenesis, which as a wound-healing process increases the nutrient availability in the microenvironment. Increased oxygen and nutrients would result in the upregulation of proliferation and the downregulation of motility, creating the microsatellite cell clusters in Figure 9.8(d) (mouse brain) and Figure 9.11(b) (human specimen, below).

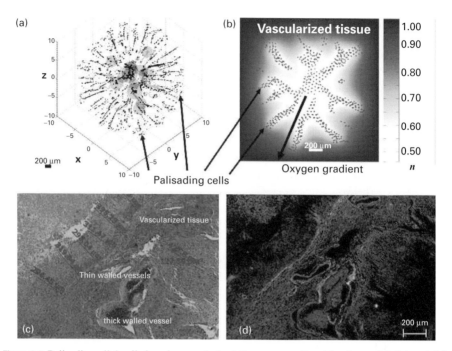

Figure 9.9 Palisading glial cells invade vascularized tissue, as predicted by the model ((a) and (b)) and as observed in histology specimens ((c) and (d)). Computer simulation showing palisading cells escaping from the perinecrotic region (dark gray) by undergoing a hypoxia-induced phenotypic change to upregulate motility and downregulate adhesion and proliferation. Cell migration occurs via chemotaxis and haptotaxis, in response to gradients in the oxygen and ECM concentration, respectively [57]. Dark dots, conducting vessels; light dots, nonconducting vessels. (b) The background shows the distribution of oxygen concentration; $n = 1$ in vascularized tissue and is lower in the tumor (the light-colored perinecrotic region). C, 4× magnification of a high-grade glioblastoma interior. (d) Corresponding fluorescent image showing the vessels. As predicted by the simulations ((a) and (b)), palisading malignant glial cells (c) invade the vascularized tissue (d), amidst a tangle of thick-walled and thin-walled large vessels and a few smaller vessels from two areas of necrosis. Also, note the distances between the necrosis and the vessels (c), corroborating the choice for the diffusion penetration length [57]. Further phenotypic changes (e.g., the downregulation of motility and the upregulation of proliferation) may occur when migrating cells reach tissue regions richer in substrates, leading to the morphologies described in Figures 9.10(d) and 9.11(a)–(d). Reprinted from Bearer *et al.* [57], with permission from the American Association for Cancer Research.

We now turn our attention to the effects of the genotype. Although there is abundant information about the biological and clinical behavior of these tumors and the genetic pathways involved [443], a quantitative link between genotype and disease progression remains elusive. Glioma progression is characterized by unscheduled glial-cell proliferation and the infiltration of normal brain tissue [466] and exhibits a complex interaction of multiple nonrandom genetic events that include the activation of proto-oncogenes and the inactivation of tumor suppressor genes [63].

For simplicity, for the simulations in [57] it was assumed that the tumor-cell population contains two transformed species, each characterized by a genotype consisting of a set of oncogenes and a set of tumor-suppressor genes. It was also assumed that a set of oncogenes can increase the uptake of nutrient to twice the normal rate and that a set of mutated tumor-suppressor genes downregulates the apoptosis rate by about 10%. These rates affect the simulated tumor-species evolution, as described in Section 5.10. Species 2 denotes the more aggressive species. In the study [57], it was assumed further that nutrient levels remain sufficiently high that the rate of cell apoptosis dominates the rate of cell necrosis. Computer simulation of a human glioma over time in the proliferative growth stage produces finger-like extensions (Figure 9.10).

The expression of oncogenes and the absence of tumor-suppressor pathways initially results in the net growth of a relatively low-grade type of morphology. After two months, the tumor has expanded to approximately 3 mm by co-opting the vasculature (not shown) while retaining a compact shape with negligible necrosis (Figure 9.10(a)). However, increased nutrient demand generates hypoxic and other substrate gradients pointing radially outwards from the tumor (not shown). A second, more proliferative, clone is generated by ongoing hypoxia-driven [347] mutations and is starting to grow (lower left corner, shaded area). Its higher cellular uptake introduces perturbations in the spatial gradients of oxygen, further enhancing local hypoxia. These gradients generate spatially heterogeneous cell proliferation and migration. After four months, this perturbation triggers a morphologic instability, which noticeably deforms the tumor mass (lower left). Hypoxia and necrosis are present within the regions where the more malignant clone grows. Shape instability causes clusters of clone 2 to protrude "finger-like" (the darker regions) into the mass of clone 1 first and into the host brain later, growing at the expense of the less proliferative clone and the host tissue. This detachment of clusters has also been observed to occur in the model (cf. Figure 9.8(c), [230]). These fingers grow away from the bulk tumor and tend to follow the substrate gradients.

In about six months' time, the aggressive, invasive, proliferation of clone 2 (the darker regions) enables it to infiltrate almost all regions of the tumor, in particular around the boundary, and leads to a higher-grade tumor. A bud-like protrusion emerges on the tumor at lower left. Hypoxic, necrotic, areas continue to expand (Figure 9.10(b)). In eight months, the glioma has aggressively infiltrated the surrounding brain tissue. Clone 1 is being confined by competition with clone 2. Extensive necrosis is present. Additional buds have appeared, and the initial (middle) bud has grown into an invasive finger. Strands and clusters of clone 2 drive the growth of the finger and buds (note the extent of the darker area). Clone 1 has been mostly eliminated from this region of the tumor, and remains stagnant. In 12 months the surrounding brain has been severely compromised. The expansion of clone 2, accompanied by continued necrosis, is now the main determinant of the tumor morphology. The tumor reaches a size of approximately 4 cm in a little over one year of simulated time, which is consistent with human glioma progression (Figure 9.10(c) shows only a portion of the tumor at one year).

Different tumors are likely to have different genomic instability factors, i.e., different types and mutation rates. The idealized tumor in Figure 9.10(a)–(c) was programmed to exhibit the progressive appearance of one highly malignant clone (clone 2). In reality

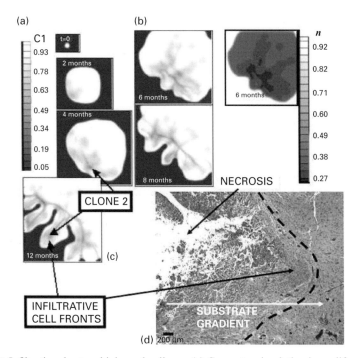

Figure 9.10 Infiltration due to a high-grade glioma. (a) Computer simulation in proliferative growth stage (the field of view is 6–10 mm). For each time snapshot, two-dimensional slices depict the spatial distribution of two different clones: lower-grade clone 1, C1, evolving to higher-grade clone 2; $C2 = 1\text{-}C1$. Gray scale: the local mass fraction of C1. The arrows pointing to darker areas in the tumor indicate C2. (b) Simulated oxygen concentration n at 6 months, indicating hypoxic gradients ($n = 1$ in normal brain and is lower in a tumor). The larger oxygen uptake of C2 enhances local hypoxia (see, e.g., the bottom left tumor corner) and leads to shape instability in which clusters of C2 cells protrude finger-like into the tumor mass of C1 first, and into the host brain later. The model predicts that C2 expansion is the main determinant of tumor morphology. (c) Tumor detail at 12 months, showing invading fingers. (d) Histology section of a tumor front (left) showing a finger invading the normal brain (bar, 200 μm). Note the clearly demarcated margin (left of the broken line) between the tumor and the more normal brain. Neovascularization at the tumor–brain interface is visible as darker spots (on the right) in the brain parenchyma, implying that substrate gradients drive collective tumor-cell infiltration into the brain. The morphology and size of the invading finger are consistent with simulation predictions ((a) and (b)), suggesting the proliferative growth stage. Reprinted from Bearer *et al.* [57], with permission from the American Association for Cancer Research.

multiple clones may arise with varying degrees of malignancy. A histology section of glioblastoma from one patient revealed the tip of a round invading finger (Figure 9.10(d)), which is consistent with the morphology and size of these infiltrative cell clusters predicted by simulations during the proliferative growth stage (cf. Figure 9.10(c) at 12 months, where the tip of one protruding cell front is about 2 mm in size). The fact that tumor cells rely on vessels beyond the protrusions – and may grow towards the blood vessels that they stimulate [50, 229, 542] – also suggests a proliferative growth stage since these vessels increase substrate availability in the microenvironment. Older tumor

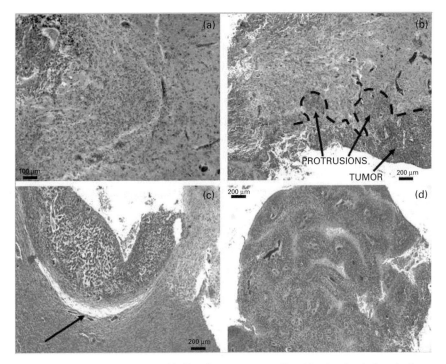

Figure 9.11 Histologic sections reveal infiltrative patterns predicted by the simulations in the proliferative growth stage. (a) Higher magnification of the invading finger shown in Figure 9.10(d) (adapted from Frieboes *et al.* [229], with permission from Elsevier). (b) Two invasive protrusions are seen emanating from the tumor mass. (c) The boundary of another protrusion from a different specimen is indicated (arrow); there also seems to be ECM degradation. (d) Morphology details of one large bulb-shaped tumor protrusion from (b) (reprinted from Frieboes *et al.* [229], with permission from Elsevier). The interior reveals clusters of viable cells surrounding blood vessels and, further away, areas depleted of cells indicating increased substrate availability in the microenvironment. Altogether these data suggest the proliferative growth stage predicted by the model for this invading morphology. Parts (b) and (d) are reprinted from Bearer *et al.* [57], with permission from the American Association for Cancer Research.

vessels may have thicker walls, which are not as permeable for nutrient and/or oxygen exchange, and they may become occluded owing to increased pressure from the tumor mass, further promoting substrate gradients.

Additional histology sections from four glioblastomas (Figure 9.11) reveal protruding fronts of cells pointing away from a necrotic area into an area of the host brain where neovascularization is evident. These invading fronts are also consistent with the tumor boundary morphology predicted by simulation in the proliferative growth regime (Figure 9.10(a)–(c)).

While infiltrative shapes were consistently observed in histological sections [57], the model predicts that their size may vary depending on the stage of growth. For example, these shapes can vary from extremely slender fingers in the hypoxic-growth regime to single rows of palisading cells migrating up substrate concentration gradients (Figure 9.9(a), (b)) and thus away from hypoxic regions, as seen in the histology (Figure 9.9(c), (d)).

9.3   Modeling outlook

The multidisciplinary tumor modeling reviewed in this chapter aims to establish a framework for the prediction and interpretation of experimental results, quantitatively linking invasive tumor morphologies to events at the microscale such as spatial variations in cell substrates, heterogeneous vascularization, and mutations in the genotype. The consequences of this variability can be highly multiscalar since morphology and oxygen and nutrient gradients at the tumoral scale and molecular interactions at the cell scale are strongly linked. We have shown applications of both single-phase and multiphase tumor models, as well as a hybrid approach that can contribute to the understanding of these multiscale phenomena by providing a means to incorporate cell-level processes in tissue-scale simulations. The results support the hypothesis, first proposed in 2003 by Cristini *et al.* [149] and developed in subsequent work [57, 145, 228–230, 437, 584, 585, 697, 722], that a heterogeneous oxygen and nutrient supply may provide a diffusional instability mechanism that drives tumor growth and invasiveness. These considerations may be relevant during chemotherapy, radiotherapy, and anti-angiogenic therapy, all of which could introduce spatial and temporal variations in critical cell substrates [227, 620, 621]. In contrast, tumors may approach a compact noninvasive morphology when cell adhesion or other stabilizing mechanical forces (e.g., tumor encapsulation) are maximized. Compact tumor morphologies may be achievable by maintaining uniform nutrient levels at the cellular scale and homogeneous microenvironmental conditions, thus suppressing instability [145].

The multidisciplinary modeling of cancer enables the prediction of cellular and molecular perturbations that may alter invasiveness and that can be measured through changes in tumor morphology. Therefore, morphologies obtained from the model could be used both to understand the underlying cellular physiology and to predict subsequent invasive behavior. For example, novel individualized therapeutic strategies could be designed in which microenvironment and cellular factors are manipulated to decrease invasiveness and promote well-defined tumor margins – an outcome that would also benefit treatment by improving local tumor control through surgery or radiation. In addition to existing strategies that act on relevant cellular behaviors (e.g., the promotion of tumor-cell adhesiveness [145, 230, 235, 528]) or that target oncogenes such as EGFR, tumor morphological stability could be enhanced by improving nutrient supply [145, 437], thus enforcing a more homogeneous microenvironment and normoxic conditions. This could be achieved through "vascular normalization" [145, 342] or uniform nanoparticle delivery [211], e.g., releasing oxygen and anti-angiogenic drugs. By providing adequate and uniform nutrient and oxygen, there would be the additional benefit that more benign clones would be maintained, helping to keep malignant clones under control by competition for oxygen and cell nutrients. Further, by maintaining microenvironmental homogeneity the effects of genetic mutations that lead to morphological instability may be minimized (e.g., [366]) without direct intervention at the genotype level.

The work presented in this chapter shows that applying biologically founded mathematical modeling to quantify the connections between the microenvironment, tumor morphology, genotype, and phenotype may direct prognosis beyond the limitations of current methodologies and may suggest new directions in the way that cancer growth and invasion are regarded. This type of modeling may be used to study system perturbations by therapeutic intervention and may aid in the design of novel clinical endpoints in therapeutic trials. By integrating the model with patient data for key tumor phenotypic and microenvironmental parameters [584], model results could be used to enhance clinical outcome prognostication. Initial conditions regarding a tumor's physical location, structure, and vasculature, e.g., as obtained from dynamic-contrast-enhanced magnetic resonance imaging (DCE–MRI), and possibly coupled with computed tomography (CT) would be translated using a computer program to the model's coordinate system [584], e.g., a finite-element computational mesh discretizing the space occupied by the tumor and host tissues [697]. Viable-region spatial information and microvasculature structure would be obtained from histopathology [229]. Vasculature-specific information could be defined from DCE–CT, yielding blood volume, flow, and microvascular permeability parameters [584]. Other input data include cell-scale parameters (e.g., proliferation rates). The model then calculates local tumor growth, angiogenesis, and the response to treatment under various conditions by solving in time and space the conservation and other equations at the tissue scale.

Invasive characteristics may strongly influence whether a tumor can be effectively treated by local resection, and may suggest specific treatment options [31, 227, 585, 621]. Observations of tumor morphology, for example, could indicate the presence of hypoxia and, therefore, the potential to respond to oxygen-dependent treatments such as radiation therapy or certain chemotherapy treatments. By quantifying the close connection between the observable morphology of the tumor boundary and cellular and molecular dynamics, the modeling of cancer using a multidisciplinary approach provides a quantitative tool for the study of tumor progression and diagnostic and prognostic applications. This connection is important because the dynamics that give rise to various tumor morphologies also control invasiveness. In particular, by describing the morphology as a function of parameters dependent on cellular and environmental phenomena [614, 667], multidisciplinary modeling quantifies, under the unifying umbrella of morphological stability analyses, the often seemingly diverse and unrelated morphologies and invasive phenotypes observed clinically and experimentally.

10 Agent-based cell modeling: application to breast cancer[1]

With P. Macklin and M. E. Edgerton

In Chapter 6 we discussed an agent-based cell model that can be applied to a variety of biological systems, with a particular emphasis on epithelial cancers. We now illustrate the model by applying it to breast cancer and demonstrating its use in obtaining theoretical biologic and clinical insights, including quantitative predictions that can be assessed using patient immunohistochemistry and histopathology data. The theoretical biologic and clinical significance is discussed.

10.1 Introduction

Ductal carcinoma in situ (DCIS) is the most prevalent precursor to invasive breast cancer (IC), the second-leading cause of death in women in the United States. The American Cancer Society predicted that 50 000 new cases of DCIS alone (excluding lobular carcinoma in situ) and 180 000 new cases of IC would be diagnosed in 2007 [24, 346]. Co-existing DCIS is expected in 80% of IC, or 144 000 cases [397]. Because DCIS is a known precursor to IC, this leads us to hypothesize that up to 75% of DCIS cases progress to invasion prior to detection by screening mammography. While DCIS itself is not life-threatening, it is a very important precursor to IC because (i) it can be treated and (ii) if left untreated it is likely to progress to IC, which is a deadly disease [368, 515, 582].

Women tend to prefer breast-conserving surgery (BCS), also known as lumpectomy, to complete mastectomy for the treatment of DCIS [617]: in the United States today, approximately two-thirds of women diagnosed with DCIS will opt for BCS over mastectomy. Women who undergo BCS face two problems. First, an estimated 38%–72% of women seeking BCS will not have their entire tumor removed in one surgery and may require up to three surgeries (called re-excisions) for complete removal of the DCIS [106, 134, 179]. Second, DCIS recurs at the same location greater than 20% of the time in patients who undergo BCS alone [525]. To combat this recurrence, women are advised to undergo radiation therapy to the breast, which induces the residual cells of DCIS to apoptose. Even in women who have been treated with surgery and radiation, DCIS recurs approximately 10% of the time [525]. Half these recurrences already show

[1] This chapter is an extension Macklin *et al.* (2009) [433], and includes work in preparation by Macklin *et al.* [434] and Edgerton *et al.* [195].

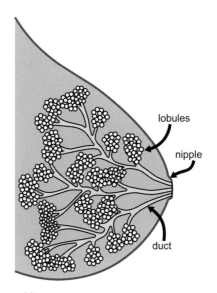

Figure 10.1 Breast duct tree architecture.

progression to invasive cancer (IC). The single most important underlying problem that contributes to both re-excisions and to recurrences is DCIS that is left inside the breast [616].

Hence, predicting the size and shape of DCIS is critical to successfully eradicating the disease in patients and preventing recurrences, which often progress to deadlier invasive carcinoma. In addition, understanding the progression from DCIS to IC is key to developing future treatments to improve patient survival. Mathematical modeling can play a role in both these tasks. In this chapter, we apply the agent-based model from Chapter 6 to DCIS. Such a model is well suited to patient-specific calibration, can be modularly extended to focus attention on specific aspects of biological interest, and can be used for generating testable scientific hypotheses. The model presented here can be incorporated into a broader, multiscale framework (such as that discussed in Chapter 7) from which patient-specific clinical predictions of DCIS outcome can be made [136, 193, 194, 434].

10.1.1 Biology of breast-duct epithelium

As an organ, the breast is organized as a system of 12–15 independent, largely parallel, duct systems: clusters of milk-producing lobules that feed into a branched duct system that terminates at the nipple [286, 472, 503, 687]. See Figure 10.1. The duct systems are separated by supporting ligaments and fatty tissue and drained by the lymphatic system (not shown) [651]. The ducts have a well-characterized microarchitecture: each duct is a tubular arrangement of epithelial cells, surrounded by myoepithelial cells (epithelial cells with muscle-like properties, such as the ability to contract the duct to transport milk) and a basement membrane (BM). The center of the duct, known as the lumen, is filled

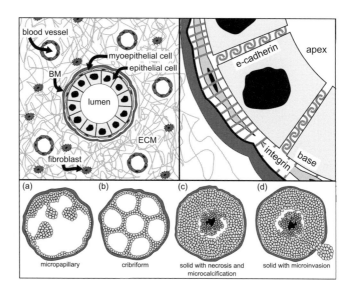

Figure 10.2 Top left: typical breast duct micro-anatomy. Top right: Breast duct epithelial cell polarization. Bottom: major DCIS types and IDC. Reprinted with permission from [434].

with either milk (during lactation) or fluid. See Figure 10.2, top left. Surrounding and supporting the duct is the stroma, a scaffolding of collagen and other fibers (collectively called the extracellular matrix, or ECM) that is secreted and maintained by fibroblasts. The stroma also contains blood vessels that supply oxygen, glucose, and growth factors to the tissue. A key aspect of this architecture is that the epithelial cells in the breast duct have no direct access to oxygen and nutrients; instead, these must diffuse into the duct through the BM.

The arrangement of the epithelial cells in the duct depends upon the polarization of the cells and the anisotropic distribution of the various surface adhesion molecules. Integrins line the cell base and adhere to several ligands (generally laminin and fibronectin) on the BM; E-cadherin molecules cover the cell surface between the base and apex and adhere to E-cadherin molecules on neighboring cells [90]. See Figure 10.2, top right. The careful orchestration of integrin-mediated cell–BM adhesion and E-cadherin-mediated cell–cell adhesion helps to determine the tissue geometry [301, 685]. While the epithelial-cell population oscillates with the menstrual cycle (e.g., [374, 375]), on average it is maintained in homeostasis by a careful balance of cell proliferation and apoptosis. Microenvironmental changes can trigger internal signaling responses in the epithelial cells that lead to either proliferation or apoptosis, as warranted by required for satisfactory maintenance of the tissue architecture. After apoptotic cells disintegrate into apoptotic bodies, they are either absorbed by surrounding epithelial cells or digested by macrophages that travel through and along the BM [369, 389].

The integrin signaling pathway allows cells to detect detachment from the BM: when integrins become adhered to their ligands on the BM, they send signals within the cell that trigger the production of survival proteins (e.g., FAK) that inhibit p53-mediated apoptosis [330, 677]. Loss of attachment to the BM therefore allows apoptosis to occur, thus preventing the overgrowth of cells into the lumen [162]. E-cadherin signaling helps the cells to detect the presence or absence of neighbors: when E-cadherin on a cell is bound to E-cadherin on neighboring cells, its intracellular domain binds to and sequesters β-catenin near the cell membrane. This prevents β-catenin from transcribing the Cyclin D1, c-myc, and Axin2 proteins associated with cell-cycle progression. (See Section 2.1.5.) As a result, cell cycling is inhibited [69, 313, 428, 597]. When a neighboring cell dies, E-cadherin signaling is reduced thereby allowing the cell cycle to progress. This results in the production of a new daughter cell to fill in the gap in the duct epithelium. The epithelial cells also respond to hormones (intercellular signaling molecules) that bind to surface receptors. Estrogen, progesterone, androgen, prolactin, and epidermal growth factor all affect epithelial-cell proliferation and apoptosis decisions, such as the decision to increase proliferation prior to lacation (in order to enlarge the breast-duct system and prepare the lobules [32]) and or the decision to increase apoptosis during breast involution (the "shutdown" process after lactation [54]).

10.1.2 Biology of DCIS

Overexpressed oncogenes and underexpressed tumor suppressor genes can disrupt the balance of epithelial cell proliferation and apoptosis, leading to cell overproliferation. This can occur either by the accumulation of DNA mutations (genetic damage) [619] or epigenetic anomalies (e.g., alterations in heritable CH_3 methyl groups that suppress key oncogenes) [8]. The transformation from regular breast epithelium to carcinoma is thought to occur in stages. For simplicity, we neglect the benign, precursor, transformations (e.g., atypical ductal hyperplasia, or ADH [619]) and focus on DCIS.

In the most well-differentiated classes of DCIS, the epithelial cells maintain their basic polarity and anisotropic surface adhesion receptor distributions, resulting in partial recapitulation of the nonpathological duct structure within the lumen. These demonstrate either finger-like growths into the lumen (micropapillary, see Figure 10.2, bottom (a)) or arrangements of duct-like structures (cribriform, see Figure 10.2, bottom (b)) [618]. The cells in solid-type DCIS lack polarity and do not develop these microstructures. Instead, they proliferate until they fill the entire lumen (Figure 10.2, bottom (c)) [162]. The proliferating cells uptake oxygen and nutrients as they diffuse into the duct through the basement membrane, leading to oxygen and nutrient gradients (due to decreasing oxygen and nutrient concentrations with distance from the BM). If the central oxygen is sufficiently depleted, the interior tumor cells die and form a necrotic core of cellular debris (comedo-type solid DCIS, see Figure 10.2, bottom (c)) [618]. These cells are typically not phagocytosed by non-apoptotic epithelial cells (since there are none nearby) or macrophages (since they are too far from the BM). Instead, they swell and burst [49] and their solid (i.e., non-water) components are slowly calcified [638]. In fact, mammograms generally rely upon these calcifications for DCIS detection [137].

While it is tempting to regard DCIS as a progression from regular epithelium to cribriform or micropapillary ("partially transformed") and then to the solid type ("fully transformed"), there is insufficient evidence to support such a linear progression and indeed the mutation pathway from noncancerous epithelium to DCIS is currently an open question [205, 565]. The dominant type of DCIS in any particular case may well depend upon which genes are mutated; for example, the cribriform DCIS microarchitecture could be due to hyperproliferation in cells whose genes regulating polarization (particularly of E-cadherin) are intact. Owing to the current difficulty in fully characterizing DCIS carcinogenesis, there is an excellent opportunity for mathematical modelers to test competing theories by generating testable quantitative hypotheses.

Ductal carcinoma in situ is a pre-malignant cancer because the BM confines it to the duct system, blocking metastasis. However, it is an important precursor stage of invasive ductal carcinoma (IDC), where further genetic or epigenetic mutations lead to tumor cell motility along the BM, the secretion of matrix metalloproteinases, which degrade the BM, and subsequent invasion of the surrounding stroma (Figure 10.2, bottom (d)) [6, 615]. An estimated three-quarters of all DCIS cases are already invasive at the time of detection [24, 346, 397]. Thus, there is a substantial risk that an undetected DCIS precursor (e.g., vasopressin (ADH)) can progress to IDC between annual mammograms [193]. Predicting the behavior of DCIS is important to understanding and, it is hoped, to preventing progression to IDC.

10.2   Adaptation of the agent model

We now adapt the agent model from Chapter 6 to the geometry and biology of solid-type non-motile DCIS. Tumor cells in the viable rim can be quiescent ($\mathcal{Q}$), apoptotic ($\mathcal{A}$), or proliferative ($\mathcal{P}$). For simplicity, we allow cells in hypoxic regions, where the oxygen level $\sigma < \sigma_H$, to bypass the hypoxic state (i.e., we neglect $\mathcal{H}$ and immediately enter the necrotic state ($\mathcal{N}$), eventually becoming calcified debris ($\mathcal{C}$).[2] Thus, $\beta_H = \infty$. Because the cells are assumed non-motile, we neglect the motile state $\mathcal{M}$ and set $\mathbf{F}_{loc}$ to zero. We assume that there is no extracellular matrix in the duct lumen, and so $\mathbf{F}_{cma} = \mathbf{0}$. Cell–cell adhesion is assumed to be homophilic between E-cadherin molecules, and cell–BM adhesion is assumed to be heterophilic between integrins and uniformly distributed ligands on the BM. In the simulations below, we neglect the presence of noncancerous epithelial cells lining the duct.

10.2.1   Oxygen and metabolism

In the DCIS model we assume that oxygen is uptaken at a rate λ_p by proliferating cells and at a rate λ_n by nonproliferating cells (including quiescent and apoptotic cells) and that the oxygen level decays at a rate λ_b in the necrotic core (containing necrotic cells and calcified debris) and in the duct lumen. In regions containing a mixture of viable and non-viable tissue and lumen, we assign a volume-averaged uptake rate. We discuss the orders of magnitude for λ_p, λ_n, and λ_b in Section 10.3.1 below.

10.2.2   Duct geometry

We denote the duct lumen by Ω and the duct boundary (the BM) by $\partial\Omega$. In this chapter, we will treat the duct as a rectangular region (a longitudinal cross section of a cylinder)

---

[2] In Chapters 6 and 10, σ and g denote the oxygen and glucose levels, which are generalized by the substrate level n in the remainder of the book. In Chapters 6 and 10, n denotes an integer.

Table 10.1. Main parameters for the agent-based model of DCIS

Parameter	Physical meaning	Data source
σ_{H}	hypoxic threshold	literature data [679]
λ_{p}	oxygen uptake rate by proliferating tumor cells	[220, 679] and analysis [434]
λ_{n}	oxygen uptake rate by non-proliferating tumor cells	[220, 679] and analysis [434]
λ_{t}	oxygen uptake rate by all tumor cells	see discussion in the text
λ_{b}	background oxygen decay rate	analysis [434]
$\langle \lambda \rangle$	mean oxygen uptake rate in viable rim	literature data [220]
L	oxygen diffusion length scale	literature data [220]
ℓ_{duct}	length of breast duct segment	set at 1 mm
r_{duct}	breast duct radius	histopathology measurement
β_{H}^{-1}	hypoxic survival time	simplified to 0
$\beta_{\mathrm{N}}^{-1}\ (=\beta_{C}^{-1})$	time to necrose and calcify	parameter study [434]
β_{P}^{-1}	mean cell cycle time	literature data [512]
β_{A}^{-1}	time to complete apoptosis	analysis [434] of lit. data [406]
α_{A}^{-1}	mean time to enter apoptosis	cleaved Caspase-3 immunostain
$\overline{\alpha}_{\mathrm{P}}^{-1}$	mean time to enter proliferative state	Ki-67 immunostain
R_{cca}	cell–cell adhesion-interaction distance	analysis [434] of cell deformation data [296]
R_{cba}	cell–BM adhesion-interaction distance	analysis [434] of cell deformation data [296]
V_{S}/V	solid fraction of individual cell	analysis [434] of literature data [440]
$n_{\mathrm{cca}}, n_{\mathrm{cba}}$	cell–cell and cell–BM adhesion powers in the potential function in Section 6.2.2	set equal to 1
$n_{\mathrm{ccr}}, n_{\mathrm{cbr}}$	cell–cell and cell–BM repulsion powers in the potential function in Section 6.2.2	set equal to 1
α_{cca}	strength of cell–cell adhesion	cell density in viable rim
α_{ccr}	strength of cell–cell repulsion	cell density in viable rim
α_{cba}	strength of cell–BM adhesion	set equal to α_{cca}
α_{cbr}	strength of cell–BM repulsion	set equal to $5\alpha_{\mathrm{cba}}$

of radius r_{duct} and length ℓ_{duct}. We terminate the left-hand side of the rectangular region with a semicircle, as an initial approximation to a lobule. See Figure 10.3 for a typical simulation view. We introduce a framework that allows us to simulate DCIS growth in an arbitrary duct geometry, such as near a branch point in the duct tree. We represent the duct wall implicitly by introducing an auxiliary signed distance function d (a *level set function*) satisfying

$$
\begin{cases}
d(\mathbf{x}) > 0, & \mathbf{x} \in \Omega, \\
d(\mathbf{x}) = 0, & \mathbf{x} \in \partial\Omega, \\
d(\mathbf{x}) < 0, & \mathbf{x} \notin \overline{\Omega} = \Omega \cup \partial\Omega, \\
|\nabla d(\mathbf{x})| \equiv 1.
\end{cases}
\tag{10.1}
$$

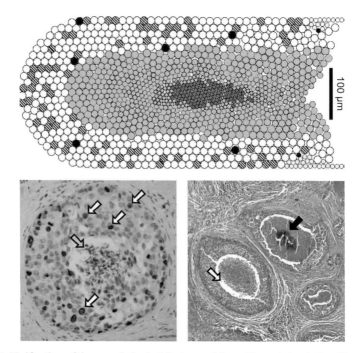

Figure 10.3 Verification of the morphological features of the calibrated simulation. Top: the simulation at time 30 days. The white cells are quiescent, the striped cells are proliferating (virtual Ki-67), the black cells are apoptotic (virtual cleaved Caspase-3), the light-gray cells are necrotic, and the central dark gray cells are calcified debris. The small cells along the BM are noncancerous epithelium. Bottom left: Ki-67 immunohistochemistry of a duct cross section. The white arrows show Ki-67-positive nuclei. The light-gray arrow shows necrotic debris. Bottom right: H&E staining showing calcification (black arrow) and the gap between the viable rim and the necrotic core (white arrow). Reprinted with permission from [434].

The gradient of the distance function gives the normal vector $\mathbf{n}$ (oriented into the lumen) to the interior duct surface. See [229, 435–439] where the method has been used to describe moving tumor boundaries.

Level set methods were first developed by Osher and Sethian [508] and have been used to study moving surfaces that experience frequent topology changes (e.g., mergers of regions or fragmentation), especially in the contexts of fluid mechanics and computer graphics; see the books by Sethian [602] and Osher and Fedkiw [507] and also [506, 508, 603]. For more information on the level set method and its application to fluid mechanics, see [4, 447, 448, 506–508, 602, 603, 646].

10.2.3   Intraductal oxygen diffusion

We model the release of oxygen by blood vessels outside the duct, its diffusion through the duct wall $\partial\Omega$ and within the duct lumen Ω, and its cellular uptake and decay (e.g., by reaction with molecules in the interstitial fluid), as follows:

$$\begin{cases} \dfrac{\partial \sigma}{\partial t} = D\nabla^2\sigma - \lambda\sigma & \text{if } \mathbf{x} \in \Omega, \\ \sigma = \sigma_B & \text{if } \mathbf{x} \in \partial\Omega, \end{cases} \tag{10.2}$$

where σ is the nondimensional oxygen level (i.e., the oxygen level scaled by the oxygen concentration σ_∞ in well-oxygenated tissue near the blood vessels in the stroma), D is the oxygen diffusion coefficient, λ is the local oxygen uptake or decay rate (generally λ averages 0.1 min^{-1} [512], for simplicity currently assumed equal for all cell types), and σ_B is the (nondimensional) oxygen level on the basement membrane.

The oxygen diffusion equation admits an intrinsic length scale $L = \sqrt{D/\overline{\lambda}}$, which we use to nondimensionalize space in Eq. (10.2); $\overline{\lambda}$ is a characteristic value of λ. From the literature, $\overline{\lambda} \approx 0.1$ min^{-1} and $L \approx 100$ μm [512].

10.2.4 Numerical method

We introduce an independent computational mesh for oxygen that discretizes the duct lumen with spacing $\Delta x = \Delta y = 0.1$ (approximately 10 μm spacing in dimensional space) in order to resolve the oxygen gradients. We use a cell interaction mesh with 1-μm spacing to avoid directly testing each cell for interaction with every other cell, hence avoiding an $\mathcal{O}\left(\# \text{cells}^2\right)$ computational cost.

We use an object-oriented C++ framework, where each cell is an instance of the class `Cell` and is endowed with instances of the class `Cell_properties` (proliferation and apoptosis parameters, initial radius and volume, etc.) and of the class `Cell_state` (cell state, position, velocity). We order the cells with a doubly linked list structure which allows us easily to delete apoptosed cells and insert new daughter cells.

The following procedure updates our agent-based model at time t to the next simulation time $t + \Delta t$.

1. Update the oxygen solution on the oxygen mesh using standard explicit forward Euler finite difference methods; see Chapter 8 and [434].
2. Iterate through all the cells to update the interaction mesh.
3. Iterate through all the cells to update their states according to Section 6.2.3. Update the necrosing cells' radii, volumes, and calcification as described above.
4. Iterate through all the cells to update their velocities as described above.
5. Iterate through all the cells to determine max $|\mathbf{v}_i|$. Use this to determine the new Δt using the stability criterion $\Delta t < 1/\max|\mathbf{v}_i|$.
6. Iterate through all the cells to update their positions according to their new velocities. We use forward Euler stepping ($\mathbf{x}_i(t + \Delta t) = \mathbf{x}_i(t) + \Delta t\, \mathbf{v}_i(t)$), although improved (e.g., Runge–Kutta) methods are straightforward.

These steps require at most cycling through all the cells. If interaction testing can be made similarly efficient then the overall simulation requires a computational effort that is linear in the number of cells.

Efficient interaction testing

With the spatial resolution given by the interaction mesh (1-μm spacing), we create an array of linked lists of interactions as follows.

1. Let $R = 2 \max_i \left\{ r_{cca}^i \right\}$.
2. Initialize the array such that each pointer is NULL.
3. For each cell i, append its memory address to the list for each mesh point within a distance R of its center $\mathbf{x}_i$.

Once complete, at any mesh point (i, j) we have a linked list of cells which are allowed to interact with a cell centered at or near (x_i, y_j).

We use this list whenever we compute a quantity of the form

$$\sum_j f\left(\text{cell}_i, \text{cell}_j\right)(x_k, y_\ell), \tag{10.3}$$

by contracting the sum to the members of the linked list at (x_k, y_ℓ). Because the number of points written to the array is fixed for each cell, this reduces the computational cost of cell–cell interaction testing to $\mathcal{O}\left(\# \text{ cells}\right)$, rather than the more typical $\mathcal{O}\left(\# \text{ cells}^2\right)$. Furthermore, this interaction data structure still allows arbitrary cell–cell interactions. Notice that this computational gain relies upon the fact that cells can only interact over finite distances.

10.3   Patient-specific calibration with patient data

To make the model predictive we must constrain the non-patient-specific parameters as much as possible (e.g., by literature searches and analysis of the mathematical model behavior across the parameter space) and calibrate the undetermined parameters using available patient-specific data. We now summarize key parameter estimates made in [434] and follow with a calibration protocol. In this discussion, we neglect hypoxia and motility and take $\beta_N = \beta_{NL} = \beta_{NS} = \beta_C$.

10.3.1   Estimating "universal" parameters

We first estimate parameters that are common to all patients, using literature searches of theoretical and experimental biology results and prior modeling efforts.

Cell cycle, apoptosis, and necrosis and calcification times

We estimated that the cell cycle time β_P^{-1} is 18 hours from the modeling literature (e.g., see [512]). We obtain the estimate $\beta_A^{-1} \approx 8.6$ hours and the estimate $\beta_N^{-1} \approx 30$ days in Section 10.4.1.

Oxygen parameters

From the literature, the mean cellular oxygen uptake rate is $\langle \lambda \rangle = 0.1 \text{ min}^{-1}$ (in the viable rim) and $L = 100$ μm. To estimate the hypoxic threshold σ_H, we examined the mitosis function $k_m(\sigma)$ used in [679], which is the basis of the breast cancer model in [220]. Ward and King [679] found that, at the step function limit, $k_m(\sigma) \propto H(\sigma - \sigma_c)$, where σ_c is the minimum, "cutoff", oxygen level for cell proliferation; they determined that $\sigma_c \approx 0.2$ experimentally when σ is nondimensionalized by σ_∞, the far-field nutrient value in well-vascularized, nonpathologic, tissue. Because the step function limit is similar to our parameter α_P, our parameter σ_H is analogous to σ_c in [679] and, as we have nondimensionalized oxygen by the nutrient value in well-vascularized, nonpathologic, breast tissue, we set $\sigma_H = 0.2$ as well. We observed in our immunohistochemical and histological images that the quiescent and proliferating viable tumor cells have the same general size; this suggests that the quiescent tumor cells are relatively metabolically active compared with the noncancerous, long-term-quiescent, cells; generally, these are smaller, with condensed nuclei (which relates to a lack of transcriptional activity), reduced mitochondrial populations [223], and less cytosol. Hence, we estimate that λ_p and λ_n are of similar orders of magnitude. In [434] a parameter study found that $\lambda_p \gg \lambda_n$ was inconsistent with the population dynamic and morphologic characteristics of DCIS observed in our immunohistochemistry and histologic data. For simplicity, we set $\lambda_p = \lambda_n = \lambda_t$ and $\lambda_t = \lambda_b$; the more general case was investigated in [434]. A statistical analysis of the viable rim thickness and tumor cell density in multiple breast ducts also supported our approximation $\lambda_p \approx \lambda_n$ [432].

Cell mechanics

We estimated the cells' solid-volume fraction V_S/V at approximately 10% by combining the published data of [440] with the assumption that the solid component is 1–10 times denser than water [433, 434]. We estimated the maximum cell–cell and cell–BM interaction distances R_{cca} and R_{cba} using published measurements of breast cancer cell deformations. Byers *et al.* [91] found the deformation of MCF-7 (an adhesive, moderately aggressive breast cancer cell line) and MCF-10A (a benign cell line) breast epithelial cells to be bounded at around 50% to 70% of the cell radius in shear flow conditions; this is an upper bound on R_{cca} and R_{cba}. Guck *et al.* [296] measured breast epithelial cell deformability (defined as additional stretched length over relaxed length) after 60 seconds of stress. Deformability was found to increase with malignant transformation: MCF10 deformed by 10.5%; MCF7 deformed by 21.4% but by 30.4% after weakening of the cytoskeleton; and MDA-MB-231 (an aggressive cancer cell line) deformed by 33.7%. Because DCIS is moderately aggressive, we used the MCF7 estimate and thus set $R_{cca}^i = R_{cba}^i = 1.214 r_i$. It is likely that the cell–cell and cell–BM adhesive forces decrease rapidly with distance, and so we used the lowest (simplest) adhesion powers that capture a smooth decrease at the maximum interaction distances, i.e., $n_{cca} = n_{cba} = 1$. For simplicity we also set $n_{ccr} = n_{cbr} = 1$.

10.3.2 Calibrating patient-specific parameters

We now present the patient-specific portion of the calibration protocol, as detailed in [434]. The following patient-specific data are available:

- the average duct radius $\langle R \rangle$ and the average viable rim thickness $\langle T \rangle$, measured directly on the IHC images;
- the average cell density $\langle \rho \rangle$ in the viable rim, measured by counting nuclei and computing the viable rim size;
- the cell confluence f in the viable rim, defined to be the area fraction of the viable region occupied by cell nuclei and cytoplasm;
- the proliferating index PI, measured by staining images for Ki-67 (a nuclear protein marker for cell cycling) and then counting the total number of Ki-67-positive nuclei versus the total number of nuclei; and
- the apoptotic index AI, measured by staining images for cleaved Caspase-3, an "executioner" caspase involved throughout most of the apoptosis process. Because Caspase-3 is a cytosolic protein, we identified cleaved Caspase-3 positive cells by comparing the whole-cell staining intensities. The apoptotic index is then computed across the viable rim, as for the PI.

Geometry
We matched the simulated duct radius to the mean measured duct radius $\langle R \rangle$. We obtained the average (equivalent) cell radius from the mean viable rim cell density $\langle \rho \rangle$ and measured the confluence f $(0 \leq f \leq 1)$ by the relation

$$f = \langle \rho \rangle \pi r^2. \tag{10.4}$$

Oxygen level
For the special case we are considering here, $\lambda_p = \lambda_n = \langle \lambda \rangle$; we assume that λ_b is stipulated as an additional constraint $\Lambda_b = \lambda_b / \langle \lambda \rangle$. The more general case is considered by separating the viable rim into fluid, proliferating cells, and non-proliferating cells and applying additional constraints to both $\lambda_n / \langle \lambda \rangle$ and $\lambda_b / \langle \lambda \rangle$ to determine the oxygen uptake rate uniquely [434].

Next, we used the mean viable rim thickness $\langle T \rangle$ as an indicator of oxygenation and thus determined the boundary oxygen value σ_B. In two dimensions (the three-dimensional results are given in Section 10.4.2), the steady-state oxygen profile away from the leading edge reduces to the simple one-dimensional equation

$$0 = \begin{cases} D\sigma'' - \langle \lambda \rangle \sigma, & 0 < x < \langle T \rangle, \\ D\sigma'' - \Lambda_b \langle \lambda \rangle \sigma, & \langle T \rangle < x < \langle R \rangle, \end{cases} \tag{10.5}$$

with the boundary and matching conditions

$$\sigma(0) = \sigma_B, \qquad \sigma(\langle T \rangle) = \sigma_H, \qquad \sigma'(\langle R \rangle) = 0, \tag{10.6}$$

$$D \lim_{x \uparrow \langle T \rangle} \sigma'(x) = D \lim_{x \downarrow \langle T \rangle} \sigma'(x). \tag{10.7}$$

Here, x is the distance from the duct wall.

After applying all the conditions except $\sigma(0) = \sigma_B$, we have, for $0 < x < \langle T \rangle$,

$$\sigma(x) = \sigma_H \left[\cosh\left(\frac{x - \langle T \rangle}{L} \right) - \sqrt{\Lambda_b} \tanh\left(\frac{\langle R \rangle - \langle T \rangle}{L/\sqrt{\Lambda_b}} \right) \sinh\left(\frac{x - \langle T \rangle}{L} \right) \right],$$

and, for $\langle T \rangle < x < \langle R \rangle$,

$$\sigma(x) = \sigma_{\mathrm{H}} \left[\cosh \left(\frac{x - \langle T \rangle}{L/\sqrt{\Lambda_{\mathrm{b}}}} \right) - \tanh \left(\frac{\langle R \rangle - \langle T \rangle}{L/\sqrt{\Lambda_{\mathrm{b}}}} \right) \sinh \left(\frac{x - \langle T \rangle}{L/\sqrt{\Lambda_{\mathrm{b}}}} \right) \right].$$

We evaluate the first expression at $x = 0$ to determine σ_{B}:

$$\sigma_{\mathrm{B}} = \sigma_{\mathrm{H}} \left[\cosh \frac{\langle T \rangle}{L} + \sqrt{\Lambda_{\mathrm{b}}} \tanh \left(\frac{\langle R \rangle - \langle T \rangle}{L/\sqrt{\Lambda_{\mathrm{b}}}} \right) \sinh \frac{\langle T \rangle}{L} \right]. \qquad (10.8)$$

Finally, we compute the mean oxygen value across the viable rim:

$$\langle \sigma \rangle = \sigma_{\mathrm{H}} \frac{L}{\langle T \rangle} \left[\sqrt{\Lambda_{\mathrm{b}}} \tanh \left(\frac{\langle R \rangle - \langle T \rangle}{L/\sqrt{\Lambda_{\mathrm{b}}}} \right) \left(\cosh \frac{\langle T \rangle}{L} - 1 \right) + \sinh \frac{\langle T \rangle}{L} \right]. \qquad (10.9)$$

Population dynamics

Using the analysis in Section 6.5, given β_{P}, β_{A}, and measurements of the PI and AI we can solve Eqs. (6.51), (6.52) in the steady state to determine $\langle \alpha_{\mathrm{P}} \rangle$ and α_{A}:

$$\langle \alpha_{\mathrm{P}} \rangle = \frac{\beta_{\mathrm{P}} \left(\mathrm{PI} + \mathrm{PI}^2 \right) - \beta_{\mathrm{A}} \mathrm{AI} \cdot \mathrm{PI}}{1 - \mathrm{AI} - \mathrm{PI}}, \qquad (10.10)$$

$$\alpha_{\mathrm{A}} = \frac{\beta_{\mathrm{A}} \left(\mathrm{AI} - \mathrm{AI}^2 \right) + \beta_{\mathrm{P}} \mathrm{AI} \cdot \mathrm{PI}}{1 - \mathrm{AI} - \mathrm{PI}}. \qquad (10.11)$$

We calibrate the functional form for α_{P} by combining the result (10.10) with the computed mean oxygen value from the previous step and then solving for $\bar{\alpha}_{\mathrm{P}}$:

$$\langle \alpha_{\mathrm{P}} \rangle = \bar{\alpha}_{\mathrm{P}} \frac{\langle \sigma \rangle - \sigma_{\mathrm{H}}}{1 - \sigma_{\mathrm{H}}}. \qquad (10.12)$$

Cell–cell mechanics

For confluent cells in solid-type DCIS ($f = 1$), we convert the mean density $\langle \rho \rangle$ to an equivalent cell spacing s (measured between cell centers) via

$$s = \sqrt{\frac{2}{\sqrt{3} \langle \rho \rangle}}, \qquad (10.13)$$

which is based upon matching the mean cell density to a hexagonal cell packing. We balance the cell–cell adhesive and repulsive forces at this equilibrium spacing. If $n_{\mathrm{cca}} = n_{\mathrm{ncr}} = 1$, $R_{\mathrm{cca}} = 1.14r$, $\langle R \rangle_{\mathrm{ccr}} = r$, and $\mathcal{E} = 1$ then

$$\frac{\alpha_{\mathrm{cca}}}{\alpha_{\mathrm{ccr}}} = \frac{\varphi'(s; 2r, n_{\mathrm{ccr}})}{\varphi'(s; 2R_{\mathrm{cca}}, n_{\mathrm{cca}})} = \frac{(1 - s/(2r))^{n_{\mathrm{ccr}}+1}}{(1 - s/(2.428r))^{n_{\mathrm{cca}}+1}}. \qquad (10.14)$$

This leaves a free parameter, since in effect the density determines the equilibrium spacing but we have not stipulated how *strictly* the density value is enforced. It may be possible to constrain the mechanics fully by matching the simulation to the variance in ρ; this is the subject of ongoing research. In the meantime, we have found that setting $\alpha_{\mathrm{ccr}} = 8$ sufficiently enforces the density value [434].

Table 10.2. Key data for a de-identified patient

Quantity	Measured mean	Units
duct radius r_{duct}	170.10	μm
viable rim thickness T	76.92	μm
PI	17.43	%
raw AI	0.638	%
corrected AI	0.831	%
cell density ρ	0.003213	cells μm^{-2}

Cell–BM mechanics

Because we have no direct data on the cell–BM mechanical interactions, we choose the parameters to prevent cells from penetrating the duct wall; $\alpha_{\text{cbr}} = 5$ suffices when $\alpha_{\text{cba}} = \alpha_{\text{cca}}$. It should be possible to constrain the parameter values further by comparing patient data with the simulated tumor propagation speed and leading-edge morphology as α_{cba} and α_{cca} are varied; such parameter studies are the topic of ongoing research [434].

10.3.3   Sample patient calibration and verification

We demonstrated the calibration protocol using IHC and histopathology material from a de-identified mastectomy patient from the M. D. Anderson Cancer Center (de-identified case number 100019). The measurements for this patient are given in Table 10.2. Because the cells are nearly confluent in the viable rim, we estimated that $f \approx 1$. By the cell–cell mechanics calibration in Eq. (10.4), $r_{\text{cell}} = \sqrt{1/(\rho\pi)} \approx 9.953$ μm. Using the above estimates of cell deformability, we set $R_{\text{cca}} = R_{\text{cba}} = 1.21r_{\text{cell}} \approx 12.0834$.

From the oxygen protocol (with $\lambda_{\text{p}} = \lambda_{\text{n}} = \lambda_{\text{b}} = 0.1$), we estimated the boundary value σ_{B} as 0.3861 (Eq. (10.8)), and $\langle \sigma \rangle$ as 0.2794 (Eq. (10.9)). For further investigations of the case $\lambda_{\text{p}} \neq \lambda_{\text{n}}$ and $\lambda_{\text{p}} \neq \lambda_{\text{b}}$ see [434].

Using the measured AI and PI values, along with $\beta_{\text{P}}^{-1} = 18$ h and $\beta_{\text{A}}^{-1} = 8.6$ h (see Section 10.4.1), we estimated the population dynamic parameters as

$$\alpha_{\text{A}}^{-1} \approx 47\,196.349 \text{ min} \quad \text{and} \quad \overline{\alpha}_{\text{p}}^{-1} \approx 434.527 \text{ min};$$

see Eqs. (10.11), (10.12).

For the mechanics, the protocol gives $s \approx 18.957$ μm (Eq. (10.13), $\alpha_{\text{ccr}} = 8$, and $\alpha_{\text{cca}} \approx 0.3915$ (Eq. (10.14)). We set $\alpha_{\text{cba}} = \alpha_{\text{cca}}$ and $\alpha_{\text{cbr}} = 5$, although we are currently investigating the effect of an imbalance between α_{cca} and α_{cba} [434].

Verification of calibration

To verify the calibration, we seeded a small section of a 1 mm virtual duct with tumor having AI and PI values matching the IHC measurements. We then ran the simulation to time $t = 30$ days and checked the model's predictions of AI, PI, viable rim thickness, and density in the viable rim. (See Section 6.6.1 for the fully dynamic simulation.)

Table 10.3. Verification of the patient-specific calibration. All figures are given as mean ± standard deviation. There is no standard deviation for the simulated cell density because it was calculated over the entire viable rim

Quantity	Patient data	Simulated
PI (%)	17.43 ± 10.48	17.193 ± 7.216
AI (%)	0.831 ± 0.572	1.447 ± 3.680
Viable rim thickness (μm)	76.92 ± 13.70	80.615 ± 4.454
Cell density (cells/μm^{-2})	0.003213 ± 6.89e-4	0.003336

We sliced the computational domain at time $t = 30$ days into slices 6 μm thick and performed virtual immunohistochemistry on those slices. We calculated the viable rim thickness in each slice and the average cell density over the entire viable tumor region. See Table 10.3. The proliferative index values match extremely well, and the apoptotic index values are within error tolerances. Because apoptosis is a rare stochastic event ($< 1\%$) in a region with fewer than 500 cells, we expect considerable noise; indeed, this is observed in the patient's AI value as well. The viable rim thickness is within the error bounds, and the cell density is in excellent agreement. Because all the numerical targets (outlined in Table 10.2) are within the error bounds, the calibration was regarded as a success.

We also compared the general tumor morphology with hematoxylin and eosin (H&E) stains from the patient (Figure 10.3, bottom right) and the spatial distribution of proliferating cells with Ki-67 immunostains from the patient (Figure 10.3, bottom left). The virtual DCIS reproduced the expected tumor microarchitecture: a viable rim closest to the duct wall, an interior necrotic core, and sporadic interior microcalcification. The simulation also recapitulated the general distribution of proliferating cells across the viable rim: in both the simulation and the Ki-67 imaging, cycling tumor cells were observed most frequently along the duct wall where oxygen is most plentiful and almost never at the perinecrotic boundary, where substrate levels are lowest. This evidence supports the dependence of α_P upon σ in our model. This theme is discussed in greater detail in Section 10.4.2.

10.4   Case studies

We now consider three case studies in which the agent model was used to facilitate predictive breast cancer research. First, we illustrate the utility of the model in estimating biological parameters that are difficult or impossible to measure experimentally. Second, we use the analytical volume-averaged behavior of the model to generate testable biological hypotheses of DCIS behavior, test those hypotheses using actual DCIS data, and use the results to refine and extend our model. Third, we demonstrate the use of the agent model in calibrating multiscale cancer simulation frameworks and compare a framework's predictions of tumor size with actual clinical data, discussing the clinical significance of the latter application and also future applications.

10.4.1   Estimating difficult physical parameters

Apoptosis time β_A^{-1}

The time course from the initial signal to commence apoptosis to final cell lysis is difficult to quantify [327]. Early reviews by key apoptosis researchers estimated that the early cellular events in apoptosis comprise a fast process on the order of minutes, with the digestion of apoptotic bodies occurring within hours of phagocytosis [369]. Hu *et al.* [327] conducted a detailed *in vivo* observation of the apoptosis of epithelial cells in the rat hippocampus, observing cells breaking up in 12–24 hours and the complete elimination of apoptotic bodies within 72 hours. Experimental work in [592] similarly observed most apoptotic processes to take place over a time on the order of hours. This provides the bound $\beta_A^{-1} \leq 24$ h. It also suggests that apoptotic bodies are absorbed by the surrounding cells in under 48 hours after cell lysis. In total, the experimental observations in the literature lead us to the estimate $\beta_A^{-1} \approx \mathcal{O}(10 \text{ h})$.

To estimate β_A for breast epithelial cells, we built on our working hypothesis that cancer cells use the same basic mechanisms of proliferation and apoptotis as noncancerous cells, only with altered frequency [300]. Hence, we postulated that β_A and β_P are the same for DCIS cell lines and noncancerous breast epithelial cells. Equation (6.50) gives us a means to estimate β_A: assuming that on average noncancerous breast epithelial tissue is in homeostasis (when averaged through the duration of the menstrual cycle), then $\dot{N} = 0$, where N is the number of cells, and we find

$$\beta_A = \beta_P \frac{P}{A} = \beta_P \frac{\text{PI}}{\text{AI}}. \tag{10.15}$$

In [406] the average proliferative and apoptotic indices of noncancerous breast epithelial cells in several hundred pre-menopausal (aged under 50 years old) women were reported from measurements to be 0.0252 ± 0.0067 and 0.0080 ± 0.0006, respectively. While AI and PI can vary considerably in time owing to hormone cycling in the menstrual cycle [492], when averaged over many women (who fall at different points in this cycle), the effects of the monthly variation should be cancelled out. On the basis of a cell cycle time $\beta_P^{-1} = 18$ h, we estimate that $\beta_A = 0.175$ h^{-1}, giving an estimated time for apoptotis β_A^{-1} of approximately 5.7 h. This is consistent with our estimated order of magnitude, 10 h.

In the same study, PI and AI were measured for several hundred post-menopausal women aged over 50 years old at 0.0138 ± 0.0069 and 0.0043 ± 0.0007, respectively. Using these figures gives a similar estimate to that above, $\beta_A \approx 0.178$ h^{-1}. The similarity of the figures in pre-menopausal and post-menopausal women supports our working hypothesis that β_A and β_P are relatively fixed for the cell type even when apoptosis and proliferation occur with differing frequencies and in different hormonal environments. We also note that conducting the same calculation with the data from [492] gives $\beta_A \approx 0.26$ h^{-1} and an estimated apoptosis time of 3.9 h. This work used a much smaller sample size, but nonetheless is generally consistent with our estimate.

We now attempt to improve our estimate to account for detection shortcomings in the immunostaining. (See the introduction in [189] for a good overview of the current methods of detecting apoptotic cells in histologic tissue cultures.) The AI measurements in [406] were obtained by TUNEL assay, which relies upon detecting DNA fragmentation. According to the detailed work on Jurkat cell apoptosis in [592], there is an approximately three-hour lag between the inducement of apoptosis (observable by rapid changes in the mitochondrial membrane potential voltage and the ratio of ATP and ADP) and the detection of DNA laddering and chromatin condensation. Cleaved Caspase-3 activity was neglibible for the first 60 minutes and steadily climbed thereafter, peaking after 180 minutes and reaching approximtely 10% of that peak in 50–60 minutes. On this basis, we would expect that TUNEL-assay-based AI figures fail to detect approximately the first three hours of apoptosis and that cleaved Caspase-3-based AI staining fail to detect the first one-to-two hours. Thus, we increase our estimate for β_A^{-1} to 8.6 hours. This gives correction factors to account for undetected apoptotic cells by TUNEL assay and cleaved-Caspase-3 immunostaining, as follows:

$$AI_{actual} \approx \frac{8.6}{5.6} AI_{TUNEL}, \tag{10.16}$$

and

$$\frac{8.6}{7.6} AI_{Caspase-3} \leq AI_{actual} \leq \frac{8.6}{6.6} AI_{Caspase-3}. \tag{10.17}$$

Calcification time β_C^{-1}

There are very few literature data available on the time taken for the completion of necrosis and calcification of breast tumor cells. The best available experimental data are generally animal time-course studies of arterial calcification; we use these to estimate the order of magnitude of β_C^{-1}. Time-course studies on *post mortem* cardiac valves in [348] reported significant tissue calcification between seven days (when there was a 10% increase in Ca incorporation) and 14 days (by which time there was a 40% increase) after injection by TGF-β1. Lee *et al.* [405] examined a related process (elastin calcification) using a rat subdermal model, demonstrating that calcification occurs gradually over the course of two to three weeks. Gadeau *et al.* [250] measured calcium accumulation in rabbit aortas following oversized balloon angioplasty injury. Calcified deposits appeared as soon as two to four days after the injury, increased over the course of eight days, and approached a steady state at between eight and 30 days. Hence, we estimate β_C^{-1} is on the order of days to weeks.

To sharpen our estimate of the calcification time parameter β_C^{-1} we conducted a parameter study using the fully dynamic model (see Section 6.6.1) that we calibrated in Section 10.3.3. We varied β_C^{-1} from 12 hours to 30 days and simulated our calibrated DCIS model to 30 days; the results are given in Table 10.4. We found that calcification times under 15 days lead to necrotic cores that are nearly entirely calcified; this is not observed in H&E image data. See Figure 10.3, bottom right, black arrow. However, a 30-day calcification time leads (as expected) to a complete absence of microcalcifications

Table 10.4. Parameter study on the calcification time

β_C^{-1}	12 hours	1 day	5 days	15 days	30 days
% of core calcified	94.0%	83.7%	51.1%	6.9%	0%

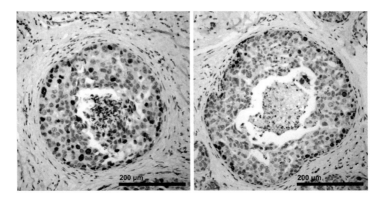

Figure 10.4 Ki-67 immunohistochemistry for ducts F3 (left) and F19 (right) for de-identified patient case 100019. Ki-67-positive nuclei stain darkly in the images. Reprinted with permission from [434].

in the core at time 30 days. Because DCIS tumors are hypothesized to grow to steady state in as little as two to three months [136, 193, 194], we expect microcalcification by this time. Hence our sharpened estimate of β_C^{-1} is 15 days, which is consistent with the literature. Parameter studies such as these are significant because they allow us to estimate physical quantities that are difficult or impossible to determine experimentally.

10.4.2   Generating and testing hypotheses

Recall that, when the agent-model behavior is averaged across the entire viable rim, we obtain a nonlinear system of two ODEs in PI and AI:

$$\dot{\text{PI}} = \langle \alpha_P \rangle (1 - \text{AI} - \text{PI}) - \beta_P \left(\text{PI} + \text{PI}^2 \right) + \beta_A \text{AI} \times \text{PI}$$
$$\dot{\text{AI}} = \alpha_A (1 - \text{AI} - \text{PI}) - \beta_A \left(\text{AI} - \text{AI}^2 \right) - \beta_P \text{AI} \times \text{PI}. \qquad (10.18)$$

As detailed earlier for fixed AI, PI, β_A, and β_P, this can be used to determine $\langle \alpha_P \rangle$, α_A, and ultimately $\bar{\alpha}_P$. If, instead, we regard α_A and $\bar{\alpha}_P$ as fixed and replace $\langle \alpha_P \rangle$ with $\alpha_P(S, \sigma, \bullet)$, we obtain a nonlinear system for AI and PI that varies with σ. If we solve this system to steady state for $\sigma_H < \sigma < 1$, we can use the model to predict the relationship between proliferation and oxygen availability. In [433] this analysis led us to hypothesize Michaelis–Menten population kinetics: for sufficient nutrient availability the proliferation saturates, indicating that oxygenation is no longer the primary growth-limiting factor.

We now test this hypothesis on the basis of a careful analysis of Ki-67 immunohistochemistry in two ducts (F3 and F19) for a DCIS patient (de-identified case 100019) [136, 193, 194]. See Figure 10.4. For each duct, we calculate the distances of all nuclei

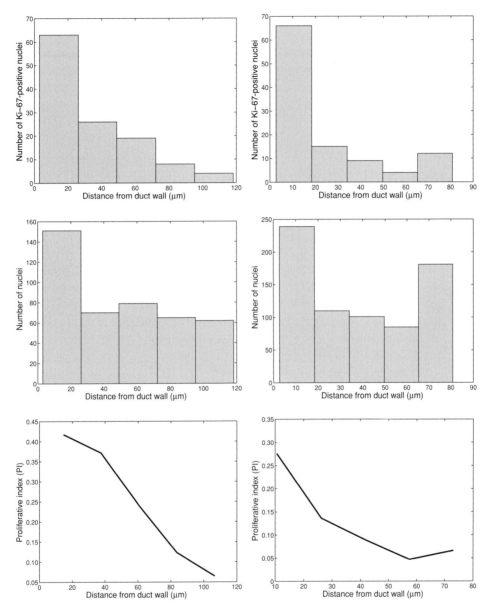

Figure 10.5 Histograms of Ki-67-positive nuclei vs. distance from the duct wall (top row) and all nuclei vs. distance from the duct wall (middle row), and plots of proliferative index vs. distance from the duct wall (bottom row). These measurements are for ducts F3 (left-hand column) and F19 (right-hand column). Reprinted with permission from [434].

and of Ki-67-positive nuclei to the duct wall, the mean distance from the duct centroid to the duct wall (i.e., the radius R), and the mean duct viable rim thickness T. Next, we create a histogram of Ki-67-positive nucleus distances to the duct wall (Figure 10.5, top row) and a histogram of the distances of all the nuclei from the duct wall *using the same "bins"* (Figure 10.5, second row), and then we divide these to obtain a graph of proliferative index PI versus distance from the duct wall (Figure 10.5, third row).

Next, we estimate the three-dimensional steady-state oxygen profile through the cylindrical ducts (assumed radially symmetric with no variation in the longitudinal direction) from the ODE

$$0 = L^2 \left(\sigma'' + \frac{1}{r} \sigma' \right) - \sigma, \qquad 0 < r < R, \tag{10.19}$$

with boundary conditions

$$\sigma(R - T) = \sigma_{\mathrm{H}}, \qquad \sigma'(0) = 0. \tag{10.20}$$

The solution is

$$\sigma(r) = \frac{\sigma_{\mathrm{H}}}{I_0 \left((R - T)/L \right)} I_0 \left(r/L \right), \tag{10.21}$$

where L is the diffusion length scale (assumed to be 100 μm in [220, 679]), I_n is the nth-order modifed Bessel function of the first kind, σ is nondimensionalized by the normoxic oxygen level in nonpathological tissue, and σ_{H} is the hypoxic threshold oxygen value (assumed to be 0.2 in [220, 679]). The mean value of the oxygen solution in the viable rim $(R - T < r < R)$ is given explicitly by

$$\langle \sigma \rangle = \frac{\sigma_{\mathrm{H}}}{I_0 \left((R - T)/L \right)} \frac{2L}{2RT - T^2} \left[R I_1 \left(R/L \right) - (R - T) I_1 \left((R - T)/L \right) \right]. \tag{10.22}$$

For the duct in F3,

$$R \approx 188.4634 \text{ μm}, \qquad T \approx 119.0256 \text{ μm}, \qquad \text{and} \qquad \langle \sigma \rangle \approx 0.282\,145,$$

and for the duct in F19,

$$R \approx 217.5548 \text{ μm}, \qquad T \approx 97.9602 \text{ μm}, \qquad \text{and} \qquad \langle \sigma \rangle \approx 0.280\,459.$$

By correlating the oxygen solutions (not shown) with the PI profiles, we estimate the relationship between the measured PI and the value of σ in the ducts. In Figure 10.6 we plot PI values for F3 (broken curve) and for F19 (dotted curve) as well as the predicted values (solid curve) from [434]. As can be seen, the theoretical prediction and measurements agree qualitatively but not quantitatively. We conclude that, while proliferation (given by PI) correlates with oxygen levels throughout the tumor, oxygenation alone cannot fully determine PI. Hence, we hypothesize that there must be additional heterogeneities in other microenvironmental factors (e.g., EGF), in gene expression, or in protein signaling across the tumor.

The next natural question is whether we can account for these heterogeneities with our current functional form by calibrating the agent model to individual ducts. If we can then this is further evidence that (i) we have chosen a suitable theoretical stochastic framework for the agent model and (ii) future work must incorporate more sophisticated gene and protein signaling models. To address this question, we next calibrate the agent model for each duct to determine α_{A} and $\overline{\alpha_{\mathrm{P}}}$. We use AI $= 0.008\,838$ for each duct, as in [434], and the measured values of PI, R, and T given above for each duct. For the duct

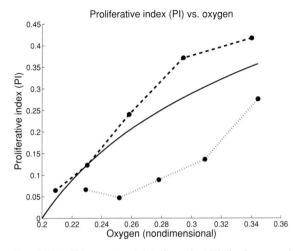

Figure 10.6 The predicted PI (solid curve) and data from duct F3 (broken curve) and from duct F19 (dotted curve), for case 100019. Reprinted with permission from [434].

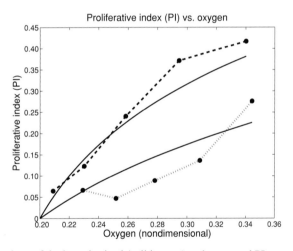

Figure 10.7 Comparison of the hypothesized (solid curve) and measured PI vs. σ curves for duct F3 (broken curve) duct F19 (dotted curve). Reprinted with permission from [434].

in F3,

$$PI = 0.281\,030, \qquad \alpha_A \approx 0.001\,624\,05\,h^{-1},$$

$$\langle \alpha_P \rangle \approx 0.027\,757\,9\,h^{-1}, \qquad \text{and} \qquad \overline{\alpha_P}(\mathcal{S}, \bullet) \approx 0.270\,331\,h^{-1};$$

and for the duct in F19,

$$PI = 0.148045, \qquad \alpha_A \approx 0.001\,290\,67\,h^{-1},$$

$$\langle \alpha_P \rangle \approx 0.011\,019\,0\,h^{-1}, \qquad \text{and} \qquad \overline{\alpha_P}(\mathcal{S}, \bullet) \approx 0.109\,562\,h^{-1}.$$

These values we generated PI versus σ curves for the individual ducts using Eq. (10.18) and compared them with the measured data; see Figure 10.7. There is generally much improved quantitative agreement between the predicted (solid) and measured (broken and dotted) curves. The difference in the predicted curves for the two ducts is due to the substantial difference in their $\bar{\alpha}_P$ values: $\bar{\alpha}_P$ is much greater for F3, which has the overall higher PI curve.

We next examine the data for the ducts in Figure 10.4, within the context of our modeling framework and the predicted PI versus σ curves, in order to generate additional biological hypotheses. Notice that the cell density is lower in F3 (Figure 10.4, left shows larger nuclei and cells, with greater spacing between cells) than in F19 (Figure 10.4, right shows smaller nuclei and cells, with less spacing between cells). These facts lead us to hypothesize that $\bar{\alpha}_P$ decreases with increasing cell density. E-cadherin/β-catenin signaling may be the physiological explanation of the phenomenon: when E-cadherin on one cell is bound to E-cadherin on a neighboring cell, β-catenin binds to the phosphorylated receptors, blocking the downstream pro-proliferative activity of the E-cadherin. (See Section 2.1.5.) For higher cell densities, more cell surfaces are in contact with one another, providing greater opportunities for E-cadherin binding; we consequently hypothesize that cell density correlates with cell cycle blockade by the E-cadherin/β-catenin pathway, resulting in the apparent relationship between cell density and $\bar{\alpha}_P$. Further evidence can be seen in duct F19 (Figure 10.4, right): the majority of the proliferation activity is in a single layer of cells along the duct wall. Because these cells adhere to the basement membrane, they present less surface for E-cadherin binding activity (relative to that of the interior cells), resulting in a reduced E-cadherin blockade of proliferation.

These hypotheses can be tested by correlating $\bar{\alpha}_P$ with the cell density in a larger number of ducts, performing an immunohistochemical (IHC) test for β-catenin activity, and correlating β-catenin-mediated transcription (indicated by the presence of β-catenin in the cell nuclei) with cell density and distance from the duct wall. One could use these data to hypothesize, calibrate, and test new functional forms for α_P such as

$$\alpha_P(\mathcal{S}, \sigma, \bullet, \circ) = \bar{\alpha}_P(\bullet, \circ) \left(1 - \mathcal{E} \langle \mathcal{E} \rangle \frac{\rho}{\rho_{\max}} \right) \left(\frac{\sigma - \sigma_H}{1 - \sigma_H} \right), \qquad (10.23)$$

where ρ is the local cell density, $\rho_{\max}$ is the density at which PI ≈ 0, $\mathcal{E}$ is the cell's (nondimensional) E-cadherin expression, and $\langle \mathcal{E} \rangle$ is the mean E-cadherin expression for the tumor. In such a formulation, $\bar{\alpha}_P(\bullet, \circ)$ determines the cell's $\mathcal{Q} \to \mathcal{P}$ transition rate in normoxic conditions with minimal E-cadherin signaling; it depends upon the cell's genetic profile $\bullet$ and potentially upon other signaling and/or microenvironmental factors $\circ$. These ideas are the subject of ongoing research by Macklin, Cristini, Edgerton, and others.

10.4.3   Calibrating multiscale modeling frameworks: preliminary results

In Chapter 7 we discussed a multiscale modeling framework in which data from various sources and scales (e.g., molecular data from IHC, cell-scale data from motility assays, and tissue-scale geometric data from MRI) are propagated throughout the framework through appropriate dynamic upscaling and downscaling between the scales. The net result is a simulator that can simulate whole three-dimensional tumors in large microenvironments while efficiently incorporating molecular- and cell-scale dynamics (e.g., hypoxic signaling and cell motility) where needed.

Recall that the overall change in the number of cells N is given by

$$\dot{N} = [\beta_\mathrm{P}\mathrm{PI}(\sigma) - \beta_\mathrm{A}\mathrm{AI}]\,N, \tag{10.24}$$

where we write $\mathrm{PI}(\sigma)$ to emphasize the dependency of PI on oxygen level, as demonstrated in Section 10.4.2. For the continuum model, the analogous form (neglecting cell transport) is given by

$$\dot{\rho} = (\lambda_\mathrm{M}\sigma - \lambda_\mathrm{A})\,\rho. \tag{10.25}$$

By averaging across a fixed volume and equating these terms, we estimate that

$$\lambda_\mathrm{M} \approx \frac{\beta_\mathrm{P}\langle\mathrm{PI}\rangle}{\langle\sigma\rangle} \quad \text{and} \quad \lambda_\mathrm{A} \approx \beta_\mathrm{A}\mathrm{AI}, \tag{10.26}$$

leading us to a preliminary upscaling between the agent and continuum models:

$$A = \frac{\lambda_\mathrm{A}}{\lambda_\mathrm{M}} = \langle\sigma\rangle\frac{\beta_\mathrm{A}\mathrm{AI}}{\beta_\mathrm{P}\langle\mathrm{PI}\rangle}, \tag{10.27}$$

or alternatively (by equating the cell proliferation when $\sigma = 1$ in both models),

$$\lambda_\mathrm{M} \approx \beta_\mathrm{P}\mathrm{PI}(1) \quad \Rightarrow \quad A \approx \frac{\beta_\mathrm{A}\mathrm{AI}}{\beta_\mathrm{P}\mathrm{PI}(1)}. \tag{10.28}$$

We applied the upscaling in Eq. (10.27) to the AI and PI data for 12 de-identified-index cases obtained from archived mastectomy-patient material at the M. D. Anderson Cancer Center. (See [195] for more information on how the cases were selected and how the patient tissues were prepared and processed to obtain AI, PI, viable rim thickness, and viable volume fraction.) The data are given in Table 10.5. Applying Eq. (10.27) to this data we obtained a patient-specific value of A (Eq. (10.27)) for each case; see the fourth column of Table 10.5.

Next, we predicted the nondimensional steady-state tumor size as a function of A by solving the model in [149] (see Chapter 3) with spherical symmetry. The resulting curve is given in Figure 10.8. To compare this predicted relationship between A and the nondimensional tumor size R for the individual cases, we must determine a patient-specific length scale L in order to nondimensionalize the patients' measured tumor sizes (column 7 in Table 10.5). The diffusional length scale used in [149] (see Section 10.2.3) was formulated for solid tumors, whereas DCIS grows in ducts that comprise only a fraction of the measured tumor volume. We modified the length scale by re-examining the nutrient transport equation. If f is the volume fraction of the breast tissue occupied by viable tumor then the nutrient equation can be altered to describe the reduced uptake in the overall tissue:

$$0 = D\nabla^2\sigma - f\lambda\sigma, \tag{10.29}$$

Table 10.5. Summary of pathological features with parameter values and predictions for index series[a]

Case ID	Subtype	Grade	A	L_0 (μm)	R (cm) (predicted)	R (cm) (geometric average of measured values[c])	Model prediction accurate
14	Cribriform	2	0.004	171.83	34.63	0.58	−
19	Mixed[b]	3	0.0247	78.87	1.72	1.14	+
8	Cribriform	2	0.0342	183.22	5.52	0.46	−
28	Solid	3	0.0368	86.58	1.33	1.47	+
13	Solid	3	0.0373	96.43	1.51	1.64	+
22	Cribriform	3	0.0441	97.08	1.30	1.04	+
18(L)	Mixed[b]	3	0.0498	111.71	1.44	1.64	+
21	Cribriform	2	0.0601	113.11	1.17	1.03	+
23	Solid	3	0.120	134.78	0.75	0.58	+
15	Cribriform	1	0.132	147.77	0.75	0.48	+
17	Mixed[b]	2	0.223	108.92	0.28	0.56	+
18(R)	Cribriform	1	0.280	116.35	0.24	0.53	+

[a] A volume density of 24.8% averaged over all cases was used for f.
[b] Mixed solid and cribriform subtypes.
[c] Geometric mean radius calculated from the measured dimensions.

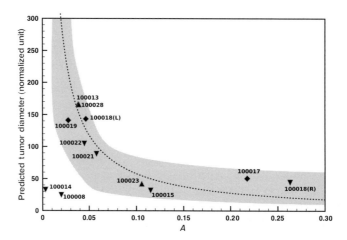

Figure 10.8 Validation of model predictions: solid-type (▲), cribriform-type (▼), and mixed-type (◆) DCIS plotted against the model prediction (dotted line) from [149]. The shaded region gives the standard deviation in A (as measured in multiple breast ducts for each individual patient) for each case.

which admits the modified length scale

$$L = \sqrt{\frac{D}{f\lambda}} = \frac{1}{\sqrt{f}}L_0. \tag{10.30}$$

This scale now accounts for the depletion of oxygen across the entire cancerous region rather than only within the solid-tumor portion. We used an average value of $f = 0.248$ across all 12 de-identified index cases [195].

We used the measured viable rim thickness L_0 (Table 10.5, column 5) for each case; this takes into account the variability from patient to patient in the vascularity of the tissue between the ducts and in the tumor cell density, as well as any differences in the cellular oxygen uptake rates. We nondimensionalized the measured tumor sizes using this length scale (Table 10.5, column 7). These predictions are presented as the labeled points in Figure 10.8. There was good qualitative and quantitative agreement between the predictions (dotted curve) and measurements (plotted points) for these de-identified index cases. To better quantify the predictions, we also estimated the measurement error (shaded region) using the standard deviations in AI and PI (horizontal breadth of the shaded region). Ten of the 12 cases (83%: labeled '+' in Table 10.5: column 8) of the cases fell within the error estimate. In the remaining two cases (17%) the tumor sizes were substantially overestimated.

Biological and modeling significance

The success of the model in predicting tumor sizes serves to validate the biological modeling hypotheses at all scales, the calibration technique, and the upscaling linking the scales. Had any of these been invalid then it is likely that the model predictions would have been much less accurate.

At the cellular scale, the successful predictions validate the fundamental dependency of proliferation upon oxygen and nutrient availability as well as the relative independence of apoptosis and nutrient availability; the latter runs counter to the clear link between hypoxia and necrosis, underscoring the importance of properly modeling the nuances in cell death at the cellular scale. This theme was explored further in Section 6.6.3.

Given this successful upscaling, the work gives credence to the functional forms for the cell proliferation and apoptosis parameters λ_M and λ_A. This provides a concrete connection between cell-scale quantities (PI and AI) and macroscopic model parameters (λ_M and λ_A), allowing a better physical interpretation of the macroscopic parameters. The parameter λ_M measures the rate of cell division in normoxic tissue, and this should be inversely proportional to the cell cycle time β_P^{-1} plus the mean waiting time between cell cycles, here functionally encapsulated in PI. Similarly, the parameter λ_A gives the mean rate of cell death, which incorporates both the time scale of apoptosis (β_A^{-1}) and the mean waiting time to apoptosis (encapsulated in AI). Both parameters implicitly involve tumor genetics and proteomics through AI and PI.

These results and a similar analytical upscaling suggest a functional form for calibrating the necrosis volume-loss parameter λ_N appearing in the earlier continuum models reported in [229, 430, 431, 435–439] and in Chapter 3: $\lambda_N \approx \beta_{NL}(1 - V_S/V)$, where β_{NL}^{-1} is the mean time for necrotic cells to lyse and lose their water content and V_S/V is the non-water fraction of each cell. In Section 6.6.3 we estimated β_{NL}^{-1} is on the order of 1 to 5 days, and in Section 10.3.1, we estimated that $V_S/V \approx 0.1$; hence, we can deduce that λ_N is in the range $0.18 - 0.9 \, \text{day}^{-1}$. This range is consistent with prior estimates using alternative approaches. In parameter studies on

λ_N/λ_M conducted in [437], $0.1\,\text{day}^{-1} \le \lambda_N \le 1.0\,\text{day}^{-1}$ gave necrotic core sizes and morphologies consistent with *in vitro* tumor spheroids such as those reported in [145]. Furthermore, calibration work on noncalcified glioblastoma multiforme in [227] gave the estimate $\lambda_N \approx 0.7\,\text{day}^{-1}$.

At the whole-tumor scale the good model predictions validate the nutrient and oxygen diffusional limit to tumor growth, even in vascularized tissue [145, 149]. In this case the diffusional-limit theory holds well once it has been appropriately adapted to growth in a sparse duct microarchitecture interspersed by well-vascularized breast stroma: at the macroscopic scale this is completely analogous to the growth of a well-vascularized tumor. The model's success also validates the modification we made to the oxygen length scale to account for the breast tissue microarchitecture and points to the likelihood of success when we integrate this microarchitecture more fully into the cell- and multicell-scale tumor behavior.

Last, this early success of the model suggests that, for short time scales or near steady-state conditions (when the parameter values are relatively constant), we can calibrate and predict tumor growth on the basis of measurable physical quantities (proliferation and apoptosis rates, etc.) alone, without the need for the precise proteomic and genetic information that ultimately determine these physical quantitites. In effect, cancer can be treated as an engineering problem determined by physical processes, without regard for the genetic and molecular basis for those processes. As was highlighted in our work in Section 10.4.2, the molecular (and hence phenotypic) characteristics of an actual tumor can vary considerably across the tumor at any given time. Hence, molecular- and cellular-scale modeling *is* required if we are to refine our modeling to predict accurately tumor morphology, motility, and other fine-scale details in patient-specific simulated tissues, as well as to understand a tumor's heterogeneous response to therapy. Indeed, this is the essence of multiscale modeling: adequately to incorporate an increasing amount of data from various modeling scales in order to improve the predictivity of the modeling framework.

Clinical implications

The fact that 10 of the 12 index cases could be accurately predicted using steady-state theory suggests that DCIS emerges quickly from an undetectable precursor state (e.g., ADH) betweeen annual mammograms. Indeed, in an exploration of the time to reach steady state, we noted that DCIS tumors reach 95% of their maximum size within three months of initiation for a physiological range of values of A, f, and L_0 (results not shown). We therefore expect that 85% of DCIS should be at steady state for women undergoing yearly mammograms. For a sample size of 12, we should therefore expect one to two (on average, 1.8) cases to be smaller than the steady-state predictions, which is fully consistent with the two overestimated cases denoted by the minius signs in column 8 of Table 10.5.

This has clear clinical implications. Given the relatively fast time scale of DCIS progression, at-risk populations (e.g., families with BRCA1 or BRCA2 mutations) may require more frequent surveillance than annual mammograms to adequately detect and treat breast cancer before it progresses to invasion. Indeed, an estimated 75% of all DCIS cases are already invasive at the time of detection [24, 346, 397]. Alternatively, low-dose chemotherapeutics could be prescribed for such high-risk groups to slow down (undetected) DCIS progression, in order to allow for adequate detection by annual mammograms.

Given the fast progression of DCIS to a steady state and the prevalence of hypoxia and necrosis when large or densely packed ducts are involved, we see that tumor cells may be subject to hypoxic stress for substantial periods of time prior to detection by annual mammogram. This is consistent with the prevalence of co-existent invasive carcinoma in newly detected DCIS cases [397]. Our simulations show that the extent of necrosis can be predicted by identifying regions of severe hypoxia. On the basis of our simulations, necrosis occurs primarily in larger ducts with densely packed DCIS. Thus, a tumor's physical location, kinetic rates of proliferation and apoptosis, and local cell density are determinant predictors of the extent of necrosis. Given that the perinecrotic rim of a tumor represents the cell population at greatest risk for evolution to invasion, these measureable quantities could be better predictors of which DCIS cases are more likely to become invasive than the grade or necrosis is today.

Our results also suggest new possible correlates for compromised margins (a predictor of tumor recurrence) and DCIS behavior. Several groups have studied the relationship between the frequency with which residual DCIS is found in a re-excision and the margin status of the previous excision. For example, in a study of core biopsy predictors of compromised margins, it was reported in [178] that surface area involvement of cores by DCIS, solid type, high grade, presence of necrosis, and presence of calcifications all correlated with compromised margins in univariate analysis; surface area involvement persisted in multivariate analysis. This is consistent with our model's primary inputs: AI and PI correlate with grade, f is determined by the surface area DCIS density (and increases with solid-type DCIS). Furthermore, in our model necrosis and calcification increase with hypoxia, which scales roughly with tumor size and likelihood of invasion. Thus the morphological characteristics that in [178] are correlated with compromised margins are histological surrogates for parameter inputs in our mechanistic model; conversely, our model should be able to use these quantifiable physical measurements to predict compromised margins more specifically and accurately. Thus, the model provides a mechanistic explanation for many of the morphological correlates that have been used to predict clinical outcomes in DCIS.

Similarly, high-grade DCIS (especially solid-type) and DCIS with comedo-type necrosis are both considered to be correlates of higher risk for subsequent invasion. From the model we conclude that the involvement of larger ducts, lower values of A, and more dense microarchitecture will result in more necrosis and hence should correspond to a higher risk of hypoxic stress and pro-invasion mutations. The interaction of physical location, growth and apoptosis rates, and local cell density are more specific predictors of the extent of hypoxia and necrosis than the gross and morphologic parameters in typical use (grade, subtype, and comedo-type necrosis). Loss of p53 activity has also been suggested as a contributor to invasive potential in high-grade DCIS (see e.g. [469]). Such a loss would decrease native rates of apoptosis and thus decrease A. For high-grade

DCIS, with its higher range of PI values in the denominator of A, the decrease in A could be even more pronounced, leading to rapid tumor growth and the evolution of extensive hypoxic stress zones. Thus, a decrease in A and the generation of more hypoxic stress could be a mechanism by which the loss of p53 activity contributes to invasive potential in high-grade DCIS.

Long-term vision: a surgical planning tool

In the longer term, a computational model of DCIS built upon these results could lead to better predictions of the tumor volume extent prior to treatment, thus providing a clinical tool to assist in (i) determining when a mastectomy is the preferred treatment over breast-conserving surgery, (ii) predicting an adequate tumor excision volume and geometry for breast-conserving surgery, and (iii) defining an optimal zone for radiation therapy. We have already demonstrated a good predictive capability of the tumor volume; if we can additionally correlate mammographic or other imaging data to the tumor morphology (or predict it in near-real time by simulating growth in the duct system), we could overlay the predicted tumor morphology on real-time imaging during surgery.

In Figure 10.9, we show a mock-up of what such a planning tool might look like. Using the shape defined by the post-mastectomy pathology specimen as a proof of concept, along with a current volume prediction from Section 10.4.3, a simulation result could be overlaid on medical imagery to plan a surgical excision. If the pathology-based shape estimate could be replaced by either a more detailed simulation geometry or better pre-surgical shape measurements, the software (the mock-up shown as largest labeled volume) could predict much more precise surgical margins than current mammographic measurements (the innermost labeled volume), potentially allowing the patient to avoid re-excision.

In comparison, the volume predicted from measuring distances between calcifications in the mammogram is known to be inadequate (see e.g. [294]). While currently the model alone does not predict the tumor shape, this information could be obtained in near-real time using an imaging modality such as MRI. Then, using the tumor shape defined by MRI along with immunohistochemical and histological inputs from the core biopsy, one could use the model to visualize the volume requiring resection rather than having to rely on viewing two-dimensional images (see e.g. [397]).

10.5   Concluding remarks

In this chapter, we adapted the agent-based model presented in Chapter 6 to ductal carcinoma in situ of the breast. After developing and testing a patient-specific calibration protocol, we surveyed several applications to test the model's predictive power.

We began by using the model to help estimate difficult biophysical parameters pertaining to cell death. By applying a volume-averaged version of the model to histopathologic data from normal breast epithelium, we were able to estimate the time duration of apoptosis as about 8.6 hours; this parameter can be difficult to observe experimentally. Furthermore, we arrived at the same estimate when using data from both pre- and

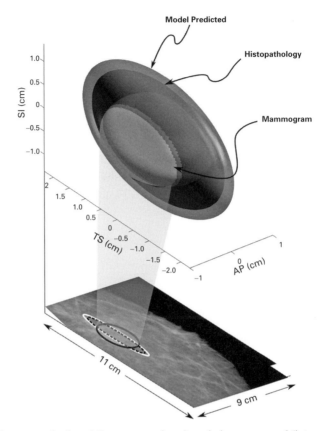

Figure 10.9 Mammography-based (innermost volume), pathology-measured (intermediate volume), and model-predicted (largest volume) excision volumes. The model-predicted excision region uses the pathology-measured shape as a proof of concept. Reprinted from the submission [195].

post-menopausal women, thereby supporting the biological hypothesis that cancerous and noncancerous cells use the same physical mechanisms (in this case apoptosis and proliferation), but with altered frequency. We applied a numerical implementation of the model to conduct a parameter study on the time duration of cell calcification, arriving at an estimate of approximately 15 days. Currently, cell calcification is difficult to study *in vitro* and only limited, indirectly available, data exist for calcification *in vivo*.

Next, we examined the ability of the model to make testable predictions on cell biology. The model predicted a Michaelis–Menten-type response of cell proliferation to oxygen availability; a subsequent analysis of patient immunohistochemistry verified the prediction with excellent quantitative agreement. However, this agreement came with an important caveat: the precise relationship between proliferation and oxygen availability can vary substantially across a single tumor even for fixed times – pointing to genetic and proteomic variation across a patient's tumor, such as in the E-cadherin/β-catenin signaling pathway.

Last, we conducted a preliminary study on the ability of the agent model to calibrate continuum-scale models (by upscaling), in the case of patient-specific predictions of breast tumor volume. We found that the agent model, as part of a larger multiscale modeling framework, had success in predicting patient-specific tumor sizes in a small group of de-identified index cases. This success points to the multiscale model's potential as a tool that could be used in conjunction with an imaging modality to construct a volume around tumor axes and midpoints in MRI and other imagery. This would help surgeons and pathologists to visualize DCIS tumors during surgery. We see the model as an important first step in understanding the physical changes that result from molecular alterations and that contribute to the development of invasive breast cancer.

11 Conclusion

Tumors are complex systems dominated by large numbers of processes with highly nonlinear dynamics spanning a wide range of dimensions. Typically, such complex systems can be understood only through complementary experimental investigation and mathematical modeling. New methodologies are needed to integrate and quantify the myriad variables and enable the prediction of outcomes, the selection of existing therapies, and the development of new treatments, possibly on a personalized, individual, basis. Mathematical modeling can provide a rigorous, precise, approach for quantifying correlations between tumor parameters, prognosis, and treatment outcomes. Integration of these elements into a multidisciplinary model of tumor progression would be an important tool to advance clinical decision-making. Thus, there is a critical need for biologically realistic and predictive multiscale and multivariate models of tumor growth and invasion and, as we have seen in this book, there has been much recent effort directed towards this goal.

The tumor models and simulation results that we have reviewed demonstrate that, while mathematical analyses still lag behind experimentation, significant progress has been made in providing important insights into the root causes of solid-tumor invasion and metastasis and in providing and assessing effective treatment strategies. For example, the results provide a means to associate quantitatively tumor growth and the effects of a limited supply of cell substrates. Inhomogeneities in the substrate availability and/or host tissue biomechanical properties can result in tumor morphological instability. These instabilities may allow a tumor to overcome diffusional limitations on growth and to grow to sizes larger than would be possible if it had a compact shape. Thus, diffusional instability could provide an additional pathway for tumor invasion.

This has important implications for therapy. Since decreasing the cell substrate levels in the microenvironment tends to increase tumor fragmentation and invasion into the surrounding tissue, this may have to be taken into consideration during anti-angiogenic therapy. Indeed, experimental studies have shown that anti-angiogenic therapies may result in the production of multifocal tumors [58, 170, 398, 570, 576, 581, 596]. However, combining anti-angiogenic and anti-invasive drugs such as Met inhibitors [48, 73, 476] or hepatocyte growth factor (HGF) antagonists [163, 468] may reduce fragmentation [58]. This is confirmed by increasing the cell–cell adhesion (or decreasing the cell mobility) in the mathematical models [145]. Conversely, increasing the cell substrate levels leads to greater morphological stability of tumors, making them more resectable. This suggests [145] that treatments that seek to normalize tumor vasculature (by selectively

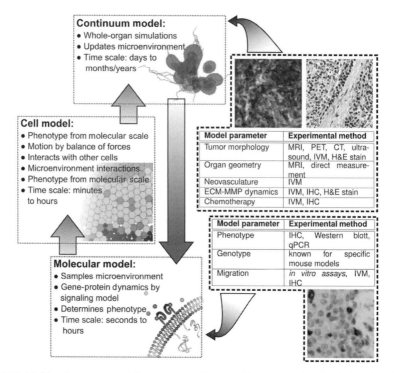

Figure 11.1 Multiscale cancer modeling incorporating data flow across the physical scales of a tumor.

"pruning" weak blood vessels with targeted anti-angiogenic therapy) may stabilize tumor morphology by providing increased access to cell substrates. Since such treatments may also increase accessibility to chemotherapeutic agents [342, 620], mathematical analyses provide additional support for the use of targeted anti-angiogenic therapy adjuvant to chemotherapy and resection.

On a broader level, the mathematical models reviewed in this book are part of an overall effort to study cancer as a complex system where there is variability and coupling among biophysical processes across a wide range of spatial and temporal scales. Taken together, the results suggest that tumor morphology and dynamics are coupled in complex, nonlinear, ways to cell phenotype and to molecular properties (e.g., genetics) and phenomena in the cell's environment such as hypoxia. These properties and phenomena act both as regulators of morphology and as determinants of invasion potential by controlling cell proliferation and migration mechanisms [235, 614, 667]. The importance of this close connection between tumor morphology and the underlying cellular and molecular scales is that it could allow observable properties of a tumor (e.g., morphology) to be used to understand the underlying cellular physiology and predict invasive behavior and response to treatment by means of mathematical modeling.

Figure 11.1 depicts the information flow between model components of various scales in a multiscale modeling framework. The tissue-scale model parameters can be dynamically obtained the upscaling the cell-scale model. At the cell scale, the pheno-type is determined by subcellular models that incorporate receptor signaling dynamics. By mechanistically tying macroscopic behavior to rigorous signaling models, new and emerging experimental pathway measurements can be incorporated and their impact can be assessed on the overall tumor behavior. The tissue-scale model is then used to model efficiently the whole-tumor evolution over longer times, while at the same time providing better modeling of continuum-scale phenomena such as tissue mechanics. As a simulation progresses, the tissue scale feeds back to recalibrate the cellular and subcellular scales, which then continue to inform the macroscopic scale in a dynamic, cyclic, and bidirectional flow of information across the multiple scales. We envision that such an approach, with its rigorous multiscale calibration protocols, would be a critical component in patient-specific tumor simulations and therapy planning.

In summary, the nonlinear multiscale modeling of solid-tumor growth has made great strides in recent years and the body of research on mathematical models promises to continue its dramatic growth in the future. Mathematical modeling and numerical sim-ulation are poised to provide a more comprehensive understanding of cellular diversity and adaptation by describing the complex interactions among tumor cells and their microenvironment [614, 667]. This approach is expected to improve current cancer-modeling efforts, because a multiscale approach links together previous work focused on specific scales and specific processes (e.g., single-cell motion) to enable large-scale simulations of *in vivo* tumors that encompass a wide range of length and time scales; thus the cancer can be modeled as a complex biological system. Further, this method-ology allows the possibility of going beyond the current understanding of invasion and migration [143, 200, 231, 233, 235, 364, 388, 568, 579, 614, 667, 699, 708], with the eventual goal of predicting disease progression and treatment response on the basis of on patient-specific tumor characteristics.

References

[1] A. C. Abajian and J. S. Lowengrub. An agent-based hybrid model for avascular tumor growth. *UCI Undergrad. Res. J.*, 11, 2008.

[2] R. Abbott, S. Forrest, and K. Pienta. Simulating the hallmarks of cancer. *Art. Life*, 12(4): 617–634, 2006.

[3] H. Acker, J. Carlsson, W. Mueller-Klieser, and R. Sutherland. Comparative po2 measurements in cell spheroids cultured with different techniques. *Br. J. Cancer*, 56: 325–327, 1987.

[4] D. Adalsteinsson and J. A. Sethian. The fast construction of extension velocities in level set methods. *J. Comput. Phys.*, 148(1): 2–22, 1999.

[5] J. Adam. General aspects of modeling tumor growth and the immune response. In J. Adam and N. Bellomo, eds., *A Survey of Models on Tumor Immune Systems Dynamics*, pp. 15–87. Birkhäuser, Boston, 1996.

[6] T. L. Adamovich and R. M. Simmons. Ductal carcinoma in situ with microinvasion. *Am. J. Surg.*, 186(2): 112–116, 2003.

[7] B. Addison-Smith, D. McElwain, and P. Maini. A simple mechanistic model of sprout spacing in tumour-associated angiogenesis. *J. Theor. Biol.*, 250: 1–15, 2008.

[8] L. Ai, W.-J. Kim, T.-Y. Kim, *et al.* Epigenetic silencing of the tumor suppressor cystatin m occurs during breast cancer progression. *Cancer Res.*, 66: 7899–909, 2006.

[9] T. Alarcón, H. Byrne, and P. Maini. A cellular automaton model for tumour growth in inhomogeneous environment. *J. Theor. Biol.*, 225: 257–274, 2003.

[10] T. Alarcón, H. Byrne, and P. Maini. A multiple scale model for tumor growth. *Multiscale Model. Sim.*, 3: 440–475, 2005.

[11] M. Alber, M. Kiskowski, J. Glazier, and Y. Jiang. On cellular automaton approaches to modeling biological cells. In R. Rosenthal and D. Gilliam, eds., *IMA Series on Mathematical Systems Theory In Biology, Communication and Finance*, vol. 142, pp. 1–40. Springer, New York, 2002.

[12] B. Alberts, A. Johnson, J. Lewis, M. Raff, K. Roberts, and P. Walter. *Molecular Biology of the Cell*. Garland Science, New York, fifth edition, 2007.

[13] J. W. Allen, S. R. Khetani, R. S. Johnson, and S. Bhatia. *In vitro* liver tissue model established from transgenic mice: role of HIF-1alpha on hypoxic gene expression. *Tissue Eng.*, 12(11): 3135–3147, 2006.

[14] D. Ambrosi, F. Bussolino, and L. Preziosi. A review of vasculogenesis models. *Comp. Math. Meth. Med.*, 6: 1–19, 2005.

[15] D. Ambrosi, A. Duperray, V. Peschetola, and C. Verdier. Traction patterns of tumor cells. *J. Math. Biol.*, 58(1–2): 163–181, 2009.

[16] D. Ambrosi, A. Gamba, and G. Serini. Cell directional and chemotaxis in vascular morphogenesis. *Bull. Math. Biol.*, 66: 1851–1873, 2004.

[17] D. Ambrosi and F. Guana. Mechanical aspects of growth in soft tissues. *Boll. Unione Mat. Ital.*, 7: 775–781, 2004.

[18] D. Ambrosi and F. Guana. Stress-modulated growth. *Math. Mech. Solids*, 12: 319–343, 2007.

[19] D. Ambrosi, A. Guillou, and E. DiMartino. Stress-modulated remodeling of a non-homogeneous body. *Biomech. Model. Mechanobiol.*, 7: 63–76, 2008.

[20] D. Ambrosi and F. Mollica. On the mechanics of a growing tumor. *Int. J. Eng. Sci.*, 40: 1297–1316, 2002.

[21] D. Ambrosi and F. Mollica. The role of stress in the growth of a multicell spheroid. *J. Math. Biol.*, 48: 477–499, 2004.

[22] D. Ambrosi and L. Preziosi. On the closure of mass balance models for tumor growth. *Math. Models Meth. Appl. Sci.*, 12: 737–754, 2002.

[23] D. Ambrosi and L. Preziosi. Cell adhesion mechanisms and elasto-viscoplastic mechanics of tumours. *Mech. Model. Mechanobiol.*, 8: 397–413, 2009. 10.1007/s10237-008-145-y.

[24] American Cancer Society. American cancer society breast cancer facts and figures 2007–2008. Atlanta: American Cancer Society, Inc., 2007. http: //www.cancer.org/downloads/ STT/BCFF-Final.pdf.

[25] A. Anderson. A hybrid mathematical model of solid tumour invasion: the importance of cell adhesion. *Math. Med. Biol.*, 22: 163–186, 2005.

[26] A. Anderson and M. Chaplain. Continuous and discrete mathematical models of tumor-induced angiogenesis. *Bull. Math. Biol.*, 60: 857–900, 1998.

[27] A. Anderson, M. Chaplain, E. Newman, R. Steele, and A. Thompson. Mathematical modeling of tumour invasion and metastasis. *J. Theor. Med.*, 2: 129–154, 2000.

[28] A. Anderson, M. Chaplain, K. Rejniak, and J. Fozard. Single-cell based models in biology and medicine. *Math. Med. Biol.*, 25(2): 185–6, 2008.

[29] A. Anderson and V. Quaranta. Integrative mathematical oncology. *Nat. Rev. Cancer*, 8: 227–244, 2008.

[30] A. Anderson, K. Rejniak, P. Gerlee, and V. Quaranta. Microenvironment driven invasion: a multiscale multimodel investigation. *J. Math. Biol.*, 58: 579–624, 2009.

[31] A. Anderson, A. Weaver, P. Commings, and V. Quaranta. Tumor morphology and phenotypic evolution driven by selective pressure from the microenvironment. *Cell*, 127: 905–915, 2006.

[32] E. Anderson. Cellular homeostasis and the breast. *Maturitas*, 48(S1): 13–17, 2004.

[33] M. Anderson, D. Srolovitz, G. Grest, and P. Sahini. Computer simulation of grain growth – i. kinetics. *Acta Metall.*, 32: 783–791, 1984.

[34] R. Araujo and D. McElwain. A history of the study of solid tumour growth: the contribution of mathematical modelling. *Bull. Math. Biol.*, 66: 1039–1091, 2004.

[35] R. Araujo and D. McElwain. A linear-elastic model of anisotropic tumor growth. *Eur. J. Appl. Math.*, 15: 365–384, 2004.

[36] R. Araujo and D. McElwain. New insights into vascular collapse and growth dynamics in solid tumors. *J. Theor. Biol.*, 228: 335–346, 2004.

[37] R. Araujo and D. McElwain. A mixture theory for the genesis of residual stresses in growing tissues I: A general formulation. *SIAM J. Appl. Math.*, 65: 1261–1284, 2005.

[38] R. Araujo and D. McElwain. A mixture theory for the genesis of residual stresses in growing tissues II: Solutions to the biphasic equations for a multicell spheroid. *SIAM J. Appl. Math.*, 66: 447–467, 2005.

[39] R. Araujo and D. McElwain. The nature of the stresses induced during tissue growth. *Appl. Math. Lett.*, 18: 1081–1088, 2005.

[40] N. Armstrong, K. Painter, and J. Sherratt. A continuum approach to modeling cell–cell adhesion. *J. Theor. Biol.*, 243: 98–113, 2006.

[41] P. Armstrong. Light and electron microscope studies of cell sorting in combinations of chick embryo and neural retina and retinal pigment epithelium. *Willhelm Roux Arch.*, 168: 125–141, 1971.

[42] S. Astanin and L. Preziosi. Multiphase models of tumour growth. In N. Bellomo, M. Chaplain, and E. DeAngelis, eds., *Selected Topics on Cancer Modelling: Genesis – Evolution – Immune Competition – Therapy*. Birkhäuser, Boston, 2007.

[43] C. Athale and T. Deisboeck. The effects of egf-receptor density on multiscale tumor growth patterns. *J. Theor. Biol.*, 238: 771–779, 2006.

[44] C. Athale, Y. Mansury, and T. Deisboeck. Simulating the impact of a molecular "decision-process" on cellular phenotype and multicellular patterns in brain tumors. *J. Theor. Biol.*, 233: 469–481, 2005.

[45] H. Augustin. Tubes, branches, and pillars: the many ways of forming a new vasculature. *Circ. Res.*, 89: 645–647, 2001.

[46] D. H. Ausprunk and J. Folkman. Migration and proliferation of endothelial cells in preformed and newly formed blood vessels during tumour angiogenesis. *Microvasc. Res.*, 14(1): 53–65, 1977.

[47] D. Balding and D. McElwain. A mathematical model of tumor-induced capillary growth. *J. Theor. Biol.*, 114: 53–73, 1985.

[48] A. Bardelli, M. Basile, E. Audero, *et al*. Concomitant activation of pathways downstream of grb2 and pi 3-kinase is required for met-mediated metastasis. *Oncogene*, 18: 1139–1146, 1999.

[49] L. F. Barros, T. Hermosilla, and J. Castro. Necrotic vol. increase and the early physiology of necrosis. *Comp. Biochem. Physiol. A. Mol. Integr. Physiol.*, 130: 401–409, 2001.

[50] U. Bartels, C. Hawkins, M. Jing, *et al*. Vascularity and angiogenesis as predictors of growth in optic pathway/hypothalamic gliomas. *J. Neurosurg.*, 104: 314–320, 2006.

[51] K. Bartha and H. Rieger. Vascular network remodeling via vessel cooption, regression and growth in tumors. *J. Theor. Biol.*, 241: 903–918, 2006.

[52] R. J. Basaraba, H. Bielefeldt-Ohmann, E. K. Eschelbach, *et al*. Increased expression of host iron-binding proteins precedes iron accumulation and calcification of primary lung lesions in experimental tuberculosis in the guinea pig. *Tuberculosis*, 88(1): 69–79, 2008.

[53] A. Bauer, T. Jackson, and Y. Jiang. A cell-based model exhibiting branching and anastomosis during tumor-induced angiogenesis. *Biophys. J.*, 92: 3105–3121, 2007.

[54] F. O. Baxter, K. Neoh, and M. C. Tevendale. The beginning of the end: death signaling in early involution. *J. Mamm. Gland Biol. Neoplas.*, 12(1): 3–13, 2007.

[55] B. Bazaliy and A. Friedman. A free boundary problem for an elliptic–parabolic system: application to a model of tumor growth. *Comm. Partial Diff. Eq.*, 28: 517–560, 2003.

[56] P. A. Beachy, S. S. Karhadkar, and D. M. Berman. Tissue repair and stem cell renewal in carcinogenesis. *Nature*, 432(7015): 324–331, 2004.

[57] E. Bearer, J. Lowengrub, *et al*. Multiparameter computational modeling of tumor invasion. *Cancer Res.*, 69: 4493–4501, 2009.

[58] L. Bello, V. Lucini, F. Costa, *et al*. Combinatorial administration of molecules that simultaneously inhibit angiogenesis and invasion leads to increased therapeutic efficacy in mouse models of malignant glioma. *Clin. Cancer Res.*, 10: 4527–4537, 2004.

[59] N. Bellomo, E. de Angelis, and L. Preziosi. Multiscale modeling and mathematical problems related to tumor evolution and medical therapy. *J. Theor. Medicine*, 5: 111–136, 2003.

[60] N. Bellomo, N. Li, and P. Maini. On the foundations of cancer modelling: selected topics, speculations, and perspectives. *Math. Models Meth. Appl. Sci.*, 4: 593–646, 2008.

[61] N. Bellomo and L. Preziosi. Modelling and mathematical problems related to tumor evolution and its interaction with the immune system. *Math. Comput. Modelling*, 32: 413–542, 2000.

[62] M. Ben-Amar and A. Goriely. Growth and instability in elastic tissues. *J. Mech. Phys. Solids*, 53: 2284–2319, 2005.

[63] R. Benjamin, J. Capparella, and A. Brown. Classification of glioblastoma multiforme in adults by molecular genetics. *Cancer J.*, 9: 82–90, 2003.

[64] J. R. Berenson, L. Rajdev, and M. Broder. Pathophysiology of bone metastases. *Cancer Biol. Ther.*, 5(9): 1078–1081, 2006.

[65] M. Berger and I. Rigoutsos. An algorithm for point clustering and grid generation. *IEEE Trans. Syst. Man. Cybern.*, 21: 1278–1286, 1991.

[66] H. Bernsen and A. van der Kogel. Antiangiogenic therapy in brain tumor models. *J. Neuro-oncology*, 45: 247–255, 1999.

[67] R. Betteridge, M. Owen, H. Byrne, T. Alarcón, and P. Maini. The impact of cell crowding and active cell movement on vascular tumour growth. *Networks Heterogen. Media*, 1: 515–535, 2006.

[68] Y. Bi, C. H. Stuelten, T. Kilts, *et al*. Extracellular matrix proteoglycans control the fate of bone marrow stromal cells. *J. Biol. Chem.*, 280: 30 481–30 489, 2005.

[69] M. Bienz and H. Clevers. Linking colorectal cancer to Wnt signaling. *Cell*, 103: 311–320, 2000.

[70] M. V. Blagosklonny and A. B. Pardee. The restriction point of the cell cycle. *Cell Cycle*, 1(2): 103–110, 2002.

[71] C. Blanpain and E. Fuchs. Epidermal stem cells of the skin. *Ann. Rev. Cell Dev. Biol.*, 22: 339–373, 2006.

[72] H. Bloemendal, T. Logtenberg, and E. Voest. New strategies in anti-vascular cancer therapy. *Euro J. Clin. Invest.*, 29: 802–809, 1999.

[73] C. Boccaccio, M. Andò, L. Tamagnone, *et al*. Induction of epithelial tubules by growth factor hgf depends on the stat pathway. *Nature*, 391: 285–288, 1998.

[74] K. Bold, Y. Zou, I. Kevrekidis, and M. Henson. An equation-free approach to analyzing heterogeneous cell population dynamics. *J. Math. Biol.*, 55: 331–352, 2007.

[75] A. Brandt. Multi-level adaptive solutions to boundary-value problems. *Math. Comput.*, 31: 333–390, 1977.

[76] B. Brandt, D. Kemming, J. Packeisen, *et al*. Expression of early placenta insulin-like growth factor in breast cancer cells provides an autocrine loop that predominantly enhances invasiveness and motility. *Endocrine-Related Cancer.*, 12(4): 823–827, 2005.

[77] A. Bredel-Geissler, U. Karbach, S. Walenta, L. Vollrath, and W. Mueller-Klieser. Proliferation-associated oxygen consumption and morphology of tumor cells in monolayer and spheroid culture. *J. Cell Phys.*, 153: 44–52, 1992.

[78] D. Bresch, T. Colin, E. Grenier, and B. Ribba. A viscoelastic model for avascular tumor growth. inria-00267292, version 2, 2008.

[79] D. Bresch, T. Colin, E. Grenier, B. Ribba, and O. Saut. Computational modeling of solid tumor growth: the avascular stage. Unpublished, 2007.

[80] C. Breward, H. Byrne, and C. Lewis. The role of cell–cell interactions in a two-phase model for avascular tumour growth. *J. Math. Biol.*, 45: 125–152, 2002.

[81] C. Breward, H. Byrne, and C. Lewis. A multiphase model describing vascular tumor growth. *Bull. Math. Biol.*, 65: 609–640, 2003.

[82] D. Brizel, S. Scully, J. Harrelson, *et al.* Tumor oxygenation predicts for the likelihood of distant metastases in human soft tissue sarcoma. *Cancer Res.*, 56: 941–943, 1996.

[83] D. Brizel, G. Sibley, L. Prosnitz, R. Scher, and M. Dewhirst. Tumor hypoxia adversely affects the prognosis of carcinoma of the head and neck. *Int. J. Radiat. Oncol. Biol. Phys.*, 38: 285–289, 1997.

[84] J. Brown and L. Lowe. Multigrid elliptic equation solver with adaptive mesh refinement. *J. Comput. Phys.*, 209: 582–598, 2005.

[85] R. K. Bruick and S. L. McKnight. A conserved family of prolyl-4-hydroxylases that modify HIF. *Science*, 294(5545): 1337–1340, 2001.

[86] H. Bueno, G. Ercole, and A. Zumpano. Asymptotic behaviour of quasi-stationary solutions of a nonlinear problem modelling the growth of tumours. *Nonlinearity*, 18: 1629–1642, 2005.

[87] E. Bullitt, D. Zeng, G. Gerig, *et al.* Vessel tortuosity and brain tumor malignance: a blinded study. *Acad. Radiol.*, 12: 1232–1240, 2005.

[88] A. Burton. Rate of growth of solid tumours as a problem of diffusion. *Growth*, 30: 157–176, 1966.

[89] F. Bussolino, M. Arese, E. Audero, *et al. Cancer Modelling and Simulation*, Chapter 1: Biological aspects of tumour angiogenesis, pp. 1–22. Chapman and Hall/CRC, London, 2003.

[90] L. M. Butler, S. Khan, G. E. Rainger, and G. B. Nash. Effects of endothelial basement membrane on neutrophil adhesion and migration. *Cell. Immun.*, 251: 56–61, 2008.

[91] S. Byers, C. Sommers, B. Hoxter, A. Mercurio, and A. Tozeren. Role of e-cadherin in the response of tumor cell aggregates to lymphatic, venous and arterial flow: measurement of cell–cell adhesion strength. *J. Cell Sci.*, 108: 2053–2064, 1995.

[92] H. Byrne. The effect of time delays on the dynamics of avascular tumor growth. *Math. Biosci.*, 144: 83–117, 1997.

[93] H. Byrne. The importance of intercellular adhesion in the development of carcinomas. *IMA J. Math. Med. Biol.*, 14: 305–323, 1997.

[94] H. Byrne. A comparison of the roles of localized and nonlocalised growth factors in solid tumour growth. *Math. Models Meth. Appl. Sci.*, 9: 541–568, 1999.

[95] H. Byrne. A weakly nonlinear analysis of a model of avascular solid tumour growth. *J. Math. Biol.*, 39: 59–89, 1999.

[96] H. Byrne, T. Alarcón, M. Owen, S. Webb, and P. Maini. Modeling aspects of cancer dynamics: a review. *Phil. Trans. R. Soc. A*, 364: 1563–1578, 2006.

[97] H. Byrne and M. Chaplain. Growth of nonnecrotic tumors in the presence and absence of inhibitors. *Mathl Biosci.*, 130: 151–181, 1995.

[98] H. Byrne and M. Chaplain. Mathematical models for tumour angiogenesis: numerical simulations and nonlinear wave solutions. *Bull. Math. Biol.*, 57: 461–486, 1995.

[99] H. Byrne and M. Chaplain. Growth of necrotic tumors in the presence and absence of inhibitors. *Mathl Biosci.*, 135: 187–216, 1996.

[100] H. Byrne and M. Chaplain. Modelling the role of cell–cell adhesion in the growth and development of carcinomas. *Mathl Comput. Modelling*, 24: 1–17, 1996.

[101] H. Byrne and M. Chaplain. Free boundary value problems associated with the growth and development of multicellular spheroids. *Eur. J. Appl. Math.*, 8: 639–658, 1997.

[102] H. Byrne and D. Drasdo. Individual-based and continuum models of growing cell populations: a comparison. *J. Math. Biol.*, 58(4–5): 657–87, 2009.

[103] H. Byrne, J. King, D. McElwain, and L. Preziosi. A two-phase model of solid tumour growth. *Appl. Math. Lett.*, 16: 567–573, 2003.

[104] H. Byrne and P. Matthews. Asymmetric growth of models of avascular solid tumors: exploiting symmetries. *IMA J. Math. Appl. Med. Biol.*, 19: 1–29, 2002.

[105] H. Byrne and L. Preziosi. Modelling solid tumour growth using the theory of mixtures. *Math. Med. Biol.*, 20: 341–366, 2003.

[106] N. Cabioglu, K. K. Hunt, A. A. Sahin, *et al.* Role for intraoperative margin assessment in patients undergoing breast-conserving surgery. *Ann. Surg. Oncol.*, 14(4): 1458–1471, 2007.

[107] J. Cahn and J. Hilliard. Free energy of a nonuniform system. i. interfacial free energy. *J. Chem. Phys.*, 28: 258–267, 1958.

[108] R. Cairns, T. Kalliomaki, and R. Hill. Acute (cyclic) hypoxia enhances spontaneous metastasis of kht murine tumors. *Cancer Res.*, 61: 8903–8908, 2001.

[109] D. Q. Calcagno, M. F. Leal, A. D. Seabra, *et al.* Interrelationship between chromosome 8 aneuploidy, C-MYC amplification and increased expression in individuals from northern Brazil with gastric adenocarcinoma. *World J. Gastroenterol.*, 12(38): 6027–211, 2006.

[110] V. Capasso and D. Morale. Stochastic modelling of tumour-induced angiogenesis. *J. Math. Biol.*, 58: 219–233, 2009.

[111] J. Carlsson and H. Acker. Relations ph, oxygen partial pressure and growth in cultured cell spheroids. *Int. J. Cancer*, 42: 715–720, 1988.

[112] P. Carmeliot and R. Jain. Angiogenesis in cancer and other diseases. *Nature*, 407: 249–257, 2000.

[113] L. Carreras, P. Macklin, J. Kim, S. Sanga, V. Cristini, and M. E. Edgerton. Oxygen uptake in quiescent versus cycling cells in a model of DCIS. In *Proc U S Canadian Acad. Path.*, 2010 Annual Meeting, 2010 (submitted).

[114] J. Casciari, S. Sotirchos, and R. Sutherland. Glucose diffusivity in multicellular tumor spheroids. *Cancer Res.*, 48: 3905–3909, 1988.

[115] J. Casciari, S. Sotirchos, and R. Sutherland. Variations in tumor cell growth rates and metabolism with oxygen concentration, glucose concentration, and extracellular ph. *J. Cell. Physiol.*, 151: 386–394, 1992.

[116] M. Castro, C. Molina-Paris, and T. Deisboeck. Tumor growth instability and the onset of invasion. *Phys. Rev. E*, 72: 041 907, 2005.

[117] P. Castro, P. Soares, L. Gusmo, R. Seruca, and M. Sobrinho-Simes. H-*RAS* 81 polymorphism is significantly associated with aneuploidy in follicular tumors of the thyroid. *Oncogene*, 25: 4620–4627, 2006.

[118] M. Chaplain. Reaction–diffusion prepatterning and its potential role in tumour invasion. *J. Biol. Sys.*, 3: 929–936, 1995.

[119] M. Chaplain. Avascular growth, angiogenesis and vascular growth in solid tumours: the mathematical modelling of the stages of tumour development. *Mathl Comput. Modelling*, 23: 47–87, 1996.

[120] M. Chaplain. Pattern formation in cancer. In M. Chaplain, G. Singh, and J. MacLachlan, eds., *On Growth and Form: Spatio-Temporal Pattern Formation in Biology*. Wiley, New York, 2000.

[121] M. Chaplain, M. Ganesh, and I. Graham. Spatio-temporal pattern formation on spherical surfaces: numerical simulation and application to solid tumour growth. *J. Math. Biol.*, 42: 387–423, 2001.

[122] M. Chaplain, L. Graziano, and L. Preziosi. Mathematical modelling of the loss of tissue compression responsiveness and its role in solid tumor development. *Math. Med. Biol.*, 23: 197–229, 2006.

[123] M. Chaplain and G. Lolas. Mathematical modeling of cancer cell invasion of tissue: the role of the urokinase plasminogen activation system. *Math. Models Meth. Appl. Sci.*, 15: 1685–1734, 2005.

[124] M. Chaplain, S. McDougall, and A. Anderson. Mathematical modeling of tumor-induced angiogenesis. *Ann. Rev. Biomed. Eng.*, 8: 233–257, 2006.

[125] M. Chaplain and B. Sleeman. Modelling the growth of solid tumours and incorporating a method for their classification using nonlinear elasticity theory. *J. Math. Biol.*, 31: 431–479, 1993.

[126] M. Chaplain and A. Stuart. A model mechanism for the chemotactic response of endothelial cells to tumor angiogenesis factor. *IMA J. Math. Appl. Med. Biol.*, 10: 149–168, 1993.

[127] A. Chauviere, T. Hillen, and L. Preziosi. Modeling cell movement in anisotropic and heterogeneous tissues. *Networks Hetero. Media*, 2: 333–357, 2007.

[128] A. Chauvier and L. Preziosi. A mathematical framework to model migration of a cell population in the extracellular matrix. In A. Chauvier, L. Preziosi, and C. Verdier, eds., *Cell Mechanics: From Single Cell Scale-Based Models to Multiscale Modeling*. Chapman and Hall/CRC Press, Boca Raton, 2009.

[129] A. Chauviere, L. Preziosi, and T. Hillen. Modeling the motion of a cell population in the extracellular matrix. *Discrete Cont. Dyn. Syst. B*, Supp: 250–259, 2007.

[130] C. Chen, H. Byrne, and J. King. The influence of growth-induced stress from the surrounding medium on the development of multicell spheroids. *J. Math. Biol.*, 43: 191–220, 2001.

[131] L. L. Chen, L. Zhang, J. Yoon, and T. S. Deisboeck. Cancer cell motility: optimizing spatial search strategies. *Biosys.*, 95(3): 234–242, 2009.

[132] W. W. Chen, B. Schoeberl, P. J. Jasper, *et al.* Input–output behavior of ErbB signaling pathways as revealed by a mass action model trained against dynamic data. *Mol. Syst. Biol.*, 5: 239ff, 2009.

[133] X. Chen, S. Cui, and A. Friedman. A hyperbolic free boundary problem modeling tumor growth: asymptotic behavior. *Trans. Am. Math. Soc.*, 357: 4771–4804, 2005.

[134] L. Cheng, N. K. Al-Kaisi, N. H. Gordon, A. Y. Liu, F. Gebrail, and R. R. Shenk. Relationship between the size and margin status of ductal carcinoma in situ of the breast and residual disease. *J. Natl Cancer Inst.*, 89(18): 1356–1360, 1997.

[135] J. H. Choi, Y. Y. Jeong, S. S. Shin, H. S. Lim, and H. K. Kang. Primary calcified T-cell lymphoma of the urinary bladder: a case report. *Korean J. Radiol.*, 4(4): 252–254, 2003.

[136] Y.-L. Chuang, M. E. Edgerton, P. Macklin, S. Wise, J. S. Lowengrub, and V. Cristini. Modeling ductal carcinoma in situ using modified Cahn–Hilliard equations (in preparation), 2010.

[137] S. Ciatto, S. Bianchi, and V. Vezzosi. Mammographic appearance of calcifications as a predictor of intraductal carcinoma histologic subtype. *Eur. Radiology*, 4(1): 23–26, 1994.

[138] R. G. Clyde, J. L. Brown, T. R. Hupp, N. Zhelev, and J. W. Crawford. The role of modelling in identifying drug targets for diseases of the cell cycle. *J. R. Soc. Interface*, 3(10): 617–627, 2006.

[139] D. Coffey. Self-organization, complexity and chaos: the new biology for medicine. *Nat. Med.*, 4: 882, 1998.

[140] P. Colella, D. T. Graves, T. J. Ligocki, *et al.* CHOMBO software package for AMR applications: design document. Technical report, Lawrence Berkeley National Laboratory, Applied Numerical Algorithms Group; NERSC Division, Berkeley, 2003.

[141] B. Coleman and W. Noll. Thermodynamics of elastic materials with conduction and viscosity. *Arch. Rat. Mech.*, 13: 167–178, 1963.

[142] M. Conacci-Sorrell, J. Zhurinsky, and A. Ben-Zeév. The cadherin–catenin adhesion system in signaling and cancer. *J. Clin. Invest.*, 109(8): 987–991, 2002.

[143] J. Condeelis, R. Singer, and J. Segall. The great escape: when cancer cells hijack the genes for chemotaxis and motility. *Ann. Rev. Cell Dev. Biol.*, 21: 695–718, 2005.

[144] A. Coniglio, A. deCandia, S. DiTalia, and A. Gamba. Percolation and burgers' dynamics in a model of capillary formation. *Phys. Rev. E*, 69: 051 910, 2004.

[145] V. Cristini, H. Frieboes, R. Gatenby, S. Caserta, M. Ferrari, and J. Sinek. Morphologic instability and cancer invasion. *Clin. Cancer Res.*, 11: 6772–6779, 2005.

[146] V. Cristini, H. Frieboes, X. Li, *et al.* Nonlinear modeling and simulation of tumor growth. In N. Bellomo, M. Chaplain, and E. de Angelis, eds., *Modelling and Simulation in Science, Engineering and Technology*. Birkhäuser, Boston, 2008.

[147] V. Cristini, X. Li, J. Lowengrub, and S. Wise. Nonlinear simulations of solid tumor growth using a mixture model: invasion and branching. *J. Math. Biol.*, 58(4–5): 723–763, 2009.

[148] V. Cristini and J. Lowengrub. Three-dimensional crystal growth. i. linear analysis and self-similar evolution. *J. Crystal Growth*, 240: 267–276, 2002.

[149] V. Cristini, J. Lowengrub, and Q. Nie. Nonlinear simulation of tumor growth. *J. Math. Biol.*, 46: 191–224, 2003.

[150] S. Cui. Analysis of a mathematical model for the growth of tumors under the action of external inhibitors. *J. Math. Biol.*, 44: 395–426, 2002.

[151] S. Cui. Analysis of a free boundary problem modeling tumor growth. *Acta Math. Sinica*, 21: 1071–1082, 2005.

[152] S. Cui. Well-posedness of a multidimensional free boundary problem modelling the growth of nonnecrotic tumors. *J. Func. Analysis*, 245: 1–18, 2007.

[153] S. Cui. Lie group action and stability analysis of stationary solutions for a free boundary problem modelling tumor growth. *J. Diff. Eq.*, 246(5): 1845–1882, 2009.

[154] S. Cui and J. Escher. Asymptotic behaviour of solutions of a multidimensional moving boundary problem modeling tumor growth. *Comm. Partial Diff. Eq.*, 33: 636–655, 2008.

[155] S. Cui and J. Escher. Well-posedness and stability of a multi-dimensional tumor growth model. *Arch. Rat. Mech. Analysis*, 191: 173–193, 2009.

[156] S. Cui and A. Friedman. Analysis of a mathematical model of the effect of inhibitors on the growth of tumors. *Math. Biosci.*, 164: 103–137, 2000.

[157] S. Cui and A. Friedman. A free boundary problem for a singular system of differential equations: an application to a model of tumor growth. *Trans. Amer. Math. Soc.*, 255: 3537–3590, 2003.

[158] S. Cui and A. Friedman. Formation of necrotic cores in the growth of tumors: analytic results. *Acta Mat. Sci.*, 26: 781–796, 2006.

[159] S. Cui and X. Wei. Global existence for a parabolic–hyperbolic free boundary problem modelling tumor growth. *Acta Math. Appl. Sinica*, 21: 597–614, 2005.

[160] S. Cui and S. Xu. Analysis of mathematical models for the growth of tumors with time delays in cell proliferation. *J. Math. Anal. Appl.*, 336: 523–541, 2007.

[161] J. Dallon and H. Othmer. How cellular movement determines the collective force generated by the dictyostelium discoideum slug. *J. Theor. Biol.*, 231: 299–306, 2004.

[162] C. G. Danes, S. L. Wyszomierski, J. Lu, C. L. Neal, W. Yang, and D. Yu. 14-3-3ζ down-regulates p53 in mammary epithelial cells and confers luminal filling. *Cancer Res.*, 68: 1760–1767, 2008.

[163] K. Date, K. Matsumoto, K. Kuba, H. Shimura, M. Tanaka, and T. Nakamura. Inhibition of tumor growth and invasion by a four-kringle antagonist (hgf/nk4) for hepatocyte growth factor. *Oncogene*, 17: 3045–3054, 1998.

[164] E. De Angelis and L. Preziosi. Advection–diffusion models for solid tumour evolution in vivo and related free boundary problem. *Math. Models Meth. Appl. Sci.*, 10: 379–407, 2000.

[165] C. R. De Potter, I. Eeckhout, A.-M. Schelfhout, M.-L. Geerts, and H. J. Roelsh. Keratinocyte induced chemotaxis in the pathogenesis of Paget's disease of the breast. *Histopath.*, 24(4): 349–56, 1994.

[166] J. Debnath and J. Brugge. Modelling glandular epithelial cancers in three-dimensional cultures. *Nat. Rev. Cancer*, 5: 675–688, 2005.

[167] J. Debnath, K. Mills, N. Collins, M. Reginato, S. Muthuswarmy, and J. Brugge. The role of apoptosis in creating and maintaining luminal space within normal and oncogene-expressing mammary acini. *Cell*, 111: 29–40, 2002.

[168] T. Deisboeck, M. Berens, A. Kansal, S. Torquato, A. Stemmer-Rachamimov, and E. Chiocca. Pattern of self-organization in tumour systems: complex growth dynamics in a novel brain tumour spherical model. *Cell Proliferation*, 34: 115–134, 2001.

[169] T. Deisboeck, L. Zhang, J. Yoon, and J. Costa. In silico cancer modeling: is it ready for prime time? *Nat. Clin. Practice Oncol.*, 6(1): 34–42, 2009.

[170] K. DeJaeger, M. Kavanagh, and R. Hill. Relationship of hypoxia to metastatic ability in rodent tumors. *Br. J. Cancer*, 84: 1280–1285, 2001.

[171] H. D. Dell. Milestone 1 (1889) Seed and soil hypothesis: observations from a ploughman. *Nat. Rev. Cancer*, 6: S7, 1989. http: //www.nature.com/milestones/milecancer/index.html.

[172] R. Demicheli, G. Pratesi, and R. Foroni. The exponential-gompertzian growth model: data from six tumor cell lines in vitro and in vivo. Estimate of the transition point from exponential to gompertzian growth and potential clinical applications. *Tumori*, 77: 189–195, 1991.

[173] B. Desai, T. Ma, and M. A. Chellaiah. Invadopodia and matrix degradation, a new property of prostate cancer cells during migration and invasion. *J. Biol. Chem.*, 283(20): 13 856–13 866, 2008.

[174] A. Deutsch and S. Dormann. *Cellular Automaton Modeling of Biological Pattern Formation*. Birkhäuser, New York, 2005.

[175] A. E. DeWitt, J. Y. Dong, H. S. Wiley, and D. A. Lauffenburger. Quantitative analysis of the EGF receptor autocrine system reveals cryptic regulation of cell response by ligand capture. *J. Cell. Sci.*, 114(12): 2301–2313, 2001.

[176] J. Diaz and J. Tello. On the mathematical controllability in a simple growth tumors model by the internal localized action of inhibitors. *Nonlinear Anal.*, 4: 109–125, 2003.

[177] A. DiCarlo and S. Quiligotti. Growth and balance. *Mech. Res. Commun.*, 29: 449–456, 2002.

[178] M. F. Dillon, A. A. Maguire, E. W. McDermott, *et al*. Needle core biopsy characteristics identify patients at risk of compromised margins in breast conservation surgery. *Mod. Pathol.*, 21(1): 39–45, 2008.

[179] M. F. Dillon, E. W. McDermott, A. O'Doherty, C. M. Quinn, A. D. Hill, and N. O'Higgins. Factors affecting successful breast conservation for ductal carcinoma in situ. *Ann. Surg. Oncol.*, 14(5): 1618–28, 2007.

[180] R. Dillon, M. Owen, and K. Painter. A single-cell based model of multicellular growth using the immersed interface method. In B. Khoo, Z. Li, and P. Lin, eds., *Contemporary Mathematics: Moving Interface Problems and Applications in Fluid Dynamics*, vol. 466, Chapter 1, pp. 1–15. American Mathematical Society, Providence, 2008.

[181] S. Dormann and A. Deutsch. Modeling of self-organized avascular tumor growth with a hybrid cellular automaton. *In Silico Biol.*, 2: 393–406, 2002.

[182] D. Drasdo. On selected individual-based approaches to the dynamics of multicellular systems. In W. Alt, M. Chaplain, and M. Griebel, eds., *Multiscale Modeling*. Birkhäuser, Basel, 2003.

[183] D. Drasdo. Coarse graining in simulated cell populations. *Adv. Complex Syst.*, 8: 319–363, 2005.

[184] D. Drasdo and S. Höhme. Individual-based approaches to birth and death in avascular tumors. *Math. Comput. Modelling*, 37: 1163–1175, 2003.

[185] D. Drasdo and S. Höhme. A single-scale-based model of tumor growth in vitro: monolayers and spheroids. *Phys. Biol.*, 2: 133–147, 2005.

[186] D. Drasdo and S. Höhme. On the role of physics in the growth and pattern of multicellular systems: what we learn from individual-cell based models? *J. Stat. Phys.*, 128: 287–345, 2007.

[187] D. Drasdo, R. Kree, and J. McCaskill. Monte-Carlo approach to tissue cell populations. *Phys. Rev. E*, 52: 6635–6657, 1995.

[188] F. Drees, S. Pokutta, S. Yamada, W. J. Nelson, and W. T. Weis. Alpha-catenin is a molecular switch that binds E-cadherin-beta-catenin and regulates actin-filament assembly. *Cell*, 123: 903–915, 2005.

[189] W. R. Duan, D. S. Garner, S. D. Williams, C. L. Funckes-Shippy, I. S. Spath, and E. A. G. Blomme. Comparison of immunohistochemistry for activated caspase-3 and cleaved cytokeratin 18 with the TUNEL method for quantification of apoptosis in histological sections of PC-3 subcutaneous xenografts. *J. Pathol.*, 199(2): 221–228, 2003.

[190] M. Ducasse and M. A. Brown. Epigenetic aberrations and cancer. *Mol. Cancer*, 5: 60, 2006.

[191] H. S. Earp, T. L. Dawson, X. Li, and H. Yu. Heterodimerization and functional interaction between EGF receptor family members: a new signaling paradigm with implications for breast cancer research. *Breast Cancer Res. Treat.*, 35(1): 115–132, 1995.

[192] D. A. Eberhard, B. E. Johnson, L. C. Amler, *et al.* Mutations in the epidermal growth factor receptor and in KRAS are predictive and prognostic indicators in patients with non-small-cell lung cancer treated with chemotherapy alone and in combination with erlotinib. *J. Clin. Oncol.*, 23: 5900–5909, 2005.

[193] M. Edgerton, Y.-L. Chuang, J. Kim, *et al.* Using mathematical models to understand the time dependence of the growth of ductal carcinoma in situ. In *31st Annual San Antonio Breast Cancer Symp.*, Supplement to vol. 68(24): Abstract 1165, 2008.

[194] M. Edgerton, Y.-L. Chuang, P. Macklin, *et al.* Using mathematical models to understand the time-dependence of the growth of ductal carcinoma in situ. *Cancer Research*, 69 (Suppl. 2): 1165, 2009.

[195] M. Edgerton, P. Macklin, Y.-L. Chuang, *et al.* A multiscale mathematical modeling framework of ductal carcinoma in situ. *Cancer Res.*, 2010 (in preparation).

[196] A. W. El-Kareh and T. W. Secomb. A mathematical model for cisplatin cellular pharmaco-dynamics. *Neoplasia*, 5(2): 161–169, 2003.

[197] A. W. El-Kareh and T. W. Secomb. Two-mechanism peak concentration model for cellular pharmacodynamics of doxorubicin. *Neoplasia*, 7(7): 705–713, 2005.

[198] C. Elliot and S. Luckhaus. A generalized diffusion equation for phase separation of a multi-component mixture with interfacial free energy. Technical report, University Sussex and University Bonn, 1991. Inst. Math. Appl., report 887.

[199] Y. I. Elshimali and W. W. Grody. The clinical significance of circulating tumor cells in the peripheral blood. *Diagn. Mol. Pathol.*, 15(4): 187–194, 2006.

[200] P. Elvin and A. Garner. Tumour invasion and metastasis: challenges facing drug discovery. *Curr. Opin. Pharmacol.*, 5: 374–381, 2005.

[201] S. Enam, M. Rosenblum, and K. Edvardsen. Role of extracellular matrix in tumor invasion: migration of glioma cells along fibronectinpositive mesenchymal cell processes. *Neuro-surgery*, 42: 599–608, 1998.

[202] R. Erban, I. Kevrekidis, and H. Othmer. An equation-free computational approach for extracting population-level behavior from individual-based models of biological dispersal. *Physica D*, 215: 1–24, 2006.

[203] R. Erban and H. Othmer. From individual to collective behavior in bacterial chemotaxis. *SIAM J. Appl. Math.*, 65: 361–391, 2004.

[204] R. Erban and H. Othmer. From signal transduction to spatial pattern formation in *E. coli*: a paradigm for multi-scale modeling in biology. *Multiscale Model. Simul.*, 3: 362–394, 2005.

[205] B. Erbas, E. Provenzano, J. Armes, and D. Gertig. The natural history of ductal carcinoma *in situ* of the breast: a review. *Breast Canc. Res. Treat.*, 97(2): 135–44, 2006.

[206] J. Erler, K. Bennewith, M. Nicolau, *et al.* Lysyl oxidase is essential for hypoxia-induced metastasis. *Nature*, 440: 1222–1226, 2006.

[207] M. Esteban and P. Maxwell. HIF, a missing link between metabolism and cancer. *Nat. Med.*, 11: 1047–1048, 2005.

[208] J. Fang, R. Gillies, and R. Gatenby. Adaptation to hypoxia and acidosis in carcinogenesis and tumor progression. *Semin. Cancer Biol.*, 18: 330–337, 2008.

[209] A. Fasano, A. Bertuzzi, and A. Gandolfi. Chapter Mathematical modelling of tumour growth and treatment, In *Complex Systems in Biomedicine*, pp. 71–108. Springer, Milan, 2006.

[210] L. Ferrante, S. Bompadre, L. Possati, and L. Leone. Parameter estimation in a gompertzian stochastic model for tumor growth. *Biometrics*, 56: 1076–1081, 2000.

[211] M. Ferrari. Cancer nanotechnology: opportunities and challenges. *Nat. Rev. Cancer*, 5: 161–171, 2005.

[212] R. Filion and A. Popel. A reaction–diffusion model of basic fibroblast growth factor inter-actions with cell surface receptors. *Annals Biomed. Eng.*, 32: 645–663, 2004.

[213] I. Fischer, J.-P. Gagner, M. Law, E. W. Newcomb, and D. Zagzag. Angiogenesis in gliomas: biology and molecular pathophysiology. *Brain Pathol.*, 15(4): 297–310, 2005.

[214] A. B. Fisher, S. Chien, A. I. Barakat, and R. M. Nerem. Endothelial cellular response to altered shear stress. *Am. J. Physiol. Heart Circ. Physiol.*, 281(3): L529–L533, 2001.

[215] B. Flaherty, J. P. McGarry, and P. E. McHugh. Mathematical models of cell motility. *Cell. Biochem. Biophys.*, 49(1): 14–28, 2007.

[216] G. B. Fogarty, N. M. Conus, J. Chu, and G. McArthur. Characterization of the expression and activation of the epidermal growth factor receptor in squamous cell carcinoma of the skin. *Brit. J. Derm.*, 156(1): 92–98, 2007.

[217] J. Folkman. Angiogenesis in cancer, vascular, rheumatoid and other disease. *Nat. Med.*, 1: 27–30, 1995.

[218] J. Forsythe, B. Jiang, N. Iyer, S. L. F. Agani, R. Koos, and G. Semenza. Activation of vascular endothelial growth factor gene transcription by hypoxia-inducible factor 1. *Mol. Cell. Biol.*, 16: 4604–4613, 1996.

[219] L. M. Franks and M. A. Knowles. What is cancer? In Knowles and Selby [384], Chapter 1, pp. 1–24.

[220] S. Franks, H. Byrne, J. King, J. Underwood, and C. Lewis. Modeling the early growth of ductal carcinoma in situ of the breast. *J. Math. Biol.*, 47: 424–452, 2003.

[221] S. Franks, H. Byrne, H. Mudhar, J. Underwood, and C. Lewis. Mathematical modeling of comedo ductal carcinoma in situ of the breast. *Math. Med. Biol.*, 20: 277–308, 2003.

[222] S. Franks and J. King. Interactions between a uniformly proliferating tumor and its surrounding uniform material properties. *Math. Med. Biol.*, 20: 47–89, 2003.

[223] J. Freyer. Decreased mitochondrial function in quiescent cells isolated from multicellular tumor spheroids. *J. Cell. Physiol.*, 176(1): 138–49, 1998.

[224] J. Freyer and R. Sutherland. Determination of diffusion constants for metabolites in multicell tumor spheroids. *Adv. Exp. Med. Biol.*, 159: 463–475, 1983.

[225] J. Freyer and R. Sutherland. A reduction in the in situ rates of oxygen and glucose consumption of cells in emt6/ro spheroids during growth. *J. Cell. Physiol.*, 124: 516–524, 1985.

[226] J. Freyer and R. Sutherland. Regulation of growth saturation and development of necrosis in emt6/ro multicellular spheroids by the glucose and oxygen supply. *Cancer Res.*, 46: 3504–3512, 1986.

[227] H. Frieboes, M. Edgerton, J. Fruehauf, *et al.* Prediction of drug response in breast cancer using integrative experimental/computational modeling. *Cancer Res.*, 69: 4484–4492, 2009.

[228] H. Frieboes, J. Fang, Y.-L. Chuang, S. Wise, J. Lowengrub, and V. Cristini. Nonlinear simulations of three-dimensional multispecies tumor growth – ii: tumor invasion and angiogenesis. *J. Theor. Biol.*, in press.

[229] H. Frieboes, J. Lowengrub, S. Wise, *et al.* Computer simulation of glioma growth and morphology. *NeuroImage*, 37: S59–S70, 2007.

[230] H. Frieboes, X. Zheng, C.-H. Sun, B. Tromberg, R. Gatenby, and V. Cristini. An integrated computational/experimental model of tumor invasion. *Cancer Res.*, 66: 1597–1604, 2006.

[231] P. Friedl. Prespecification and plasticity: shifting mechanisms of cell migration. *Curr. Opin. Cell Biol.*, 16: 14–23, 2004.

[232] P. Friedl, E. Brocker, and K. Zanker. Integrins, cell matrix interactions and cell migration strategies: fundamental differences in leukocytes and tumor cells. *Cell Adhes. Commun.*, 6: 225–236, 1998.

[233] P. Friedl, Y. Hegerfeldt, and M. Tilisch. Collective cell migration in morphogenesis and cancer. *Int. J. Dev. Biol.*, 48: 441–449, 2004.

[234] P. Friedl, P. Noble, P. Walton, *et al.* Migration of coordinated cell clusters in mesenchymal and epithelial cancer explants in vitro. *Cancer Res.*, 55: 4557–4560, 1995.

[235] P. Friedl and A. Wolf. Tumor cell invasion and migration: diversity and escape mechanisms. *Nat. Rev. Cancer*, 3: 362–374, 2003.

[236] A. Friedman. A hierarchy of cancer models and their mathematical challenges. *Discrete Cont. Dyn. Syst. Ser. B*, 4: 147–159, 2004.

[237] A. Friedman, N. Bellomo, and P. Maini. Mathematical analysis and challenges arising from models of tumor growth. *Math. Models Meth. Appl. Sci.*, 17: 1751–1772, 2007.

[238] A. Friedman and B. Hu. Asymptotic stability for a free boundary problem arising in a tumor model. *J. Diff. Eq.*, 227: 598–639, 2005.

[239] A. Friedman and B. Hu. Bifurcation from stability to instability for a free boundary problem arising in a tumor model. *Arch. Rat. Mech. Anal.*, 180: 293–330, 2006.

[240] A. Friedman and B. Hu. Bifurcation from stability to instability for a free boundary problem modeling tumor growth by Stokes equation. *J. Math. Anal. Appl.*, 327: 643–664, 2007.

[241] A. Friedman and F. Reitich. Analysis of a mathematical model for the growth of tumors. *J. Math. Biol.*, 38: 262–284, 1999.

[242] A. Friedman and F. Reitich. On the existence of spatially patterned dormant malignancies in a model for the growth of non-necrotic vascular tumors. *Math. Models Meth. Appl. Sci.*, 11: 601–625, 2001.

[243] A. Friedman and F. Reitich. Symmetry-breaking bifurcation of analytic solutions to free boundary problems: an application to a model of tumor growth. *Trans. Am. Math. Soc.*, 353: 1587–1634, 2001.

[244] S. H. Friend, R. Bernards, S. Rogelj, *et al.* A human DNA segment with properties of the gene that predisposes to retinoblastoma and osteosarcoma. *Nature*, 323: 643–6, 1986.

[245] S. M. Frisch and H. Francis. Disruption of epithelial cell–matrix interactions induces apoptosis. *J. Cell Biol.*, 124: 619–26, 1994.

[246] R. Fukuda, H. Zhang, J.-W. Kim, L. Shimoda, C. V. Dang, and G. L. Semenza. HIF-1 regulates cytochrome oxidase subunits to optimize efficiency of respiration in hypoxic cells. *Cell*, 129(1): 111–122, 2007.

[247] D. Fukumura and R. Jain. Tumor microenvironment abnormalities: causes, consequences, and strategies to normalize. *J. Cell. Biochem.*, 101: 937–949, 2007.

[248] Y. Fung. *Biomechanics: Motion, Flow, Stress and Growth*. Springer, New York, 1990.

[249] Y. Fung. *Biomechanics: Material Properties of Living Tissues*. Springer, New York, 1993.

[250] A.-P. Gadeau, H. Chaulet, D. Daret, M. Kockx, J.-M. Daniel-Lamazié, and C. Desgranges. Time course of osteopontin, osteocalcin, and osteonectin accumulation and calcification after acute vessel wall injury. *J. Histochem. Cytochem.*, 49: 79–86, 2001.

[251] D. Galaris, A. Barbouti, and P. Korantzopoulos. Oxidative stress in hepatic ischemia-reperfusion injury: the role of antioxidants and iron chelating compounds. *Curr. Pharma. Design*, 12: 2875–2890, 2006.

[252] J. Galle, G. Aust, G. Schaller, T. Beyer, and D. Drasdo. Individual cell-based models of the spatial temporal organization of multicellular systems – achievements and limitations. *Cytometry*, 69A: 704–710, 2006.

[253] J. Galle, M. Hoffmann, and G. Aust. From single cells to tissue architecture – a bottom-up approach to modeling the spatio-temporal organization of complex multicellular systems. *J. Math. Biol.*, 58: 261–283, 2009.

[254] J. Galle, M. Loeffler, and D. Drasdo. Modeling the effect of deregulated proliferation and apoptosis on the growth dynamics of epithelial cell populations in vitro. *Biophys. J.*, 88: 62–75, 2005.

[255] J. Galle, L. Preziosi, and A. Tosin. Contact inhibition of growth described using a multiphase model and an individual cell based model. *Appl. Math. Lett.*, 22: 1483–90, 2009.

[256] A. Gamba, D. Ambrosi, A. Coniglio, *et al.* Percolation, morphogenesis and burgers dynamics in blood vessels formation. *Phys. Rev. Lett.*, 90: 11 8101, 2003.

[257] J. Ganghoffer. Some issues related to growth and goal functions for continuum biological systems. *Phil. Mag.*, 85: 4353–4391, 2005.

[258] M. Garbey and G. Zouridakis. Modeling tumor growth: from differential deformable models to growth prediction of tumors detected in pet images. *Eng. Med. Biol. Soc.*, 3: 2687–2690, 2003.

[259] H. Garcke, M. Rumpf, and U. Weikard. The Cahn–Hilliard equation with elasticity: finite element approximation and qualitative studies. *Interfaces Free Bound.* 10: 101–118, 2001.

[260] M. Gardner. The fantastic combinations of John Conway's new solitaire game "Life". *Scientific American*, 223: 120–123, October 1970.

[261] K. Garikipati, E. Arruda, K. Grosh, H. Narayanan, and S. Calve. A continuum treatment of growth in biological tissue: the coupling of mass transport and mechanics. *J. Mech. Phys. Solids*, 52: 1595–1625, 2004.

[262] L. Garner, Y. Lau, T. Jackson, M. Uhler, D. Jordan, and R. Gilgenbach. Incorporating spatial dependence into a multicellular tumor spheroid growth model. *J. Appl. Phys.*, 98: 124 701, 2005.

[263] R. Gatenby and E. Gawlinski. The glycolytic phenotype in carcinogenesis and tumor invasion: insights through mathematical models. *Cancer Res.*, 63: 3847–3854, 2003.

[264] R. Gatenby, E. Gawlinski, A. Gmitro, B. Kaylor, and R. Gillies. Acid-mediated tumor invasion: a multidisciplinary study. *Cancer Res.*, 66: 5216–5223, 2006.

[265] R. Gatenby and E. Gawlinsky. Mathematical models of tumour invasion mediated by transformation-induced alteration of microenvironmental pH. In *The Tumour Microenvironment*, pp. 85–99. Wiley, London, 2003.

[266] R. Gatenby and R. Gillies. A microenvironmental model of carcinogenesis. *Nat. Rev. Cancer*, 8: 56–61, 2008.

[267] R. Gatenby, K. Smallbone, P. Maini, *et al*. Cellular adaptations to hypoxia and acidosis during somatic evolution of breast cancer. *Br. J. Cancer*, 97: 646–653, 2007.

[268] R. Gatenby and T. Vincent. An evolutionary model of carcinogenesis. *Cancer Res.*, 63: 6212–6220, 2003.

[269] C. W. Gear and I. G. Kevrekidis. Projective methods for stiff differential equations: problems with gaps in the eigenvalue spectrum. *SIAM J. Sci. Comput.*, 24: 1091–1106, 2003.

[270] C. W. Gear, I. G. Kevrekidis, and C. Theodoropoulos. "Coarse" integration/bifurcation analysis via microscopic simulators: micro-Galerkin methods. *Comput. Chem. Eng.*, 26: 941–963, 2002.

[271] A. Gerisch and M. Chaplain. Mathematical modelling of cancer cell invasion of tissue: local and non-local models and the effect of adhesion. *J. Theor. Biol.*, 250: 684–704, 2008.

[272] P. Gerlee and A. Anderson. An evolutionary hybrid cellular automaton model of solid tumor growth. *J. Theor. Biol.*, 246: 583–603, 2007.

[273] P. Gerlee and A. Anderson. Stability analysis of a hybrid cellular automaton model of cell colony growth. *Phys. Rev. E*, 75: 051 911, 2007.

[274] P. Gerlee and A. Anderson. A hybrid cellular automaton model of clonal evolution in cancer: the emergence of the glycolytic phenotype. *J. Theor. Biol.*, 250: 705–722, 2008.

[275] J. Gevertz and S. Torquato. Modeling the effects of vasculature evolution on early brain tumor growth. *J. Theor. Biol.*, 243: 517–531, 2006.

[276] F. G. Giancotti and E. Ruoslahti. Integrin signaling. *Science*, 285(5430): 1028–1032, 1999.

[277] J. W. Gibbs. Fourier series. *Nature*, 59: 200, 1898.

[278] M. Z. Gilcrease. Integrin signaling in epithelial cells. *Cancer Lett.*, 247(1): 1–25, 2007.

[279] R. Gillies and R. Gatenby. Adaptive landscapes and emergent phenotypes: why do cancers have high glycolysis? *J. Bioenerg. Biomem.*, 39: 251–257, 2007.

[280] R. Gillies and R. Gatenby. Hypoxia and adaptive landscapes in the evolution of carcinogenesis. *Cancer Metastasis Rev.*, 26: 311–317, 2007.

[281] R. Gillies, Z. Liu, and Z. Bhujwalla. 31p-mrs measurements of extracellular ph of tumors using 3-aminopropylphosphonate. *Am. J. Physiol.*, 267: 195–203, 1994.

[282] R. Gillies, I. Robey, and R. Gatenby. Causes and consequences of increased glucose metabolism of cancers. *J. Nuclear Med.*, 49: 24S–42S, 2008.

[283] M. Gimbrone, R. Cotran, S. Leapman, and J. Folkman. Tumor growth and neovascularization: an experimental model using the rabbit cornea. *J. Nat. Cancer Inst.*, 52: 413–427, 1974.

[284] J. Glazier and F. Graner. Simulation of the differential adhesion driven rearrangement of biological cells. *Phys. Rev. E*, 47: 2128–2154, 1993.

[285] R. Godde and H. Kurz. Structural and biophysical simulation of angiogenesis and vascular remodeling. *Dev. Dyn.*, 220(4): 387–401, 2001.

[286] J. J. Going and T. J. Mohun. Human breast duct anatomy, the "sick lobe" hypothesis and intraductal approaches to breast cancer. *Breast. Cancer. Res. Treat.*, 97(3): 285–291, 2006.

[287] J. D. Gordan, J. A. Bertout, C.-J. Hu, J. A. Diehl, and M. C. Simon. HIF-2α promotes hypoxic cell proliferation by enhancing c-Myc transcriptional activity. *Cancer Cell*, 11(4): 335–347, 2007.

[288] V. Gordon, M. Valentine, M. Gardel, *et al.* Measuring the mechanical stress induced by an expanding multicellular tumor system: a case study. *Exp. Cell Res.*, 289: 58–66, 2003.

[289] T. Graeber, C. Osmanian, T. Jacks, *et al.* Hypoxia-mediated selection of cells with diminished apoptotic potential in solid tumors. *Nature*, 379: 88–91, 1996.

[290] F. Graner and J. Glazier. Simulation of biological cell sorting using a two-dimensional extended Potts model. *Phys. Rev. Lett.*, 69: 2013–2016, 1992.

[291] L. Graziano and L. Preziosi. Mechanics in tumor growth. In F. Mollica, L. Preziosi, and K. Rajagopal, eds., *Modeling of Biological Materials*, pp. 267–328. Birkhäuser, New York, 2007.

[292] H. Greenspan. Models for the growth of a solid tumor by diffusion. *Stud. Appl. Math.*, 51: 317–340, 1972.

[293] H. Greenspan. On the growth and stability of cell cultures and solid tumors. *J. Theor. Biol.*, 56: 229–242, 1976.

[294] A. Grin, G. Horne, M. Ennis, and F. P. O'Malley. Measuring extent of ductal carcinoma in situ in breast excision specimens: a comparison of four methods. *Arch. Pathol. Lab. Med.*, 133: 31–37, 2009.

[295] K. Groebe, S. Erz, and W. Mueller-Klieser. Glucose diffusion coefficients determined from concentration profiles in emt6 tumor spheroids incubated in radioactively labeled l-glucose. *Adv. Exp. Med. Biol.*, 361: 619–625, 1994.

[296] J. Guck, S. Schinkinger, B. Lincoln, *et al.* Optical deformability as an inherent cell marker for testing malignant transformation and metastatic competence. *Biophys. J.*, 88(5): 3689–3698, 2005.

[297] C. Guiot, P. D. Santo, and T. Deisboeck. Morphological instability and cancer invasion: a "splashing water drop" analogy. *Theor. Biol. Med. Model.*, 4: 4, 2007.

[298] G. P. Gupta and J. Massagu. Cancer metastasis: building a framework. *Cell*, 127(4): 679–695, 2006.

[299] G. Hamilton. Multicellular spheroids as an in vitro tumor model. *Cancer Lett.*, 131: 29–34, 1998.

[300] D. Hanahan and R. Weinberg. The hallmarks of cancer. *Cell*, 100: 57–70, 2000.

[301] R. K. Hansen and M. J. Bissell. Tissue architecture and breast cancer: the role of extracellular matrix and steroid hormones. *Endocrine-Related Cancer*, 7(2): 95–113, 2000.

[302] B. D. Harms, G. M. Bassi, A. R. Horwitz, and D. A. Lauffenburger. Directional persistence of EGF-induced cell migration is associated with stabilization of lamellipodial protrusions. *Biophys. J.*, 88(2): 1479–1488, 2005.

[303] A. Harris. Hypoxia – a key regulatory factor in tumor growth. *Nat. Rev. Cancer*, 2: 38–47, 2002.

[304] A. Harten, B. Engquist, S. Osher, and S. R. Chakravarthy. Uniformly high order accurate essentially non-oscillatory schemes, III. *J. Comput. Phys.*, 71: 231–303, 1987.

[305] H. Hashizume, P. Baluk, S. Morikawa, *et al.* Openings between defective endothelial cells explain tumor vessel leakiness. *Am. J. Pathol.*, 156: 1363–1380, 2000.

[306] H. Hatzikirou, L. Brusch, and A. Deutsch. From cellular automaton rules to an effective macroscopic mean-field description. *Acta Physica Polonica B Proc. Suppl.*, 3(2): 399–416, 2010.

[307] H. Hatzikirou, A. Deutsch, C. Schaller, M. Simon, and K. Swanson. Mathematical modeling of glioblastoma tumour development: a review. *Math. Models Meth. Appl. Sci.*, 15: 1779–1794, 2005.

[308] M. A. Hayat. *Methods of Cancer Diagnosis, Therapy, and Prognosis: Liver Cancer.* Springer, New York, fifth edition, 2009.

[309] Y. Hegerfeldt, M. Tusch, E. Brocker, and P. Friedl. Collective cell movement in primary melanoma explants: plasticity of cell–cell interaction, 1-integrin function, and migration strategies. *Cancer Res.*, 62: 2125–2130, 2002.

[310] G. Helmlinger, P. Netti, H. Lichtenbeld, R. Melder, and R. Jain. Solid stress inhibits the growth of multicellular tumor spheroids. *Nat. Biotech.*, 15: 778–783, 1997.

[311] G. Helmlinger, F. Yuan, M. Dellian, and R. Jain. Interstitial ph and po2 gradients in solid tumors in vivo: high-resolution measurements reveal a lack of correlation. *Nat. Med.*, 3: 177–182, 1997.

[312] R. S. Herbst. Review of epidermal growth factor receptor biology. *Int. J. Rad. Oncol. Biol. Phys.*, 59(2(S1)): S21–S26, 2004.

[313] S.-I. Hino, C. Tanji, K. I. Nakayama, and A. Kikuchi. Phosphorylation of β-catenin by cyclic AMP-dependent protein kinase stabilizes β-catenin through inhibition of its ubiquitination. *Molec. Cell. Biol.*, 25(20): 9063–72, 2005.

[314] S. Hiratsuka, K. Nakamura, S. Iwai, *et al.* MMP9 induction by vascular endothelial growth factor receptor-1 is involved in lung-specific metastasis. *Cancer Cell*, 2(4): 289–300, 2002.

[315] M. Höckel, K. Schlenger, B. Aral, M. Mitze, U. Schaffer, and P. Vaupel. Association between tumor hypoxia and malignant progression in advanced cancer of the uterine cervix. *Cancer Res.*, 56: 4509–4515, 1996.

[316] M. Höckel, K. Schlenger, S. Hoeckel, and P. Vaupel. Hypoxic cervical cancers with low apoptotic index are highly aggressive. *Cancer Res.*, 59: 4525–4528, 1999.

[317] M. Höckel, K. Schlenger, and P. Vaupel. Tumor hypoxia: definitions and current clinical, biologic, and molecular aspects. *J. Natl Cancer Inst.*, 93: 266–276, 2001.

[318] C. Hogea, B. Murray, and J. Sethian. Simulating complex tumor dynamics from avascular to vascular growth using a general level-set method. *J. Math. Biol.*, 53: 86–134, 2006.

[319] P. Hogeweg. Evolving mechanisms of morphogenesis: on the interplay between differential adhesion and cell-differentiation. *J. Theor. Biol.*, 203: 317–333, 2000.

[320] J. Holash, P. Maisonpierre, D. Compton, *et al.* Vessel cooption, regression, and growth in tumors mediated by angiopoietins and vegf. *Science*, 284: 1994–1998, 1999.

[321] J. Holash, S. Wiegand, and G. Yancopoulos. New model of tumor angiogenesis: dynamic balance between vessel regression and growth mediated by angiopoietins and vegf. *Oncogene*, 18: 5356–5362, 1999.

[322] M. Holmes and B. Sleeman. A mathematical model of tumor angiogenesis incorporating cellular traction and viscoelastic effects. *J. Theor. Biol.*, 202: 95–112, 2000.

[323] J. M. Horowitz, D. W. Yandell, S.-H. Park, *et al.* Point mutational inactivation of the retinoblastoma antioncogene. *Science*, 243(4893): 937–940, 1989.

[324] D. Horstmann, K. Painter, and H. Othmer. Aggregation under local reinforcement: from lattice to continuum. *Eur. J. Appl. Math.*, 15: 545–576, 2004.

[325] K. B. Hotary, E. D. Allen, P. C. Brooks, N. S. Datta, M. W. Long, and S. J. Weiss. Membrane type 1 matrix metalloproteinase usurps tumour growth control imposed by the three-dimensional extracellular matrix. *Cell*, 114(1): 33–45, 2003.

[326] M. Hu, J. Yao, L. Cai, K. E. Bachman, F. van den Brle, V. Velculescu, and K. Polyak. Distinct epigenetic changes in the stromal cells of breast cancers. *Nat. Genet.*, 37(8): 899–905, 2005.

[327] Z. Hu, K. Yuri, H. Ozawa, H. Lu, and M. Kawata. The *in vivo* time course for elimination of adrenalectomy-induced apoptotic profiles from the granule cell layer of the rat hippocampus. *J. Neurosci.*, 17(11): 3981–3989, 1997.

[328] J. Humphrey. Continuum biomechanics of soft biological tissues. *Proc. Roy. Soc. London A*, 459: 3–46, 2003.

[329] J. Humphrey and K. Rajagopal. A constrained mixture model for growth and remodeling of soft tissues. *Math. Models Meth. Appl. Sci.*, 12: 407–430, 2002.

[330] D. Ilic, E. A. Almeida, D. D. Schlaepfer, P. Dazin, S. Aizawa, and C. H. Damsky. Extracellular matrix survival signals transduced by focal adhesion kinase suppress p53-mediated apoptosis. *J. Cell Biol.*, 143: 547–560, 1998.

[331] J. H. Irving and J. G. Kirkwood. The statistical mechanical theory of transport processes. IV. The equations of hydrodynamics. *J. Chem. Phys.*, 18: 817–829, 1950.

[332] N. Ishii, D. Maier, A. Merlo, M. Tada, Y. Sawamura, A. Diserens, and E. V. Meir. Frequent co-alterations of tp53, p16/cdkn2a, p14arf, pten tumor suppressor genes in human glioma cell lines. *Brain Pathol.*, 9: 469–479, 1999.

[333] T. Ishii, J. Murakami, K. Notohara, *et al.* Oeophageal squamous cell carcinoma may develop within a background of accumulating DNA methylation in normal and dysplastic mucosa. *Gut*, 56(1): 13–19, 2007.

[334] T. Ishikawa, Y. Kobayashi, A. Omoto, *et al.* Calcification in untreated non-Hodgkin's lymphoma of the jejunum. *Acta Haematol.*, 102(4): 185–189, 1999.

[335] T. Jackson. Intracellular accumulation and mechanism of action of doxorubicin in a spatio-temporal tumor model. *J. Theor. Biol.*, 220: 201–213, 2003.

[336] T. Jackson. A mathematical investigation of the multiple pathways to recurrent prostate cancer: comparison with experimental data. *Neoplasia*, 6: 697–704, 2004.

[337] T. Jackson. A mathematical model of prostate tumor growth and androgen-independent relapse. *Disc. Cont. Dyn. Sys. B*, 4: 187–201, 2004.

[338] T. Jackson and H. Byrne. A mechanical model of tumor encapsulation and transcapsular spread. *Math. Biosci.*, 180: 307–328, 2002.

[339] R. Jain. Determinants of tumor blood flow: a review. *Cancer Res.*, 48: 2641–2658, 1988.

[340] R. Jain. Physiological barriers to delivery of monoclonal antibodies and other macromolecules in tumors. *Cancer Res.*, 50: 814s–819s, 1990.

[341] R. Jain. Delivery of molecular medicine to solid tumors: lessons from in vivo imaging of gene expression and function. *J. Control. Release*, 74: 7–25, 2001.

[342] R. Jain. Normalizing tumor vasculature with anti-angiogenic therapy: a new paradigm for combination therapy. *Nat. Med.*, 7: 987–989, 2001.

[343] R. Jain. Molecular regulation of vessel maturation. *Nat. Med.*, 9: 685–693, 2003.

[344] R. Jain. Normalization of tumor vasculature: an emerging concept in antiangiogenic therapy. *Science*, 307: 58–62, 2005.

[345] K. A. Janes and D. A. Lauffenburger. A biological approach to computational models of proteomic networks. *Curr. Opin. Chem. Biol.*, 10(1): 73–80, 2006.

[346] A. Jemal, R. Siegel, E. Ward, T. Murray, J. Xu, and M. J. Thun. Cancer statistics, 2007. *CA Cancer J. Clin.*, 57(1): 43–66, 2007.

[347] R. Jensen. Hypoxia in the tumorigenesis of gliomas and as a potential target for therapeutic measures. *Neurosurg. Focus*, 20: E24, 2006.

[348] B. Jian, N. Narula, Q.-Y. Li, E. R. Mohler III, and R. J. Levy. Progression of aortic valve stenosis: TGF-β1 is present in calcified aortic valve cusps and promotes aortic valve interstitial cell calcification via apoptosis. *Ann. Thoracic Surg.*, 75(2): 457–465, 2003.

[349] G.-S. Jiang and C.-W. Shu. Efficient implementation of weighted ENO schemes. *J. Comput. Phys.*, 126: 202–228, 1996.

[350] T. X. Jiang and C. M. Chuong. Mechanism of skin morphogenesis I: analyses with antibodies to adhesion molecules tenascin, NCAM, and integrin. *Dev. Biol.*, 150: 82–98, 1992.

[351] Y. Jiang, J. Pjesivac-Grbovic, C. Cantrell, and J. Freyer. A multiscale model for avascular tumor growth. *Biophys. J.*, 89: 3884–3894, 2005.

[352] A. Jones, H. Byrne, J. Gibson, and J. Dold. Mathematical model for the stress induced during avascular tumor growth. *J. Math. Biol.*, 40: 473–499, 2000.

[353] P. Jones and B. Sleeman. Angiogenesis – understanding the mathematical challenge. *Angiogenesis*, 9: 127–138, 2006.

[354] P. A. Jones and S. B. Baylin. The fundamental role of epigenetic events in cancer. *Nat. Rev. Genet*, 3(6): 415–428, 2002.

[355] P. A. Jones and P. W. Laird. Cancer epigenetics comes of age. *Nat. Genet.*, 21(2): 163–7, 1999.

[356] R. E. Jr., K. O'Connor, D. Lacks, D. Schwartz, and R. Dotson. Dynamics of spheroid self-assembly in liquid-overlay culture of du 145 human prostate cancer cells. *Biotech. Bioeng.*, 72: 579–591, 2001.

[357] K. Kaibuchi, S. Kuroda, and M. Amano. Regulation of the cytoskeleton and cell adhesion by the Rho family GTPases in mammalian cells. *Ann. Rev. Biochem.*, 68: 459–486, 1999.

[358] F. Kallinowski, P. Vaupel, S. Runkel, *et al.* Glucose uptake, lactate release, ketone body turnover, metabolic milieu and ph distributions in human cancer xenografts in nude rats. *Cancer Res.*, 48: 7264–7272, 1988.

[359] K. Kaneko, K. Satoh, and A. Masamune. T. myosin light chain kinase inhibitors can block invasion and adhesion of human pancreatic cancer cell lines. *Pancreas*, 24: 34–41, 2002.

[360] R. N. Kaplan, S. Rafii, and D. Lyden. Preparing the "soil": the premetastatic niche. *Cancer Res.*, 66(23): 11 089–11 093, 2006.

[361] P. I. Karecla, S. J. Green, S. J. Bowden, J. Coadwell, and P. J. Kilshaw. Identification of a binding site for integrin $\alpha e \beta 7$ in the N-terminal domain of E-cadherin. *J. Biol. Chem.*, 271: 30 909–30 915, 1996.

[362] B. Kaur, F. Khwaja, E. Severson, S. Matheny, D. Brat, and E. VanMeir. Hypoxia and the hypoxia-inducible-factor pathway in glioma growth and angiogenesis. *Neuro-Oncol.*, 7: 134–153, 2005.

[363] D. Kay and R. Welford. A multigrid finite element solver for the Cahn–Hilliard equation. *J. Comput. Phys.*, 212: 288–304, 2006.

[364] P. Keller, F. Pampaloni, and E. Stelzer. Life sciences require the third dimension. *Curr. Op. Cell Biol.*, 18: 117–124, 2006.

[365] T. Kelly, Y. Yan, R. L. Osborne, *et al*. Proteolysis of extracellular matrix by invadopodia facilitates human breast cancer cell invasion and is mediated by matrix metalloproteinases. *Clin. Exp. Metastasis*, 16(6): 501–12, 1998.

[366] P. Kenny, G. Lee, and M. Bissell. Targeting the tumor microenvironment. *Front. Biosci.*, 12: 3468–3474, 2007.

[367] K. Keren, Z. Pincus, G. M. Allen, *et al*. Mechanism of shape determination in motile cells. *Nature*, 453(7194): 475–480, 2008.

[368] K. Kerlikowske, A. Molinaro, I. Cha, *et al*. Characteristics associated with recurrence among women with ductal carcinoma in situ treated by lumpectomy. *J. Natl Cancer Inst.*, 95(22): 1692–1702, 2003.

[369] J. F. R. Kerr, C. M. Winterford, and B. V. Harmon. Apoptosis. Its significance in cancer and cancer therapy. *Cancer*, 73(8): 2013–2026, 1994.

[370] I. Kevrekidis, , C. Gear, J. Hyman, P. Kevrekidis, O. Runborg, and K. Theodoropoulos. Equation-free, coarse-grained multiscale computation: enabling microscopic simulators to perform system-level analysis. *Commun. Math. Sci.*, 1: 715–762, 2003.

[371] P. Kevrekidis, N. Whitaker, D. Good, and G. Herring. Minimal model for tumor angiogenesis. *Phys. Rev. E*, 73: 061 926, 2006.

[372] E. Khain and L. Sander. Generalized Cahn-Hilliard equation for biological applications. *Phys. Rev. E*, 77: 051 129, 2008.

[373] E. Khain, L. Sander, and C. Schneider-Mizell. The role of cell–cell adhesion in wound healing. *J. Stat. Phys.*, 128: 209–218, 2007.

[374] S. Khan, M. Rogers, K. Khurana, M. Meguid, and P. Numann. Estrogen receptor expression in benign breast epithelium and breast cancer risk. *J. Natl Canc. Inst.*, 90: 37–42, 1998.

[375] S. Khan, A. Sachdeva, S. Naim, *et al*. The normal breast epithelium of women with breast cancer displays an aberrant response to estradiol. *Canc. Epidemiol. Biomarkers Prev.*, 8: 867–872, 1999.

[376] S. Kharait, S. Hautaniemi, S. Wu, A. Iwabu, D. A. Lauffenburger, and A. Wells. Decision tree modeling predicts effects of inhibiting contractility signaling on cell motility. *BMC Syst. Biol.*, 1: 9ff, 2007.

[377] B. N. Kholodenko, O. V. Demin, G. Moehren, and J. B. Hoek. Quantification of short term signaling by the epidermal growth factor receptor. *J. Biol. Chem.*, 274(42): 30 169–30 181, 1999.

[378] J. Kim. Three-dimensional tissue culture models in cancer biology. *J. Biomol. Screening*, 15: 365–377, 2005.

[379] J. Kim, K. Kang, and J. Lowengrub. Conservative multigrid methods for Cahn–Hilliard fluids. *J. Comput. Phys.*, 193: 511–543, 2003.

[380] J. Kim, K. Kang, and J. Lowengrub. Conservative multigrid methods for ternary Cahn–Hilliard systems. *Commun. Math. Sci.*, 2: 53–77, 2004.

[381] J. Kim and J. Lowengrub. Phase field modeling and simulation of three phase flows. *Int. Free Bound.*, 7: 435–466, 2005.

[382] Y. Kim, M. Stolarska, and H. Othmer. A hybrid model for tumor spheroid growth in vitro i: theoretical development and early results. *Math. Meth. App. Sci.*, 17: 1773–1798, 2007.

[383] R. Kloner and R. Jennings. Consequences of brief ischemia: stunning, preconditioning, and their clinical implications: part 1. *Circulation*, 104: 2981–2989, 2001.

[384] M. Knowles and P. Selby, eds. *Introduction to the Cellular and Molecular Biology of Cancer*. Oxford University Press, Oxford, UK, fourth edition, 2005.

[385] K. A. Knudsen, A. P. Soler, K. R. Johnson, and M. J. Wheelock. Interaction of α-actin with the cadherin/catenin cell–cell adhesion complex via α-catenin. *J. Cell. Biol.*, 130: 66–77, 1995.

[386] A. G. Knudson. Mutation and cancer: statistical study of retinoblastoma. *Proc. Natl Acad. Sci. USA*, 68(4): 820–823, 1971.

[387] A. G. Knudson. Two genetic hits (more or less) to cancer. *Nat. Rev. Cancer*, 1(2): 157–162, 2001.

[388] L. Kopfstein and G. Christofori. Metastasis: cell-autonomous mechanisms versus contributions by the tumor microenvironment. *Cell. Mol. Life Sci.*, 63: 449–468, 2006.

[389] D. V. Krysko, T. V. Berghe, K. D'Herde, and P. Vandenabeele. Apoptosis and necrosis: detection, discrimination and phagocytosis. *Methods*, 44: 205–221, 2008.

[390] E. Kuhl and G. Holzapfel. A continuum model for remodeling in living structures. *J. Mater. Sci.*, 42: 8811–8823, 2007.

[391] R. Kuiper, J. Schellens, G. Blijham, J. Beijnen, and E. Voest. Clinical research on antiangiogenic therapy. *Pharmacol. Res.*, 37: 1–16, 1998.

[392] P. Kunkel, U. Ulbricht, P. Bohlen, *et al*. Inhibition of glioma angiogenesis and growth in vivo by systemic treatment with a monoclonal antibody against vascular endothelial growth factor receptor-2. *Cancer Res.*, 61: 6624–6628, 2001.

[393] L. Kunz-Schughart, J. P. Freyer, F. Hofstaedter, and R. Ebner. The use of 3-d cultures for high-throughput screening: the multicellular spheroid model. *J. Biomol. Screening*, 9: 273–285, 2004.

[394] R. Küppers and R. Dalla-Favera. Mechanisms of chromosomal translocations in B cell lymphomas. *Oncogene*, 20(40): 5580–5594, 2001.

[395] A. Lal, C. Glazer, H. Martinson, *et al*. Mutant epidermal growth factor receptor up-regulates molecular effectors of tumor invasion. *Cancer Res.*, 62: 3335–3339, 2002.

[396] C. R. Lamb. Diagnosis of calcification on abdominal radiographs. *Vet. Rad. Ultrasound*, 32(5): 211–220, 1991.

[397] O. T. Lampejo, D. M. Barnes, P. Smith, and R. R. Millis. Evaluation of infiltrating ductal carcinomas with a DCIS component: correlation of the histologic type of the in situ component with grade of the infiltrating component. *Semin. Diagn. Pathol.*, 11(3): 215–222, 1994.

[398] K. Lamszus, P. Kunkel, and M. Westphal. Invasion as limitation to anti-angiogenic glioma therapy. *Acta Neurochir. Suppl.*, 88: 169–177, 2003.

[399] K. Landman and C. Please. Tumour dynamics and necrosis: surface tension and stability. *IMA J. Math. Appl. Med. Biol.*, 18: 131–158, 2001.

[400] M. C. Lane, M. A. Koehl, F. WIlt, and R. Keller. A role for regulated secretion of apical extracellular matrix during epithelial invagination in the sea urchin. *Development*, 117(3): 1049–1060, 1993.

[401] H. Larjava, T. Salo, K. Haapasalmi, R. H. Kramer, and J. Heino. Expression of integrins and basement membrane components by wound keratinocytes. *J. Clin. Invest.*, 92(3): 1425–1435, 1993.

[402] D. A. Lauffenburger. Cell signaling pathways as control modules: complexity for simplicity? *Proc. Natl Acad. Sci. USA*, 97(10): 5031–5033, 2000.

[403] C. Le Clainche and M. F. Carlier. Regulation of actin assembly associated with protrusion and adhesion in cell migration. *Physiol. Rev.*, 88(2): 489–513, 2008.

[404] D. Lee, H. Rieger, and K. Bartha. Flow correlated percolation during vascular remodeling in growing tumors. *Phys. Rev. Lett.*, 96: 058 104, 2006.

[405] J. S. Lee, D. M. Basalyga, A. Simionescu, J. C. Isenburg, D. T. Simionescu, and N. R. Vyavahare. Elastin calcification in the rate subdermal model is accompanied by up-regulation of degradative and osteogenic cellular responses. *Am. J. Pathol.*, 168: 490–498, 2006.

[406] S. Lee, S. K. Mohsin, S. Mao, S. G. Hilsenbeck, D. Medina, and D. C. Allred. Hormones, receptors, and growth in hyperplastic enlarged lobular units: early potential precursors of breast cancer. *Breast Cancer Res.*, 8(1): R6, 2006.

[407] S. Lehoux and A. Tedgui. Signal transduction of mechanical stresses in the vascular wall. *Hypertension*, 32(2): 338–345, 1998.

[408] J. Less, T. Skalak, E. Sevick, and R. Jain. Microvascular architecture in a mammary carcinoma: branching patterns and vessel dimensions. *Cancer Res.*, 51: 265–273, 1991.

[409] H. Levine and M. Nilsen-Hamilton. Angiogenesis – a biochemical/mathematical perspective. *Tutorials in Math. Biosci. III*, 1872: 23–76, 2006.

[410] H. Levine, S. Pamuk, B. Sleeman, and M. Nilsen-Hamilton. Mathematical modeling of capillary formation and development in tumor angiogenesis: penetration into the stroma. *Bull. Math. Biol.*, 63: 801–863, 2001.

[411] H. Levine and B. Sleeman. Modelling tumour-induced angiogenesis. In L. Preziosi, ed., *Cancer Modelling and Simulation*, pp. 147–184. Chapman&Hall/CRC, Boca Raton, Florida, 2003.

[412] H. Levine, B. Sleeman, and M. Nilsen-Hamilton. Mathematical modeling of the onset of capillary formation initiating angiogenesis. *J. Math. Biol.*, 42: 195–238, 2001.

[413] H. Levine, M. Smiley, A. Tucker, and M. Nilsen-Hamilton. A mathematical model for the onset of avascular tumor growth in response to the loss of p53 function. *Cancer Informatics*, 2: 163–188, 2006.

[414] H. Levine, A. Tucker, and M. Nilsen-Hamilton. A mathematical model for the role of cell signal transduction in the initiation and inhibition of angiogenesis. *Growth Factors*, 20: 155–175, 2002.

[415] J. Li, P. Kevrekidis, C. Gear, and I. Kevrekidis. Deciding the nature of the coarse equation through microscopic simulations: the baby–bathwater scheme. *SIAM Rev.*, 49: 469–487, 2007.

[416] X. Li, V. Cristini, Q. Nie, and J. Lowengrub. Nonlinear three-dimensional simulation of solid tumor growth. *Disc. Dyn. Contin. Dyn. Syst. B*, 7: 581–604, 2007.

[417] S.-Y. Lin, W. Xia, J. C. Wang, K. Y. Kwong, and B. Spohn. β-Catenin, a novel prognostic marker for breast cancer: its roles in Cyclin D1 expression and cancer progression. *Proc. Natl Acad. Sci. USA*, 97(8): 4262–4266, 2000.

[418] L. Liotta and E. Kohn. The microenvironment of the tumour–host interface. *Nature*, 411: 375–379, 2001.

[419] A. Lipton. Pathophysiology of bone metastases: how this knowledge may lead to Therapeutic intervention. *J. Support. Oncol.*, 2(3): 205–220, 2004.

[420] X. D. Liu, S. Osher, and T. Chan. Weighted essentially non-oscillatory schemes. *J. Comput. Phys.*, 115: 200–212, 1994.

[421] B. Lloyd, D. Szczerba, M. Rudin, and G. Szekely. A computational framework for modeling solid tumour growth. *Phil. Trans. Roy. Soc. A*, 366: 3301–3318, 2008.

[422] B. Lloyd, D. Szczerba, and G. Szekely. A coupled finite element model of tumor growth and vascularization. In N. Ayache, S. Ourselin, and A. Maeder, eds., *Medical Image Computing and Computer-Assisted Intervention, Proc. MICCA 2007: 10th international conf.*, vol. 4792 of *Lecture Notes in Computer Science*, pp. 874–881. Springer, New York, 2007.

[423] J. Lotem and L. Sachs. Epigenetics and the plasticity of differentiation in normal and cancer stem cells. *Oncogene*, 25(59): 7663–7672, 2006.

[424] R. M. B. Loureiro and P. A. D'Amore. Transcriptional regulation of vascular endothelial growth factor in cancer. *Cytokine Growth Factor Rev.*, 16(1): 77–89, 2005.

[425] J. Lowengrub, H. Frieboes, F. Jin, *et al*. Nonlinear modeling of cancer: bridging the gap between cells and tumors. *Nonlinearity*, 23: R1–R91, 2010.

[426] V. Lubarda and A. Hoger. On the mechanics of solids with a growing mass. *Int. J. Solids Structures*, 39: 4627–4664, 2002.

[427] P. J. Lucio, M. T. Faria, A. M. Pinto, *et al*. Expression of adhesion molecules in chronic B-cell lymphoproliferative disorders. *Haematologica*, 83(2): 104–11, 1998.

[428] B. Lustig, B. Jerchow, M. Sachs, *et al*. Negative feedback loop of Wnt signaling through upregulation of conductin/axin2 in colorectal and liver tumors. *Mol. Cell. Biol.*, 22: 1184–193, 2002.

[429] B. MacArthur and C. Please. Residual stress generation and necrosis formation in multi-cell tumour spheroids. *J. Math. Biol.*, 49: 537–552, 2004.

[430] P. Macklin. Numerical simulation of tumor growth and chemotherapy. M.S. thesis, University of Minnesota School of Mathematics, September 2003.

[431] P. Macklin. Toward computational oncology: nonlinear simulation of centimeter-scale tumor growth in complex, heterogeneous tissues. Ph.D. dissertation, University of California, Irvine Department of Mathematics, June 2007.

[432] P. Macklin, L. Carreras, J. Kim, S. Sanga, V. Cristini, and M. E. Edgerton. Mathematical analysis of histopathology indicates comparable oxygen uptake rates for quiescent and proliferating breast cancer cells in DCIS (in preparation), 2010.

[433] P. Macklin, J. Kim, G. Tomaiuolo, M. E. Edgerton, and V. Cristini. Agent-based modeling of ductal carcinoma in situ: application to patient-specific breast cancer modeling. In T. Pham, ed., *Computational Biology: Issues and Applications in Oncology*, Chapter 4, pp. 77–112. Springer, New York, 2009.

[434] P. Macklin, *et al*. A composite agent-based cell model, with application to cancer, Parts I and II. *J. Theor. Biol.* (in preparation).

[435] P. Macklin and J. Lowengrub. Evolving interfaces via gradients of geometry-dependent interior Poisson problems: application to tumor growth. *J. Comput. Phys.*, 203: 191–220, 2005.

[436] P. Macklin and J. Lowengrub. An improved geometry-aware curvature discretization for level set methods: application to tumor growth. *J. Comput. Phys.*, 215: 392–401, 2006.

[437] P. Macklin and J. Lowengrub. Nonlinear simulation of the effect of microenvironment on tumor growth. *J. Theor. Biol.*, 245: 677–704, 2007.

[438] P. Macklin and J. Lowengrub. A new ghost cell/level set method for moving boundary problems: application to tumor growth. *J. Sci. Comp.*, 35(2–3): 266–299, 2008.

[439] P. Macklin, S. McDougall, A. Anderson, M. Chaplain, V. Cristini, and J. Lowengrub. Multiscale modeling and nonlinear simulation of vascular tumour growth. *J. Math. Biol.*, 58(4–5): 765–798, 2009.

[440] A. D. C. Macknight, D. R. DiBona, A. Leaf, and M. C. Mortimer. Measurement of the composition of epithelial cells from the toad urinary bladder. *J. Membrane Biol.*, 6(2): 108–126, 1971.

[441] S. Madsen, E. Angell-Petersen, S. Spetalen, S. Carper, S. Ziegler, and H. Hirschberg. Photodynamic therapy of newly implanted glioma cells in the rat brain. *Lasers Surg. Med.*, 38: 540–548, 2006.

[442] S. Maggelakis and J. Adam. Mathematical model of prevascular growth of a spherical carcinoma. *Math. Comput. Modelling*, 13: 23–38, 1990.

[443] E. Maher, F. Furnari, R. Bachoo, *et al*. Malignant glioma: genetics and biology of a grave matter. *Genes Dev.*, 15: 1311–1333, 2001.

[444] G. Majno and I. Joris. *Cells, Tissues, and Disease: Principles of General Pathology*. Oxford University Press, New York, second edition, 2004.

[445] A. G. Makeev, D. Maroudas, and I. G. Kevrekidis. "Coarse" stability and bifurcation analysis using stochastic simulators: kinetic Monte Carlo examples. *J. Chem. Phys.*, 116: 10 083–10 091, 2002.

[446] A. G. Makeev, D. Maroudas, A. Z. Panagiotopoulos, and I. G. Kevrekidis. Coarse bifurcation analysis of kinetic Monte Carlo simulations: a lattice-gas model with lateral interactions. *J. Chem. Phys.*, 117: 8229–8240, 2002.

[447] R. Malladi, J. A. Sethian, and B. C. Vemuri. Shape modeling with front propagation: a level set approach. *IEEE Trans. Pattern Anal. Mach. Intell.*, 17(2), 1995.

[448] R. Malladi, J. A. Sethian, and B. C. Vemuri. A fast level set based algorithm for topology-independent shape modeling. *J. Math. Imaging Vision*, 6(2–3): 269–289, 1996.

[449] M. Malumbres and M. Barbacis. RAS oncogenes: the first 30 years. *Nat. Rev. Cancer*, 3(6): 459–465, 2001.

[450] L. Malvern. *Introduction of the Mechanics of a Continuous Medium*. Prentice Hall, Englewood Cliffs, 1969.

[451] D. Manoussaki, S. Lubkin, R. Vernon, and J. Murray. A mechanical model for the formation of vascular networks in vitro. *Acta Biotheor.*, 44: 271–282, 1996.

[452] N. Mantzaris, S. Webb, and H. Othmer. Mathematical modeling of tumor-induced angiogenesis. *J. Math. Biol.*, 49: 111–187, 2004.

[453] B. Marchant, J. Norbury, and J. A. Sherratt. Travelling wave solutions to a haptotaxisdominated model of malignant invasion. *Nonlinearity*, 14: 1653–1671, 2001.

[454] A. F. Maree, A. Jilkine, A. Dawes, V. A. Grieneisen, and L. Edelstein-Keshet. Polarization and movement of keratocytes: a multiscale modelling approach. *Bull. Math. Biol.*, 68(5): 1169–1211, 2006.

[455] D. Martin and P. Colella. A cell-centered adaptive projection method for the incompressible Euler equations. *J. Comput. Phys.*, 163: 271–312, 2000.

[456] M. Marusic, Z. Baizer, J. Freyer, and S. Vuk-Pavlovic. Analysis of growth of multicellular tumour spheroids by mathematical models. *Cell Prolif.*, 27: 73–94, 1994.

[457] K. Matsumoto and T. Nakamura. Hepatocyte growth factor and the Met system as a mediator of tumor-stromal interactions. *Int. J. Cancer*, 119(3): 477–483, 2006.

[458] S. McDougall, A. Anderson, and M. Chaplain. Mathematical modelling of dynamic adaptive tumour-induced angiogenesis: clinical applications and therapeutic targeting strategies. *J. Theor. Biol.*, 241: 564–589, 2006.

[459] S. McDougall, A. Anderson, M. Chaplain, and J. Sherratt. Mathematical modelling of flow through vascular networks: implications for tumour-induced angiogenesis and chemotherapy strategies. *Bull. Math. Biol.*, 64: 673–702, 2002.

[460] D. McElwain and L. Morris. Apoptosis as a vol. loss mechanism in mathematical models of solid tumor growth. *Math. Biosci.*, 39: 147–157, 1978.

[461] C. Medrek, G. Landberg, T. Andersson, and K. Leandersson. Wnt-5a-CKIα signaling promotes β-catenin/E-cadherin complex formation and intercellular adhesion in human breast epithelial cells. *J. Biol. Chem.*, 284: 10 968–10 979, 2009.

[462] A. Menzel. Modelling of anisotropic growth in biological tissues – a new approach and computational aspects. *Biomech. Model. Mechanobiol.*, 3: 147–171, 2005.

[463] R. Merks, S. Brodsky, M. Goligorksy, S. Newman, and J. Glazier. Cell elongation is key to in silico replication of in vitro vasculogenesis and subsequent remodeling. *Dev. Biol.*, 289: 44–54, 2006.

[464] R. Merks and J. Glazier. Dynamic mechanisms of blood vessel growth. *Nonlinearity*, 19: C1–C10, 2006.

[465] R. Merks, E. P. A. Shirinifard, and J. Glazier. Contact-inhibited chemotaxis in de novo and sprouting blood-vessel growth. *PloS Comp. Biol.*, 4: e1000163, 2008.

[466] A. Merlo. Genes and pathways driving glioblastomas in humans and murine disease models. *Neurosurg. Rev.*, 26: 145–158, 2003.

[467] N. Metropolis, A. Rosenbluth, M. Rosenbluth, A. Teller, and E. Teller. Equation of state calculations by fast computing machines. *J. Chem. Phys.*, 21: 1087–1092, 1953.

[468] P. Michieli, C. Basilico, S. Pennacchietti, *et al.* Mutant met mediated transformation is ligand-dependent and can be inhibited by hgf antagonists. *Oncogene*, 18: 5221–5231, 1999.

[469] L. P. Middleton, G. Vlastos, N. Q. Mirza, S. Eva, and A. A. Sahin. Multicentric mammary carcinoma: evidence of monoclonal proliferation. *Cancer*, 94(7): 1910–1916, 2002.

[470] F. Milde, M. Bergdorf, and P. Koumoutsakos. A hybrid model for three-dimensional simulations of sprouting angiogenesis. *Biophys. J.*, 95: 3146–3160, 2008.

[471] S. Mitran. BEARCLAW – a code for multiphysics applications with embedded boundaries: user's manual. Department of Mathematics, University of North Carolina, http://www.amath.unc.edu/Faculty/mitran/bearclaw.html, 2006.

[472] D. F. Moffat and J. J. Going. Three dimensional anatomy of complete duct systems in human breast: pathological and developmental implications. *J. Clin. Pathol.*, 49: 48–52, 1996.

[473] A. Mogilner and L. Edelstein-Keshet. Regulation of actin dynamics in rapidly moving cells: a quantitative analysis. *Biophys. J.*, 83(3): 1237–1258, 2002.

[474] R. Montesano, K. Matsumoto, T. Nakamura, and L. Orci. Identification of a fibroblast-derived epithelial morphogen as hepatocyte growth factor. *Cell*, 67: 901–908, 1991.

[475] J. Moreira and A. Deutsch. Cellular automaton models of tumor development: a critical review. *Adv. Complex Syst.*, 5: 247–267, 2002.

[476] A. Morotti, S. Mila, P. Accornero, E. Tagliabue, and C. Ponzetto. K252a inhibits the oncogenic properties of met, the hgf receptor. *Oncogene*, 21: 4885–4893, 2002.

[477] K. Morton and D. Mayers. *Numerical Solution of Partial Differential Equations*. Cambridge University Press, second edition, 2005.

[478] B. Mosadegh, W. Saadi, S. J. Wang, and N. L. Jeon. Epidermal growth factor promotes breast cancer cell chemotaxis in CXCL12 gradients. *Biotech. Bioeng.*, 100(6): 1205–1213, 2008.

[479] W. Mueller-Klieser. Multicellular spheroids: a review on cellular aggregates in cancer research. *J. Cancer Res. Clin. Oncol.*, 113: 101–122, 1987.

[480] W. Mueller-Klieser. Three-dimensional cell cultures: from molecular mechanisms to clinical applications. *Am. J. Physiol. Cell Physiol.*, 273: C1109–C1123, 1997.

[481] W. Mueller-Klieser, J. Freyer, and R. Sutherland. Influence of glucose and oxygen supply conditions on the oxygenation of multicellular spheroids. *Br. J. Cancer*, 53: 345–353, 1986.

[482] G. Müller and J.-J. Métois. *Crystal Growth: From Fundamentals to Technology*. Elsevier, 2004.

[483] W. Mullins and R. Sekerka. Morphological instability of a particle growing by diffusion or heat flow. *J. Appl. Phys.*, 34: 323–329, 1963.

[484] G. R. Mundy. Metastasis to bone: causes, consequences and therapeutic opportunities. *Nat. Rev. Cancer*, 2(8): 584–93, 2002.

[485] J. Murray and G. Oster. Cell traction models for generation of pattern and form in morphogenesis. *J. Math. Biol.*, 33: 489–520, 1984.

[486] V. R. Muthukkaruppan, L. Kubai, and R. Auerbach. Tumor-induced neovascularization in the mouse eye. *J. Natl Cancer Inst.*, 69(3): 699–705, 1982.

[487] K. Nabeshima, T. Moriyama, Y. Asada, *et al.* Ultrastructural study of TPA-induced cell motility: human well-differentiated rectal adenocarcinoma cells move as coherent sheets via localized modulation of cell–cell adhesion. *Clin. Exp. Med.*, 13(6): 499–508, 1995.

[488] M. Nagane, A. Levitzki, A. Gazit, W. Cavenee, and H. Huang. Drug resistance of human glioblastoma cells conferred by a tumor-specific mutant epidermal growth factor receptor through modulation of bcl-x-l and caspase-3-like proteases. *Proc. Natl Acad. Sci. USA*, 95: 5724–5729, 1998.

[489] H. Naganuma, R. Kimurat, A. Sasaki, A. Fukamachi, H. Nukui, and K. Tasaka. Complete remission of recurrent glioblastoma multiforme following local infusions of lymphokine activated killer cells. *Acta Neurochir.*, 99: 157–160, 1989.

[490] J. Nagy. The ecology and evolutionary biology of cancer: a review of mathematical models of necrosis and tumor cell diversity. *Math. Biosci. Eng.*, 2: 381–418, 2005.

[491] M. N. Nakatsu, R. C. A. Sainson, J. N. Aoto, *et al.* Angiogenic sprouting and capillary lumen formation modeled by human umbilical vein endothelial cells (HUVEC) in fibrin gels: the role of fibroblasts and angiopoietin-1. *Microvasc. Res.*, 66: 102–112, 2003.

[492] M. A. H. Navarrete, C. M. Maier, R. Falzoni, L. G. de. A. Quadros, E. C. Baracat, and A. C. P. Nazário. Assessment of the proliferative, apoptotic, and cellular renovation indices of the human mammary epithelium during the follicular and luteal phases of the menstrual cycle. *Breast Cancer Res.*, 7: R306–13, 2005.

[493] C. Nelson and M. Bissell. Of extracellular matrix, scaffolds, and signaling: tissue architecture regulates development, homeostasis, and cancer. *Ann. Rev. Cell Dev. Biol.*, 22: 287–309, 2006.

[494] P. Netti, L. Baxter, Y. Boucher, R. Skalak, and R. Jain. Time dependent behavior of interstitial fluid pressure in solid tumors: implications for drug delivery. *Cancer Res.*, 55: 5451–5458, 1995.

[495] A. Neville, P. Matthews, and H. Byrne. Interactions between pattern formation and domain growth. *Bull. Math. Biol.*, 68: 1975–2003, 2006.

[496] G. Ngwa and P. Maini. Spatio-temporal patterns in a mechanical model for mesenchymal morphogenesis. *J. Math. Biol.*, 33: 489–520, 1995.

[497] M. Nichols and T. Foster. Oxygen diffusion and reaction kinetics in the photodynamic therapy of multicell tumour spheroids. *Phys. Med. Biol.*, 39: 2161–2181, 1994.

[498] R. Nishikawa, X. Ji, R. Harmon, C. Lazar, G. Gill, W. Cavenee, and H. Huang. A mutant epidermal growth factor receptor common in human glioma confers enhanced tumorigenicity. *Proc. Natl Acad. Sci. USA*, 91: 7727–7731, 1994.

[499] J. Nor, J. Christensen, J. Liu, *et al*. Up-regulation of bcl-2 in microvascular endothelial cells enhances intratumoral angiogenesis and accelerates tumor growth. *Cancer Res.*, 61: 2183–2188, 2001.

[500] M. A. Nowak, N. L. Komarova, A. Sengupta, J. V. Prasad, I.-M. Shih, B. Vogelstein, and C. Lengauer. The role of chromosomal instability in tumor initiation. *Proc. Natl Acad. Sci. USA*, 99(25): 16 226–16 231, 2002.

[501] J. O'Connor, A. Jackson, G. Parker, and G. Jayson. Dce-mri biomarkers in the clinical evaluation of antiangiogenic and vascular disrupting agents. *Br. J. Cancer*, 96: 189–195, 2007.

[502] K. Oda, Y. Matsuoka, A. Funahashi, and H. Kitano. A comprehensive pathway map of epidermal growth factor receptor signaling. *Mol. Syst. Biol.*, 1, 2005.

[503] T. Ohtake, I. Kimijima, T. Fukushima, *et al*. Computer-assisted complete three-dimensional reconstruction of the mammary ductal/lobular systems. *Cancer*, 91: 2263–2272, 2001.

[504] B. Øksendal. *Stochastic Differential Equations: An Introduction with Applications*. Springer, New York, sixth edition, 2007.

[505] M. Orme and M. Chaplain. Two-dimensional models of tumour angiogenesis and anti-angiogenesis strategies. *Math. Med. Biol.*, 14: 189–205, 1997.

[506] S. Osher and R. Fedkiw. Level set methods: an overview and some recent results. *J. Comput. Phys.*, 169(2): 463–502, 2001.

[507] S. Osher and R. Fedkiw. *Level Set Methods and Dynamic Implicit Surfaces*. Springer, New York, 2002.

[508] S. Osher and J. Sethian. Fronts propagating with curvature-dependent speed: algorithms based on Hamilton–Jacobi formulation. *J. Comput. Phys.*, 79: 12, 1988.

[509] H. G. Othmer, S. R. Dunbar, and W. Alt. Models of dispersal in biological systems. *J. Math. Biol.*, 26: 263–298, 1988.

[510] H. G. Othmer and A. Stevens. Aggregration, blowup, and collapse: the abc's of taxis in reinforced random walks. *Siam. J. Appl. Math.*, 57: 1044–1081, 1997.

[511] M. R. Owen, T. Alarcón, P. Maini, and H. Byrne. Angiogenesis and vascular remodeling in normal and cancerous tissues. *J. Math. Biol.*, 58: 689–721, 2009.

[512] M. R. Owen, H. M. Byrne, and C. E. Lewis. Mathematical modelling of the use of macrophages as vehicles for drug delivery to hypoxic tumour sites. *J. Theor. Biol.*, 226(4): 377–391, 2004.

[513] T. Padera, B. Stoll, J. Tooredman, D. Capen, E. di Tomaso, and R. Jain. Cancer cells compress intratumour vessels. *Nature*, 427: 695, 2004.

[514] D. L. Page, T. Anderson, and G. Sakamoto. *Diagnostic Histopathology of the Breast*. Churchill Livingstone, New York, 1987.

[515] D. L. Page, W. D. Dupont, L. W. Rogers, and M. Landenberger. Intraductal carcinoma of the breast: follow-up after biopsy only. *Cancer*, 49(4): 751–758, 1982.

[516] S. Paget. The distribution of secondary growths in cancer of the breast. *Lancet*, 133(3421): 571–573, 1889.

[517] S. Paku. First step of tumor-related angiogenesis. *Lab. Invest.*, 65: 334–346, 1991.

[518] D. Palmieri, C. E. Horak, J.-H. Lee, D. O. Halverson, and P. S. Steeg. Translational approaches using metastasis suppressor genes. *J. Bioenerg. Biomembr.*, 38(3–4): 151–161, 2006.

[519] E. Palsson and H. Othmer. A model for individual and collective cell movement in dictyostelium discoideum. *Proc. Natl Acad. Sci. USA*, 97: 10 338–10 453, 2000.

[520] S. Pamuk. Qualitative analysis of a mathematical model for capillary formation in tumor angiogenesis. *Math. Models Meth. Appl. Sci.*, 13: 19–33, 2003.

[521] P. Panorchan, M. S. Thompson, K. J. Davis, Y. Tseng, K. Konstantopoulos, and D. Wirtz. Single-molecule analysis of cadherin-mediated cell–cell adhesion. *J. Cell Sci.*, 119: 66–74, 2006.

[522] W. Pao, T. Y. Wang, G. J. Riely, *et al*. KRAS mutations and primary resistance of lung adenocarcinomas to gefitinib or erlotinib. *PLoS Med.*, 2: e17, 2005.

[523] S. Parnuk. A mathematical model for capillary formation and development in tumor angiogenesis: a review. *Chemotherapy*, 52: 35–37, 2006.

[524] S. Patan, S. Tanda, S. Roberge, R. Jones, R. Jain, and L. Munn. Vascular morphogenesis and remodeling in a human tumor xenograft: blood vessel formation and growth after ovariectomy and tumor implantation. *Circ. Res.*, 89: 732–739, 2001.

[525] N. Patani, B. Cutuli, and K. Mokbel. Current management of DCIS: a review. *Breast Cancer Res. Treat.*, 111(1): 1–10, 2008.

[526] A. Patel, E. Gawlinski, S. Lemieux, and R. Gatenby. A cellular automaton model of early tumor growth and invasion: the effects of native tissue vascularity and increased anaerobic tumor metabolism. *J. Theor. Biol.*, 213: 315–331, 2001.

[527] N. Paweletz and M. Knierim. Tumor-related angiogenesis. *Crit. Rev. Oncol. Hematol.*, 9: 197–242, 1989.

[528] S. Pennacchietti, P. Michieli, M. Galluzzo, S. Giordano, and P. Comoglio. Hypoxia promotes invasive growth by transcriptional activation of the met protooncogene. *Cancer Cell*, 3: 347–361, 2003.

[529] C. Peskin. The immersed boundary method. *Acta Numer.*, 11: 479–517, 2002.

[530] C. S. Peskin. Flow patterns around heart valves: a numerical method. *J. Comput. Phys.*, 10(2): 252–271, 1972.

[531] J. Peterson, G. Carey, D. Knezevic, and B. Murray. Adaptive finite element methodology for tumour angiogenesis modelling. *Int. J. Num. Meth. Eng.*, 69: 1212–1238, 2007.

[532] G. Pettet, C. Please, M. Tindall, and D. McElwain. The migration of cells in multicell tumor spheroids. *Bull. Math. Biol.*, 63: 231–257, 2001.

[533] S. Pierce. Computational and mathematical modeling of angiogenesis. *Microcirculation*, 15(8): 739–751, 2008.

[534] M. Plank and B. Sleeman. A reinforced random walk model of tumour angiogenesis and anti-angiogenic strategies. *Math. Med. Biol.*, 20: 135–181, 2003.

[535] M. Plank and B. Sleeman. Lattice and non-lattice models of tumour angiogenesis. *Bull. Math. Biol.*, 66: 1785–1819, 2004.

[536] C. Please, G. Pettet, and D. McElwain. A new approach to modeling the formation of necrotic regions in tumors. *Appl. Math. Lett.*, 11: 89–94, 1998.

[537] C. Please, G. Pettet, and D. McElwain. Avascular tumour dynamics and necrosis. *Math. Models Appl. Sci.*, 9: 569–579, 1999.

[538] N. Poplawski, U. Agero, J. Gens, M. Swat, J. Glazier, and A. Anderson. Front instabilities and invasiveness of simulated avascular tumors. *Bull. Math. Biol.*, 71: 1189–1227, 2009.

[539] L. Postovit, M. Adams, G. Lash, J. Heaton, and C. Graham. Oxygen-mediated regulation of tumor cell invasiveness. Involvement of a nitric oxide signaling pathway. *J. Biol. Chem.*, 277: 35 730–35 737, 2002.

[540] J. Pouysségur, F. Dayan, and N. Mazure. Hypoxia signalling in cancer and approaches to enforce tumour regression. *Nature*, 441: 437–443, 2006.

[541] F. Prall. Tumour budding in colorectal carcinoma. *Histopathology*, 50: 151–162, 2007.

[542] M. Preusser, H. Heinzl, E. Gelpi, *et al.* Histopathologic assessment of hot-spot microvessel density and vascular patterns in glioblastoma: poor observer agreement limits clinical utility as prognostic factors: a translational research project of the European organization for research and treatment of cancer brain tumor group. *Cancer*, 107: 162–170, 2006.

[543] L. Preziosi. *Cancer Modelling and Simulation*. Chapman and Hall/CRC, London, 2003.

[544] L. Preziosi and S. Astanin. Modelling the formation of capillaries. In A. Quarteroni, L. Formaggia, and A. Veneziani, eds., *Complex Systems in Biomedicine*. Springer, Milan, 2006.

[545] L. Preziosi and A. Tosin. Multiphase modeling of tumor growth and extracellular matrix interaction: mathematical tools and applications. *J. Math. Biol.*, 58: 625–656, 2009.

[546] A. Pries, B. Reglin, and T. Secomb. Structural adaptation and stability of microvascular networks: functional roles of adaptive responses. *Am. J. Physiol. Heart Circ. Physiol.*, 281: H1015–H1025, 2001.

[547] A. Pries, B. Reglin, and T. Secomb. Structural adaptation of vascular networks: the role of pressure response. *Hypertension*, 38: 1476–1479, 2001.

[548] A. Pries and T. Secomb. Control of blood vessel structure: insights from theoretical models. *Am. J. Physiol. Heart Circ. Physiol.*, 288: 1010–1015, 2005.

[549] A. Pries and T. Secomb. Modeling structural adaptation of microcirculation. *Microcirculation*, 15(8): 753–64, 2008.

[550] A. Pries, T. Secomb, and P. Gaehtgens. Structural adaptation and stability of microvascular networks: theory and simulations. *Am. J. Physiol. Heart Circ. Physiol.*, 275: H349–H360, 1998.

[551] W. C. Prozialeck, P. C. Lamar, and D. M. Appelt. Differential expression of E-cadherin, N-cadherin and beta-catenin in proximal and distal segments of the rat nephron. *BMC Physiol.*, 4(10), 2004.

[552] V. Quaranta, K. Rejniak, P. Gerlee, and A. Anderson. Invasion emerges from cancer cell adaptation to competitive microenvironments: quantitative predictions from multiscale mathematical models. *Sem. Cancer Biol.*, 18(5): 338–348, 2008.

[553] V. Quaranta, A. Weaver, P. Cummings, and A. Anderson. Mathematical modeling of cancer: the future of prognosis and treatment. *Clinica Chimica Acta*, 357: 173–179, 2005.

[554] C. M. Quick, W. L. Young, E. F. Leonard, S. Joshi, E. Gao, and T. Hashimoto. Model of structural and functional adaptation of small conductance vessels to arterial hypotension. *Am. J. Physiol. Heart Circ. Physiol.*, 279(4): H1645–H1653, 2000.

[555] K. C. Quon and A. Berns. Haplo-insufficiency? Let me count the ways. *Genes Dev.*, 15(22): 2917–2921, 2001.

[556] A. Ramanathan, C. Wang, and S. Schreiber. Perturbational profiling of a cell-line model of tumorigenesis by using metabolic measurements. *PNAS*, 102: 5992–5997, 2005.

[557] I. Ramis-Conde, M. Chaplain, and A. Anderson. Mathematical modelling of cancer cell invasion of tissue. *Math. Comput. Model.*, 47: 533–545, 2008.

[558] I. Ramis-Conde, D. Drasdo, A. Anderson, and M. Chaplain. Modeling the influence of the e-cadherin-beta-catenin pathway in cancer cell invasion: a multiscale approach. *Biophys. J.*, 95: 155–165, 2008.

[559] A. Rätz, A. Ribalta, and A. Voigt. Surface evolution of elastically stressed films under deposition by a diffuse interface model. *J. Comput. Phys.*, 214: 187–208, 2006.

[560] K. Rejniak. A single-cell approach in modeling the dynamics of tumor microregions. *Math. Biosci. Eng.*, 2: 643–655, 2005.

[561] K. Rejniak. An immersed boundary framework for modeling the growth of individual cells: an application to the early tumour development. *J. Theor. Biol.*, 247: 186–204, 2007.

[562] K. Rejniak and A. Anderson. A computational study of the development of epithelial acini: I. Sufficient conditions for the formation of a hollow structure. *Bull. Math. Biol.*, 70: 677–712, 2008.

[563] K. Rejniak and A. Anderson. A computational study of the development of epithelial acini: II. necessary conditions for structure and lumen stability. *Bull. Math. Biol.*, 70: 1450–1479, 2008.

[564] K. Rejniak and R. Dillon. A single cell-based model of the ductal tumor microarchitecture. *Comput. Math. Meth. Med.*, 8(1): 51–69, 2007.

[565] K. Rennstam and I. Hedenfalk. High-throughput genomic technology in research and clinical management of breast cancer. Molecular signatures of progression from benign epithelium to metastatic breast cancer. *Breast Cancer. Res.*, 8(4): 213ff, 2006.

[566] B. Ribba, T. Alarcón, K. Marron, P. Maini, and Z. Agur. The use of hybrid cellular automaton models for improving cancer therapy. In P. Sloot, B. Chopard, and A. Hoekstra, eds., *ACRI, LNCS*, pp. 444–453. Springer, Berlin, 2004.

[567] B. Ribba, O. Saut, T. Colin, D. Bresch, E. Grenier, and J. P. Boissel. A multiscale mathematical model of avascular tumor growth to investigate the therapeutic benefit of anti-invasive agents. *J. Theor. Biol.*, 243(4): 532–541, 2006.

[568] A. Ridley, M. Schwartz, K. Burridge, *et al.* Cell migration: integrating signals from front to back. *Science*, 302: 1704–1709, 2003.

[569] E. Robinson, K. Zazzali, S. Corbett, and R. Foty. α5b1 integrin mediates strong tissue cohesion. *J. Cell. Sci.*, 116: 377–386, 2003.

[570] E. Rofstad and E. Halsør. Hypoxia-associated spontaneous pulmonary metastasis in human melanoma xenographs: involvement of microvascular hotspots induced in hypoxic foci by interleukin. *Br. J. Cancer*, 86: 301–308, 2002.

[571] E. Rofstad, H. Rasmussen, K. Galappathi, B. Mathiesen, K. Nilsen, and B. Graff. Hypoxia promotes lymph node metastasis in human melanoma xenografts by up-regulating the urokinase-type plasminogen activator receptor. *Cancer Res.*, 62: 1847–1853, 2002.

[572] J. Rohzin, M. Sameni, G. Ziegler, and B. Sloane. Pericellular ph affects distribution and secretion of cathepsin b in malignant cells. *Cancer Res.*, 54: 6517–6625, 1994.

[573] T. Roose, S. J. Chapman, and P. Maini. Mathematical models of avascular tumor growth. *SIAM Rev.*, 49: 179–208, 2007.

[574] T. Roose, P. Netti, L. Munn, Y. Boucher, and R. Jain. Solid stress generated by spheroid growth using a linear poroelastic model. *Microvascular Res.*, 66: 204–212, 2003.

[575] B. Rubenstein and L. Kaufman. The role of extracellular matrix in glioma invasion: a cellular potts model approach. *Biophys. J.*, 95: 5661–5680, 2008.

[576] J. Rubenstein, J. Kim, T. Ozawa, *et al.* Anti-vegf antibody treatment of glioblastoma prolongs survival but results in increased vascular cooption. *Neoplasia*, 2: 306–314, 2000.

[577] K. Rygaard and M. Spang-Thomsen. Quantitation and gompertzian analysis of tumor growth. *Breast Cancer Res. Treat.*, 46: 303–312, 1997.

[578] Y. Saad and M. Schultz. Gmres: a generalized minimal residual algorithm for solving nonsymmetric linear systems. *SIAM J. Sci. Stat. Comput.*, 7: 856–869, 1986.

[579] E. Sahai. Mechanisms of cancer cell invasion. *Curr. Opin. Genet. Dev.*, 15: 87–96, 2005.

[580] T. Sairanen, R. Szepesi, M.-L. Karjalainen-Lindsberg, *et al.* Neuronal caspase-3 and PARP-1 correlate differentially with apoptosis and necrosis in ischemic human stroke. *Acta Neuropathologica*, 118(4): 541–552, 2009.

[581] G. Sakamoto. Infiltrating carcinoma: major histological types. In D. Page and T. Anderson, eds., *Diagnostic Histopathology of the Breast*. Churchill-Livingstone, London, 1987.

[582] M. E. Sanders, P. A. Schuyler, W. D. Dupont, and D. L. Page. The natural history of low-grade ductal carcinoma in situ of the breast in women treated by biopsy only revealed over 30 years of long-term follow-up. *Cancer*, 103(12): 2481–2484, 2005.

[583] S. Sanga, M. E. Edgerton, P. Macklin, and V. Cristini. From receptor dynamics to directed cell motion: a predictive multiscale model of cell motility in complex microenvironments. In preparation, 2010.

[584] S. Sanga, H. Frieboes, X. Zheng, R. Gatenby, E. Bearer, and V. Cristini. Predictive oncology: a review of multidisciplinary, multiscale in silico modeling linking phenotype, morphology and growth. *NeuroImage*, 37: S120–S134, 2007.

[585] S. Sanga, J. Sinek, H. Frieboes, M. Ferrari, J. Fruehauf, and V. Cristini. Mathematical modeling of cancer progression and response to chemotherapy. *Expert Rev. Anticancer Ther.*, 6: 1361–1376, 2006.

[586] B. Sansone, P. D. Santo, M. Magnano, and M. Scalerandi. Effects of anatomical constraints on tumor growth. *Phys. Rev. E*, 64: 21 903, 2002.

[587] B. Sansone, M. Scalerandi, and C. Condat. Emergence of taxis and synergy in angiogenesis. *Phys. Rev. Lett.*, 87: 128 102, 2001.

[588] M. Santini, G. Rainaldi, and P. Indovina. Apoptosis, cell adhesion and the extracellular matrix in three-dimensional growth of multicellular tumor spheroids. *Crit. Rev. Oncol. Hematol.*, 36: 75–87, 2000.

[589] M. Sarntinoranont, F. Rooney, and M. Ferrari. Interstitial stress and fluid pressure within a growing tumor. *Ann. Biomed. Eng.*, 31: 327–335, 2003.

[590] J. Satulovsky, R. Lui, and Y. L. Wang. Exploring the control circuit of cell migration by mathematical modeling. *Biophys. J.*, 94(9): 3671–3683, 2008.

[591] N. Savill and P. Hogeweg. Modeling morphogenesis: from single cells to crawling slugs. *J. Theor. Biol.*, 184: 229–235, 1997.

[592] J. L. Scarlett, P. W. Sheard, G. Hughes, E. C. Ledgerwood, H.-K. Ku, and M. P. Murphy. Changes in mitochondrial membrane potential during staurosporine-induced apoptosis in Jurkat cells. *FEBS Lett.*, 475(3): 267–272, 2000.

[593] J. Schlessinger. Ligand-induced, receptor-mediated dimerization and activation of EGF receptor. *Cell*, 110(6): 669–672, 2002.

[594] K. Schmeichel, V. Weaver, and M. Bissel. Structural cues from the tissue microenvironment are essential determinants of the human mammary epithelial cell phenotype. *J. Mammary Gland Biol. Neoplasia*, 3: 201–213, 1998.

[595] L. S. Schulman and P. E. Seiden. Statistical mechanics of a dynamical system based on Conway's game of life. *J. Stat. Phys.*, 19(3): 293–314, 1978.

[596] E. Seftor, P. Meltzer, D. Kirshmann, *et al.* Molecular determinants of human uveal melanoma invasion and metastasis. *Clin. Exp. Metastasis*, 19: 233–246, 2002.

[597] M. J. Seidensticker and J. Behrens. Biochemical interactions in the wnt pathway. *Biochim. Biophys. Acta*, 1495: 168–182, 2000.

[598] B. Selam, U. A. Kayisli, J. A. Garcia-Velasco, and A. Arici. Extracellular matrix-dependent regulation of FAS ligand expression in human endometrial stromal cells. *Biol. Reprod.*, 66(1): 1–5, 2002.

[599] G. L. Semenza. HIF-1, O_2, and the 3 PHDs: how animal cells signal hypoxia to the nucleus. *Cell*, 107(1): 1–3, 2001.

[600] G. Serini, D. Ambrosi, E. Giraudo, A. Gamba, L. Preziosi, and F. Bussolino. Modeling the early stages of vascular network assembly. *EMBO J.*, 22: 1771–1779, 2003.

[601] S. Setayeshgar, C. Gear, H. Othmer, and I. Kevrekidis. Application of coarse integration to bacterial chemotaxis. *SIAM Multiscale Model. Sim.*, 4: 307–327, 2005.

[602] J. A. Sethian. *Level Set Methods and Fast Marching Methods*. Cambridge University Press, New York, 1999.

[603] J. A. Sethian and P. Smereka. Level set methods for fluid interfaces. *Ann. Rev. Fluid Mech.*, 35(1): 341–372, 2003.

[604] M. Shannon and B. Rubinsky. The effect of tumor growth on the stress distribution in tissue. *Adv. Biol. Heat Mass Transfer*, 231: 35–38, 1992.

[605] N. Sharifi, B. T. Kawasaki, E. M. Hurt, and W. L. Farrar. Stem cells in prostate cancer: resolving the castrate-resistant conundrum and implications for hormonal therapy. *Cancer Biol. Ther.*, 5(8): 910–906, 2006.

[606] C. J. Sherr. Cancer cell cycles. *Science*, 274(5293): 1672–1677, 1996.

[607] J. Sherratt. Traveling wave solutions of a mathematical model for tumor encapsulation. *SIAM J. Appl. Math.*, 60: 392–407, 1999.

[608] J. Sherratt and M. Chaplain. A new mathematical model for avascular tumour growth. *J. Math. Biol.*, 43: 291–312, 2001.

[609] A. N. Shiryaev. *Probability*. Springer, New York, second edition, 1995.

[610] B. I. Shraiman. Mechanical feedback as a possible regulator of tissue growth. *Proc. Natl Acad. Sci. USA*, 102(9): 3318–3323, 2005.

[611] C.-W. Shu and S. Osher. Efficient implementation of essentially non-oscillatory shock-capturing schemes. *J. Comput. Phys.*, 77: 439–471, 1988.

[612] C.-W. Shu and S. Osher. Efficient implementation of essentially non-oscillatory shock capturing schemes, II. *J. Comput. Phys.*, 83: 32–78, 1989.

[613] D. Shweiki, A. Itin, D. Soffer, and E. Keshet. Vascular endothelial growth factor induced by hypoxia may mediate hypoxia-initiated angiogenesis. *Nature*, 359: 843–845, 1992.

[614] A. Sierra. Metastases and their microenvironments: linking pathogenesis and therapy. *Drug Resist. Updates*, 8: 247–257, 2005.

[615] S. A. Silver and F. A. Tavassoli. Ductal carcinoma in situ with microinvasion. *Breast J.*, 4(5): 344–348, 1998.

[616] M. J. Silverstein. Predicting residual disease and local recurrence in patients with ductal carcinoma in situ. *J. Natl Cancer Inst.*, 89(18): 1330–1331, 1997.

[617] M. J. Silverstein. Recent advances: diagnosis and treatment of early breast cancer. *BMJ*, 314(7096): 1736ff, 1997.

[618] M. J. Silverstein. Ductal carcinoma in situ of the breast. *Ann. Rev. Med.*, 51: 17–32, 2000.

[619] P. T. Simpson, J. S. Reis-Filho, T. Gale, and S. R. Lakhani. Molecular evolution of breast cancer. *J. Pathol.*, 205(2): 248–254, 2005.

[620] J. Sinek, H. Frieboes, X. Zheng, and V. Cristini. Two-dimensional chemotherapy simulations demonstrate fundamental transport and tumor response limitations involving nanoparticles. *Biomedical Microdevices*, 6: 297–309, 2004.

[621] J. Sinek, S. Sanga, X. Zheng, H. Frieboes, M. Ferrari, and V. Cristini. Predicting drug pharmacokinetics and effect in vascularized tumors using computer simulation. *J. Math. Biol.*, 58: 485–510, 2009.

[622] S. Skinner. Microvascular architecture of experimental colon tumors in the rat. *Cancer Res.*, 50: 2411–2417, 1990.

[623] V. I. F. Slettenaar and J. L. Wilson. The chemokine network: a target in cancer biology? *Adv. Drug Deliv. Rev.*, 58(8): 962–974, 2006.

[624] K. Smallbone, R. Gatenby, R. Gillies, P. Maini, and D. Gavaghan. Metabolic changes during carcinogenesis: potential impact on invasiveness. *J. Theor. Biol.*, 244: 703–713, 2007.

[625] K. Smallbone, R. Gatenby, and P. Maini. Mathematical modelling of tumour acidity. *J. Theor. Biol.*, 255: 106–112, 2008.

[626] K. Smallbone, D. Gavaghan, R. Gatenby, and P. Maini. The role of acidity in solid tumor growth and invasion. *J. Theor. Biol.*, 235: 476–484, 2005.

[627] K. Smallbone, D. Gavaghan, P. Maini, and J. M. Brady. Quiescence as a mechanism for cyclical hypoxia and acidosis. *J. Math. Biol.*, 55: 767–779, 2007.

[628] S. A. A. Sohaib and R. H. Reznek. MR imaging in ovarian cancer. *Canc. Imag.*, 7(Special Issue A): S119–S129, 2007.

[629] V. Spencer, R. Xu, and M. Bissell. Extracellular matrix, nuclear and chromatin structure, and gene expression in normal tissues and malignant tumors: a work in progress. *Adv. Cancer Res.*, 97: 275–294, 2007.

[630] T. A. Springer. Adhesion receptors of the immune system. *Nature*, 346(6283): 425–434, 1990.

[631] P. Steeg. Angiogenesis inhibitors: motivators of metastasis? *Nature Med.*, 9: 822–823, 2003.

[632] A. Stein, T. Demuth, D. Mobley, M. Berens, and L. Sander. A mathematical model of glioblastoma tumor spheroid invasion in a three-dimensional in vitro experiment. *Biophys. J.*, 92: 356–365, 2007.

[633] M. S. Steinberg and M. Takeichi. Experimental specification of cell sorting, tissue spreading, and specific spatial patterning by quantitative differences in cadherin expression. *Proc. Natl Acad. Sci. USA*, 91: 206–209, 1994.

[634] A. Stephanou, S. McDougall, A. Anderson, and M. Chaplain. Mathematical modelling of flow in 2d and 3d vascular networks: applications to anti-angiogenic and chemotherapeutic drug strategies. *Math. Comput. Modelling*, 41: 1137–1156, 2005.

[635] A. Stephanou, S. McDougall, A. Anderson, and M. Chaplain. Mathematical modelling of the influence of blood rheological properties upon adaptative tumour-induced angiogenesis. *Math. Comput. Modelling*, 44: 96–123, 2006.

[636] J. Stewart, P. Broadbridge, and J. Goard. Symmetry analysis and numerical modelling of invasion by malignant tumour tissue. *Nonlinear Dyn.*, 28: 175–193, 2002.

[637] C. Stokes and D. Lauffenburger. Analysis of the roles of microvessel endothelial cell random motility and chemotaxis in angiogenesis. *J. Theor. Biol.*, 152: 377–403, 1991.

[638] P. C. Stomper and F. R. Margolin. Ductal carcinoma in situ: the mammographer's perspective. *Am. J. Roentgenology*, 162: 585–591, 1994.

[639] E. Stott, N. Britton, J. Glazier, and M. Zajac. Simulation of benign avascular tumour growth using the Potts model. *Math. Comput. Modelling*, 30: 183–198, 1999.

[640] D. Stupack and D. Cheresh. Get a ligand, get a life: Integrins, signaling and cell survival. *J. Cell. Sci.*, 115: 3729–3738, 2002.

[641] C. Sun and L. Munn. Lattice-Boltzmann simulation of blood flow in digitized vessel networks. *Comput. Math. Appl.*, 55: 1594–1600, 2008.

[642] S. Sun, M. Wheeler, M. Obeyesekere, and C. Patrick Jr. A deterministic model of growth factor-induced angiogenesis. *Bull. Math. Biol.*, 67: 313–337, 2005.

[643] S. Sun, M. Wheeler, M. Obeyesekere, and C. Patrick Jr. Multiscale angiogenesis modeling using mixed finite element methods. *Multiscale Model. Simul.*, 4: 1137–1167, 2005.

[644] X.-F. Sun and H. Zhang. Clinicopathological significance of stromal variables: angiogenesis, lymphangiogenesis, inflammatory infiltration, MMP and PINCH in colorectal carcinomas. *Mol. Cancer*, 5: 43, 2006.

[645] K. Sundfor, H. Lyng, and E. Rofstad. Tumour hypoxia and vascular density as predictors of metastasis in squamous cell carcinoma of the uterine cervix. *Br. J. Cancer*, 78: 822–827, 1998.

[646] M. Sussman, P. Smereka, and S. Osher. A level set approach for computing solutions to incompressible two-phase flow. *J. Comput. Phys.*, 114(1): 146–159, 1994.

[647] R. Sutherland. Cell and environment interactions in tumor microregions: the multicell spheroid model. *Science*, 240: 177–184, 1988.

[648] R. Sutherland, J. Carlsson, R. Durand, and J. Yuhas. Spheroids in cancer research. *Cancer Res.*, 41: 2980–2994, 1981.

[649] K. Swanson, C. Bridge, J. Murray, and J. Alvord Virtual and real brain tumors: using mathematical modeling to quantify glioma growth and invasion. *J. Neuro. Sci.*, 216: 1–10, 2003.

[650] L. A. Taber. An optimization principle for vascular radius including the effects of smooth muscle tone. *Biophys. J.*, 74(1): 109–114, 1998.

[651] P. J. Tannis, O. E. Nieweg, R. A. Valdés Olmos, and B. B. R. Kroon. Anatomy and physiology of lymphatic drainage of the breast from the perspective of sentinel node biopsy. *J. Am. Coll. Surg.*, 192(3): 399–409, 2001.

[652] Y. Tao and M. Chen. An elliptic–hyperbolic free boundary problem modelling cancer therapy. *Nonlinearity*, 19: 419–440, 2006.

[653] Y. Tao, N. Yoshida, and Q. Guo. Nonlinear analysis of a model of vascular tumour growth and treatment. *Nonlinearity*, 17: 867–895, 2004.

[654] M. J. Terol, M. Tormo, J. A. Martinez-Climent, *et al*. Soluble intercellular adhesion molecule-1 (s-ICAM-1/s-CD54) in diffuse large B-cell lymphoma: association with clinical characteristics and outcome. *Ann. Oncol.*, 14(3): 467–474, 2003.

[655] R. Thomlinson and L. Gray. The histological structure of some human lung cancers and the possible implications of radiotherapy. *Br. J. Cancer*, 9: 539–549, 1955.

[656] B. Thorne, A. Bailey, and S. Pierce. Combining experiments with multi-cell agent-based modeling to study biological tissue patterning. *Briefings in Bioinformatics*, 8: 245–257, 2007.

[657] M. Tindall, C. Please, and M. Peddie. Modelling the formation of necrotic regions in avascular tumours. *Math. Biosci.*, 211: 34–55, 2008.

[658] S. Tong and F. Yuan. Numerical simulations of angiogenesis in the cornea. *Microvasc. Res.*, 61: 14–27, 2001.

[659] A. Tosin. Multiphase modeling and qualitative analysis of the growth of tumor cords. *Networks Heterogen. Media*, 3: 43–84, 2008.

[660] A. Tosin, D. Ambrosi, and L. Preziosi. Mechanics and chemotaxis in the morphogenesis of vascular networks. *Bull. Math. Biol.*, 68: 1819–1836, 2006.

[661] P. Tracqui. Biophysical models of tumor growth. *Rep. Prog. Phys.*, 72: 056 701, 2009.

[662] U. Trottenberg, C. Oosterlee, and A. Schüller. *Multigrid*. Academic Press, New York, 2005.

[663] C. Truesdell and R. Toupin. Classical field theories. In S. Flugge, ed., *Handbuch der Physik*, vol. 3, part I. Springer-Verlag, Berlin, 1960.

[664] S. Turner and J. Sherratt. Intercellular adhesion and cancer invasion: a discrete simulation using the extended Potts model. *J. Theor. Biol.*, 216: 85–100, 2002.

[665] B. Tysnes and R. Mahesparan. Biological mechanisms of glioma invasion and potential therapeutic targets. *J. Neurooncol.*, 53: 129–147, 2001.

[666] P. Vajkoczy, M. Farhadi, A. Gaumann, *et al.* Microtumor growth initiates angiogenic sprouting with simultaneous expression of vegf, vegf receptor-2, and angiopoietin-2. *J. Clin. Invest.*, 109: 777–785, 2002.

[667] L. van Kempen, D. Ruiter, G. van Muijen, and L. Coussens. The tumor microenvironment: a critical determinant of neoplastic evolution. *Eur. J. Cell. Biol.*, 82: 539–548, 2003.

[668] I. van Leeuwen, C. Edwards, M. Ilyas, and H. Byrne. Towards a multiscale model of colorectal cancer. *World Gastroenterol.*, 13: 1399–1407, 2007.

[669] V. V. Vasko and M. Saji. Molecular mechanisms involved in differentiated thyroid cancer invasion and metastasis. *Curr. Opin. Oncol.*, 19(1): 11–17, 2007.

[670] P. Vaupel, H. Haugland, T. Nicklee, A. Morrison, and D. Hedley. Hypoxia-inducible factor-1 alpha is an intrinsic marker for hypoxia in cervical cancer xenografts. *Cancer Res.*, 61: 7394–7398, 2001.

[671] R. Vernon, J. Angello, M. Iruela-Arispe, and T. Lane. Reorganization of basement membrane matrices by cellular traction promotes the formation of cellular networks in vitro. *Lab. Invest.*, 66: 536–547, 1992.

[672] E. Villa-Cuesta, E. Gonz'alez-P'erez, and J. Modolell. Apposition of *iroguois* expressing and non-expressing cells leads to cell sorting and fold formation in *Drosiphila* imaginal wing disc. *BMC Devel. Biol.*, 7(106), 2007.

[673] B. Vollmayr-Lee and A. Rutenberg. Stresses in growing soft tissues. *Acta Biomaterialia*, 2: 493–504, 2006.

[674] J. von Neumann. *Theory of Self-Replicating Automata*. University of Illinois Press, 1966. Edited by Arthur W. Burks.

[675] C. Walker and G. Webb. Global existence of classical solutions for a haptotaxis model. *SIAM J. Math. Anal.*, 38(5): 1694–1713, 2007.

[676] T. Walles, M. Weimer, K. Linke, J. Michaelis, and H. Mertsching. The potential of bioartificial tissues in oncology research and treatment. *Onkologie*, 30: 388–394, 2007.

[677] R. Wang, L. Jinming, K. Lyte, N. K. Yashpal, F. Fellows, and C. G. Goodyer. Role for $\beta 1$ integrin and its associated $\alpha 3$, $\alpha 5$, and $\alpha 6$ subunits in development of the human fetal pancreas. *Diabetes*, 54: 2080–9, 2005.

[678] Z. Wang, L. Zhang, J. Sagotsky, and T. Deisboeck. Simulating non-small cell lung cancer with a multiscale agent-based model. *Theor. Biol. Med. Model.*, 4: 50, 2007.

[679] J. Ward and J. King. Mathematical modelling of avascular tumour growth. *IMA J. Math. Appl. Med. Biol.*, 14: 36–69, 1997.

[680] J. Ward and J. King. Mathematical modelling of avascular-tumour growth ii: modelling growth saturation. *Math. Med. Biol.*, 16: 171–211, 1999.

[681] J. Ward and J. King. Modelling the effect of cell shedding on avasacular tumour growth. *J. Theor. Med.*, 2: 155–174, 2000.

[682] J. Ward and J. King. Mathematical modelling of drug transport in tumour multicell spheroids and monolayer cultures. *Math. Biosci.*, 181: 177–207, 2003.

[683] R. Wcislo and W. Dzwinel. Particle based model of tumor progression stimulated by the process of angiogenesis. In J. Adam and N. Bellomo, eds., *Computational Science, Proc. ICCS 2008*, pp. 177–186. Springer, Heidelberg, 2008.

[684] A. M. Weaver. Invadopodia: specialized cell structures for cancer invasion. *Clin. Exp. Metastasis*, 23(2): 97–105, 2006.

[685] C. Wei, M. Larsen, M. P. Hoffman, and K. M. Yamada. Self-organization and branching morphogenesis of primary salivary epithelial cells. *Tissue Eng.*, 13(4): 721–735, 2007.

[686] O. D. Weiner, W. A. Marganski, L. F. Wu, S. J. Altschuler, and M. W. Kirschner. An actin-based wave generator organizes cell motility. *PLoS Biol*, 5(9): e221, 08 2007.

[687] S. R. Wellings, H. M. Jensen, and R. G. Marcum. An atlas of subgross pathology of the human breast with special reference to possible precancerous lesions. *J. Natl Cancer Inst.*, 55(2): 231–273, 1975.

[688] A. Wells, B. Harms, A. Iwabu,*et al*. Motility signaled from the EGF receptor and related systems. *Meth. Mol. Biol.*, 327: 159–177, 2006.

[689] A. Wells, J. Kassis, J. Solava, T. Turner, and D. A. Lauffenburger. Growth factor-induced cell motility in tumor invasion. *Acta Oncol.*, 41(2): 124–130, 2002.

[690] M. Welter, K. Bartha, and H. Rieger. Emergent vascular network inhomogeneities and resulting blood flow patterns in a growing tumor. *J. Theor. Biol.*, 250: 257–280, 2008.

[691] M. Welter, K. Bartha, and H. Rieger. Hot spot formation in tumor vasculature during tumor growth in an arterio–venous-network environment. arXiv.org ¿ q-bio ¿ arXiv: 0801.0654v2, 2008.

[692] N. Wentzensen, S. Vinokurova, and M. von Knebel Doeberitz. Systematic review of genomic integration sites of human papillomavirus genomes in epithelial dysplasia and invasive cancer of the female lower genital tract. *Cancer. Res.*, 64(11): 3878–3884, 2004.

[693] K. Wiesenfeld and F. Moss. Stochastic resonance and the benefits of noise: from ice ages to crayfish and squids. *Nature*, 373: 33, 1995.

[694] H. S. Wiley, S. Y. Shvartsman, and D. A. Lauffenburger. Computational modeling of the EGF-receptor system: a paradigm for systems biology. *Trends Cell. Biol.*, 13(1): 43–50, 2003.

[695] S. Wise, J. Kim, and J. Lowengrub. Solving the regularized, strongly anisotropic Cahn–Hilliard equation by an adaptive nonlinear multigrid method. *J. Comput. Phys.*, 226: 414–446, 2007.

[696] S. Wise, J. Lowengrub, and V. Cristini. An adaptive algorithm for simulating solid tumor growth using mixture models (in review), 2010.

[697] S. Wise, J. Lowengrub, H. Frieboes, and V. Cristini. Three-dimensional multispecies nonlinear tumor growth. i. model and numerical method. *J. Theor. Biol.*, 253: 524–543, 2008.

[698] E. K. Wolf, A. C. Smidt, and A. E. Laumann. Topical sodium thiosulfate therapy for leg ulcers with dystrophic calcification. *Arch. Dermatol.*, 144(12): 1560–1562, 2008.

[699] K. Wolf and P. Friedl. Molecular mechanisms of cancer cell invasion and plasticity. *Br. J. Dermatology*, 154: 11–15, 2006.

[700] K. Wolf, R. Müller, S. Borgmann, E.-B. Bröcker, and P. Friedl. Amoeboid shape change and contact guidance: T-lymphocyte crawling through fibrillar collagen is independent of matrix remodeling by MMPs and other proteases. *Blood*, 102(9): 3262–3269, 2003.

[701] J. Wu and S. Cui. Asymptotic behavior of solutions of a free boundary problem modeling the growth of tumors with Stokes equations. *Discr. Contin. Dyn. Sys.*, 24(2): 625–651, 2009.

[702] J. Wu, F. Zhou, and S. Cui. Simulation of microcirculation in solid tumors. In *Proc. IEEE/ICME Int. Conf. on Complex Med. Eng.*, pp. 1555–1563, 2007.

[703] M. Wurzel, C. Schaller, M. Simon, and A. Deutsch. Cancer cell invasion of brain tissue: guided by a prepattern? *J. Theor. Med.*, 6: 21–31, 2005.

[704] Y. Xiong, P. Rangamani, B. Dubin-Thaler, M. Sheetz, and R. Iyengar. A three-dimensional stochastic spatio-temporal model of cell spreading. *Nat. Proc.*, 2007. Available from Nature Proceedings: http: //10.1038/npre.2007.62.2.

[705] R. Xu, V. Spencer, and M. Bissell. Extracellular matrix-regulated gene expression requires cooperation of swi/snf and transcription factors. *J. Biol. Chem.*, 282: 14 992–14 999, 2007.

[706] S. Xu. Hopf bifurcation of a free boundary problem modeling tumor growth with two time delays. *Chaos Solitons Fractals*, 41(5): 2491–2494, 2009.

[707] Y. Xu and R. Gilbert. Some inverse problems raised from a mathematical model of ductal carcinoma in situ. *Math. Comp. Model.*, 49(3–4): 814–828, 2009.

[708] H. Yamaguchi, J. Wyckoff, and J. Condeelis. Cell migration in tumors. *Curr. Op. Cell Biol.*, 17: 559–564, 2005.

[709] K. Yamauchi, M. Yang, P. Jiang, *et al.* Development of real-time subcellular dynamic multicolor imaging of cancer-cell trafficking in live mice with a variable-magnification whole-mouse imaging system. *Cancer. Res.*, 66: 4028–4214, 2006.

[710] S. Young and R. Hill. Effects of reoxygenation of cells from hypoxic regions of solid tumors: anticancer drug sensitivity and metastatic potential. *J. Natl Cancer Inst.*, 82: 338–339, 1990.

[711] S. Young, R. Marshall, and R. Hill. Hypoxia induces dna overreplication and enhances metastatic potential of murine tumor cells. *Proc. Natl Acad. Sci. USA*, 85: 9533–9537, 1988.

[712] J. Yu, J. Rak, B. Coomber, D. Hicklin, and R. Kerbel. Effect of p53 status on tumor response to antiangiogenic therapy. *Science*, 295: 1526–1528, 2002.

[713] A. Zagorska and J. Dulak. HIF-1: the knowns and unknowns of hypoxia sensing. *Acta Biochimica Polonica*, 51(3): 563–585, 2004.

[714] D. Zagzag, R. Amirnovin, M. Greco, *et al.* Vascular apoptosis and involution in gliomas precede neovascularization: a novel concept for glioma growth and angiogenesis. *Lab. Invest.*, 80: 837–849, 2000.

[715] M. Zajac, G. Jones, and J. Glazier. Model of convergent extension in animal morphogenesis. *Phys. Rev. Lett.*, 85: 2022–2025, 2000.

[716] M. H. Zaman, R. D. Kamm, P. Matsudaira, and D. A. Lauffenburger. Computational model for cell migration in three-dimensional matrices. *Biophys. J.*, 89(2): 1389–1397, 2005.

[717] A. Zetterberg, O. Larsson, and K. G. Wilman. What is the restriction point? *Curr. Opin. Cell Biol.*, 7(6): 835–842, 1995.

[718] L. Zhang, C. Athale, and T. Deisboeck. Development of a three-dimensional multiscale agent-based tumor model: simulating gene–protein interaction profiles, cell phenotypes and multicellular patterns in brain cancer. *J. Theor. Biol.*, 244: 96–107, 2007.

[719] L. Zhang, C. Strouthos, Z. Wang, and T. Deisboeck. Simulating brain tumor heterogeneity with a multiscale agent-based model: linking molecular signatures, phenotypes and expansion rate. *Math. Comput. Modelling*, 49: 307–319, 2009.

[720] L. Zhang, Z. Wang, J. Sagotsky, and T. Deisboeck. Multiscale agent-based cancer modeling. *J. Math. Biol.*, 58(4–5): 545–559, 2009.

[721] G. Zhao, J. Wu, S. Xu, *et al.* Numerical simulation of blood flow and interstitial fluid pressure in solid tumor microcirculation based on tumor-induced angiogenesis. *Mech. Sinica*, 23: 477–483, 2007.

[722] X. Zheng, S. Wise, and V. Cristini. Nonlinear simulation of tumor necrosis, neovascularization and tissue invasion via an adaptive finite-element/level-set method. *Bull. Math. Biol.*, 67: 211–259, 2005.

[723] F. Zhou and S. Cui. Bifurcation for a free boundary problem modeling the growth of multi-layer tumors. *Nonlinear Anal. Theory Meth. Appl.*, 68: 2128–2145, 2008.

[724] F. Zhou, J. Escher, and S. Cui. Well-posedness and stability of a free boundary problem modeling the growth of multi-layer tumors. *J. Diff. Eq.*, 244: 2909–2933, 2008.

[725] D. Zipori. The mesenchyme in cancer therapy as a target tumor component, effector cell modality and cytokine expression vehicle. *Cancer Metastasis Rev.*, 25(3): 459–467, 2006.

Index